ANALYSIS OF BASIC SYSTEMS

ANALYSIS OF BASIC SYSTEMS

Saurabh Mani Tripathy

M.Tech. Fellow

B.Tech. (EEE), G.S.M. (IEEE)

Department of Electrical Engineering

M.M.M. Engineering College

Gorakhpur (U.P.)

UNIVERSITY SCIENCE PRESS

(An Imprint of Laxmi Publications Pvt. Ltd.)

BANGALORE • CHENNAI • COCHIN • GUWAHATI • HYDERABAD

JALANDHAR • KOLKATA • LUCKNOW • MUMBAI • RANCHI

NEW DELHI • BOSTON, USA

Published by :
UNIVERSITY SCIENCE PRESS
(*An Imprint of Laxmi Publications Pvt. Ltd.*)
113, Golden House, Daryaganj,
New Delhi-110002

Phone : 011-43 53 25 00
Fax : 011-43 53 25 28

www.laxmipublications.com
info@laxmipublications.com

Price : **Rs. 150.00** *Only.* *First Edition : 2009*

OFFICES

India

© **Bangalore**	080-26 61 15 61
© **Chennai**	044-24 34 47 26
© **Cochin**	0484-239 70 04
© **Guwahati**	0361-254 36 69, 251 38 81
© **Hyderabad**	040-24 65 23 33
© **Jalandhar**	0181-222 12 72
© **Kolkata**	033-222 74 384
© **Lucknow**	0522-220 95 78
© **Mumbai**	022-24 91 54 15, 24 92 78 69
© **Ranchi**	0651-221 47 64

USA

Boston

11, Leavitt Street, Hingham,
MA 02043, USA

UAB-9317-150-ANALYSIS OF BASIC SYSTEMS **C-16611 /08/10**

Typeset at : Sukuvisa Enterprises, New Delhi. ***Printed at*** : Ajit Printers, Delhi.

DEDICATED

In the Loving Memory of

Late Smt. Sampati Devi & Late Shri Jagdish Narayan Tripathi
(My Grandparents)

CONTENTS

PREFACE

The scope of real-time applications of the techniques of signal and system analysis continues to expand as engineers are confronted with new challenges involving the synthesis and analysis of complex processes. In many contexts in which signals and systems arise, we are presented with a specific system and are interested in characterizing it in detail to understand how it will respond to various input signals. This book deals with the signals and their interaction in systems.

This book comprises of six chapters *viz.* "Signals and Systems", "Fourier Series Analysis", "Fourier Transform Analysis", "Laplace Transform Analysis", "z-Transform Analysis" and "State-Variable Analysis". All six chapters are arranged with a unique style in order to deal with fundamentals of the subject keeping in mind the necessities of the students of Indian subcontinent and the same of all foreign universities. Each chapter also includes a list of references in order to assist the reader who is interested in pursuing additional and more advanced studies of the methods and applications of signal and system analysis. Much of the material in this book forms a direct bridge to other engineering subjects such as *Control Systems, Digital Signal Processing* and *Communication Systems.*

The emphasis is on concepts as well as on mathematical derivations in order to help the readers to understand the theoretical and practical aspects of the subject. Another feature of the book is the emphasis given to illustrations. A number of examples have been dealt for different cases so as to bring automatically out the basic concepts incorporated. This is also an exclusive feature of the present book. Furthermore, a sufficient number of problems have also been included in this book.

Primarily intended as an introductory text for course in **"Basic Systems Analysis"** or **"Linear Systems Analysis"** at the level of undergraduate students, I hope that this book would also be useful for the students at postgraduate level.

I lay no claim to the original research in preparing the book. Liberal use of the materials available has been made. I have tried to fashion the vast amount of material available from various sources into coherent body of description and analysis.

I have taken considerable care to ensure that this text is as error free and effective as possible and will be grateful if any error remains behind being pointed out with other suggestions for further improvement of the book.

— *AUTHOR*

ACKNOWLEDGEMENT

I wish to communicate my deep sense of gratitude to my parents Smt. Shashi Tripathi and Shri Lal Bihari Tripathi, who gave physical basis for representing me as well as to my brother Kaustubh Mani Tripathi for his valuable suggestions throughout the preparation of the manuscript of this book.

My sincere thanks go to Dr. A.K. Pandey, Assistant Professor, Department of Electrical Engineering and to Dr. Govind Pandey, Assistant Professor, Department of Civil Engineering, M.M.M. Engineering College, Gorakhpur for their encouragement and appreciation.

I would like to express my gratitude to all faculty members and supporting staff members of Electrical Engineering Department, M.M.M. Engineering College, Gorakhpur for their helps rendered.

I also thank to all my friends, colleagues and especially to my students for their cooperation, suggestions, recommendations and criticism during whole way of this publication.

— *AUTHOR*

Chapter 1

SIGNALS AND SYSTEMS

1.1 INTRODUCTION

The ideas and techniques associated with the concepts of **"signals and systems"** play a significant role in wide areas of science and technology such as communication engineering, circuit designing, biomedical engineering, energy generation, chemical process control and speech processing *etc*. The **signals,** which are functions of one or more independent variables, contain information about the behaviour or the nature of some physical phenomenon whereas the **systems** manipulate one or more signals (inputs) to extract additional information from them, thereby yielding new signals (outputs) of some desired behaviour. Voltage and current in an electrical circuit, which are functions of time, are examples of signals and a circuit itself is an example of a system, which in this case responds to applied voltage and current.

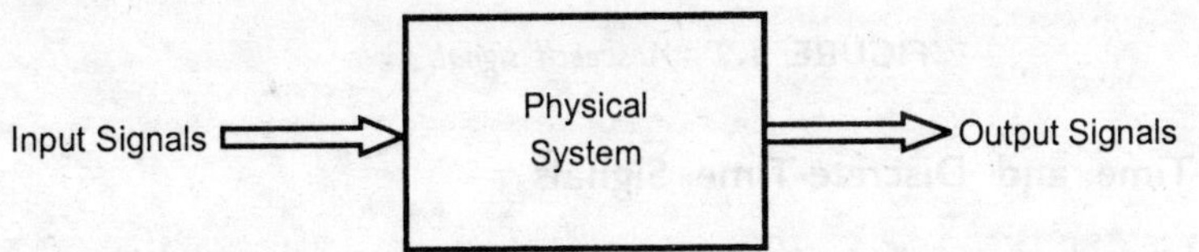

FIGURE 1.1 *Illustration of the interaction of signals in a system.*

The engineers must have a good understanding of the **behaviour** of the physical system under different working conditions before they can start **designing new systems** and / or **redesigning existing ones** in order to improve their performance. The change in the outputs and other internal variables in response to different inputs and initial conditions are called the **"behaviour"** of the physical system and the process of determining how a physical system would behave under different operating conditions is called **"system analysis"**. In order to study the behaviour of a system we must establish the **functional relationships** between different variables. These relationships are given by a set of mathematical equations called the **"mathematical model"** of the system. The interaction of signals in a system is illustrated in Figure 1.1.

1.2 CLASSIFICATION OF SIGNALS

1.2.1 One-Dimensional and Multi-Dimensional Signals

The dimension of a signal is defined as **number of independent variables** on which the value of signal depends. If a signal is a function of single variable, *i.e.*, the value of signal depends on a single variable, the signal is said to be **one-dimensional**. A speech signal is an example of a one-dimensional signal whose amplitude varies with time depending upon the spoken word and who speaks it.

On the other hand, if a signal is a function of two or more variables, *i.e.*, the value of signal depends on two or more variables, the signal is said to be **multi-dimensional**. An image with the horizontal and vertical coordinates representing the two dimensions is an example of a two-dimensional signal.

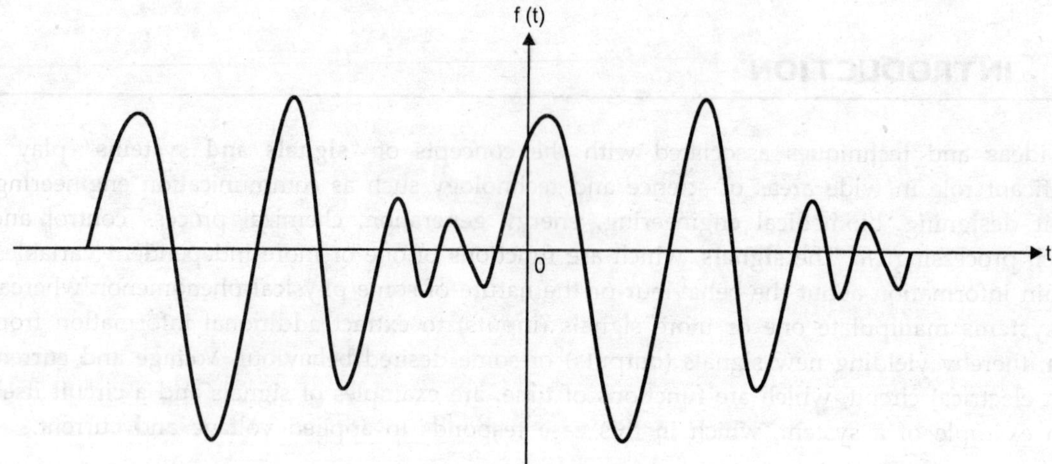

FIGURE 1.2 *A speech signal.*

1.2.2 Continuous-Time and Discrete-Time Signals

A signal which is defined at **every instant of time** is known as a **continuous-time signal** whereas a signal which is defined only at **discrete instants of time** is called a **discrete-time signal.** In case of discrete-time signals, the independent variable, thus, assumes **equally spaced discrete values** only. A discrete-time signal is often **derived** from a continuous-time signal by **sampling** it at a uniform rate. Figure 1.3 illustrates the relationship between a continuous-time signal $f(t)$ and a discrete-time signal $f(k)$. Throughout this book, we shall use the symbol t to denote time for continuous-time signals / functions and symbol k to denote time for discrete-time signals / functions.

1.2.3 Analog and Digital Signals

The terms analog and digital are **often confused** with that of continuous-time and discrete-time. In fact these two are **not the same**. A signal whose amplitude can take on **any value** in a continuous-range is known as an **analog signal**. This means that analog signal amplitude can take on an **infinite number of values**.

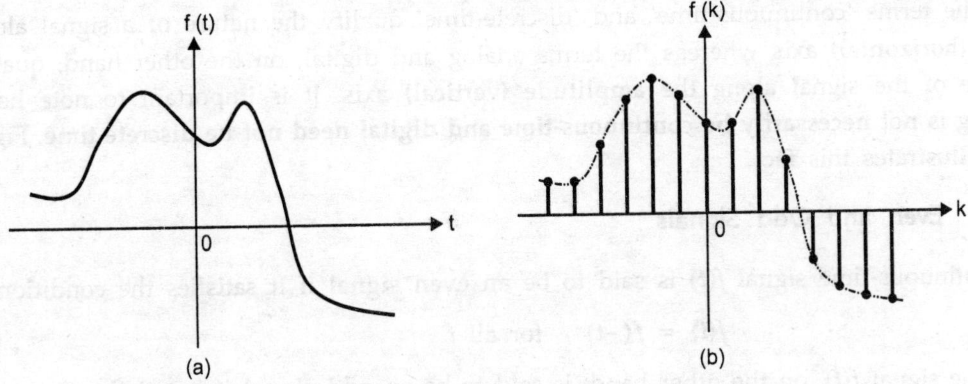

FIGURE 1.3 *Relationship between continuous-time and discrete-time signals*
(a) Continuous-time signal $f(t)$
(b) Discrete-time signal $f(k)$.

A **digital signal**, on the other hand, is one whose amplitude can take on only a **finite number of values**. Signals associated with a digital computer are **digital** because they take on only two values, *i.e.*, **binary signals**. However, for a signal to qualify as digital, the number of values **need not be restricted to two**. It can be any **finite number**.

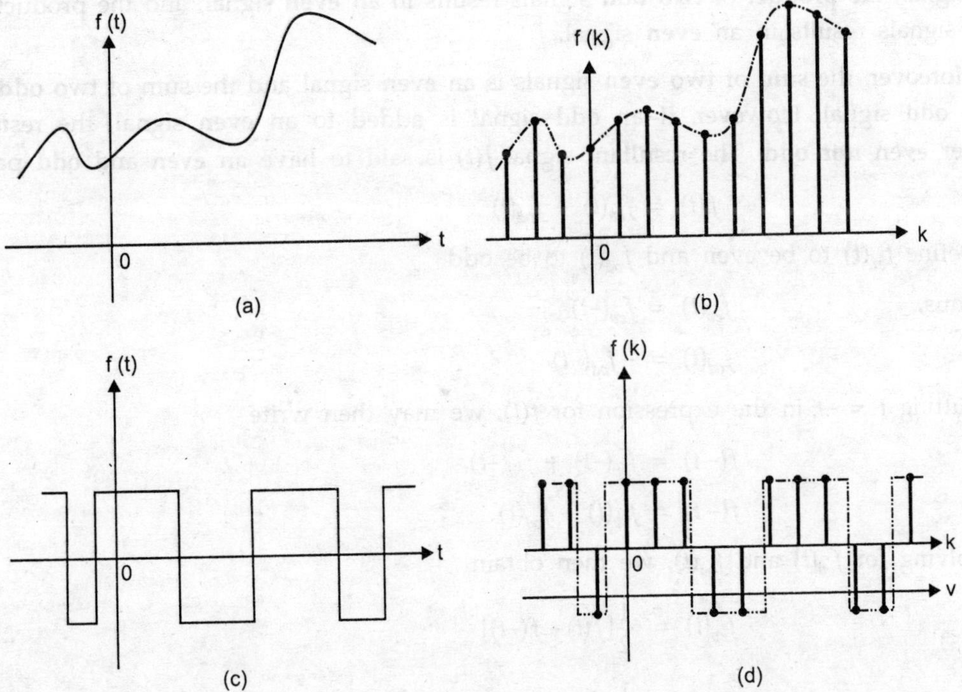

FIGURE 1.4 *Analog is not necessarily be continuous-time and digital need not be discrete-time*
(a) Continuous-time, analog signal $f(t)$ *(b) Discrete-time, analog signal $f(k)$*
(c) Continuous-time, digital signal $f(t)$ *(d) Discrete-time, digital signal $f(k)$.*

The terms 'continuous-time' and 'discrete-time' qualify the nature of a signal along the **time (horizontal) axis** whereas the terms analog and digital, on the other hand, qualify the nature of the signal along the **amplitude (vertical) axis.** It is important to note here that **analog is not necessarily be continuous-time and digital need not be discrete-time.** Figure 1.4 well illustrates this fact.

1.2.4 Even and Odd Signals

A continuous-time signal $f(t)$ is said to be an **even signal** if it satisfies the condition

$$f(t) = f(-t) \quad \text{for all } t \qquad \text{...(1.1 A)}$$

The signal $f(t)$, on the other hand, is said to be an **odd signal** if it satisfies the condition

$$f(t) = -f(-t) \quad \text{for all } t \qquad \text{...(1.1 B)}$$

In other words, we may state that **even signals** are **symmetrical** about **vertical axis** or **time-origin** whereas **odd signals** are **anti-symmetrical** about the **time-origin.** Figure 1.5 well illustrates this fact. A continuous-time, odd signal must necessarily be **zero** at $t = 0$, since equation (1.1 B) requires that $f(0) = -f(0)$.

It is important here to state that the **product** of an **even** and an **odd** signal results in an **odd** signal; the product of **two odd** signals results in an **even** signal; and the product of **two even** signals results in an **even** signal.

Moreover, the **sum** of **two even** signals is an **even** signal and the sum of **two odd** signals is an **odd** signal. However, if an **odd** signal is added to an **even** signal, the resultant is **neither even nor odd**. The resultant signal $f(t)$ is said to have an **even** and **odd** part, *i.e.,*

$$f(t) = f_{ev}(t) + f_{od}(t) \qquad \text{...(1.2)}$$

Define $f_{ev}(t)$ to be even and $f_{od}(t)$ to be odd.

Thus,
$$f_{ev}(t) = f_{ev}(-t)$$
and
$$f_{od}(t) = -f_{od}(-t)$$

Putting $t = -t$ in the expression for $f(t)$, we may then write

$$f(-t) = f_{ev}(-t) + f_{od}(-t)$$
or
$$f(-t) = f_{ev}(t) - f_{od}(t)$$

Solving for $f_{ev}(t)$ and $f_{od}(t)$, we then obtain

$$f_{ev}(t) = \frac{1}{2}[f(t)+f(-t)] \qquad \text{...(1.3 A)}$$

and
$$f_{od}(t) = \frac{1}{2}[f(t)-f(-t)] \qquad \text{...(1.3 B)}$$

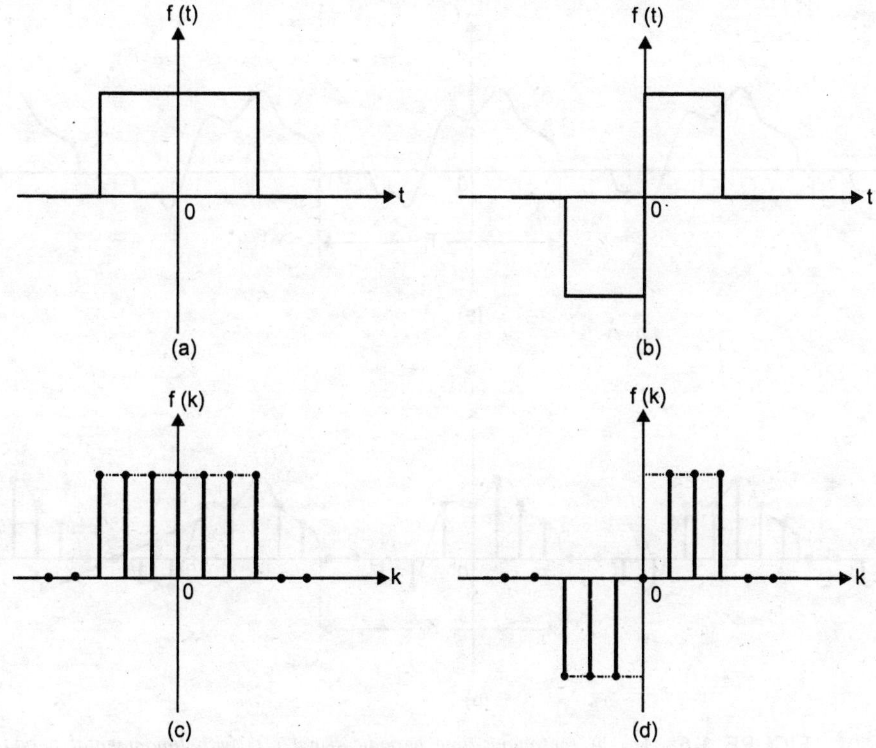

FIGURE 1.5 (*a*) *Continuous-time, even signal*
(*b*) *Continuous-time, odd signal*
(*c*) *Discrete-time, even signal*
(*d*) *Discrete-time, odd signal.*

In an analogous manner, a discrete-time signal $f(k)$ is said to be an **even signal** if it satisfies the condition

$$f(k) = f(-k) \qquad \text{for all } k \qquad \text{...(1.4 A)}$$

The signal $f(k)$, on the other hand, is said to be an **odd signal** if it satisfies the condition

$$f(k) = -f(-k) \qquad \text{for all } k \qquad \text{...(1.4 B)}$$

A discrete-time, odd signal must necessarily be **zero** at $k = 0$, since equation (1.4 B) requires that $f(0) = -f(0)$.

1.2.5 Periodic and Non-Periodic Signals

A continuous-time signal $f(t)$ is said to be **periodic** if for some positive constant T_0

$$f(t) = f(t + T_0) \qquad \text{for all } t \qquad \text{...(1.5)}$$

If, however, no value of T_0 exists to satisfy the condition above, the signal $f(t)$ is called a **non-periodic signal**. Figure 1.6 (*a*) shows an example of a **periodic continuous-time signal** with period T_0.

The smallest value of T_0 that satisfies periodicity condition (1.5) is called the **fundamental period** of $f(t)$. Accordingly, the fundamental period T_0 defines the **duration** of one complete cycle of $f(t)$.

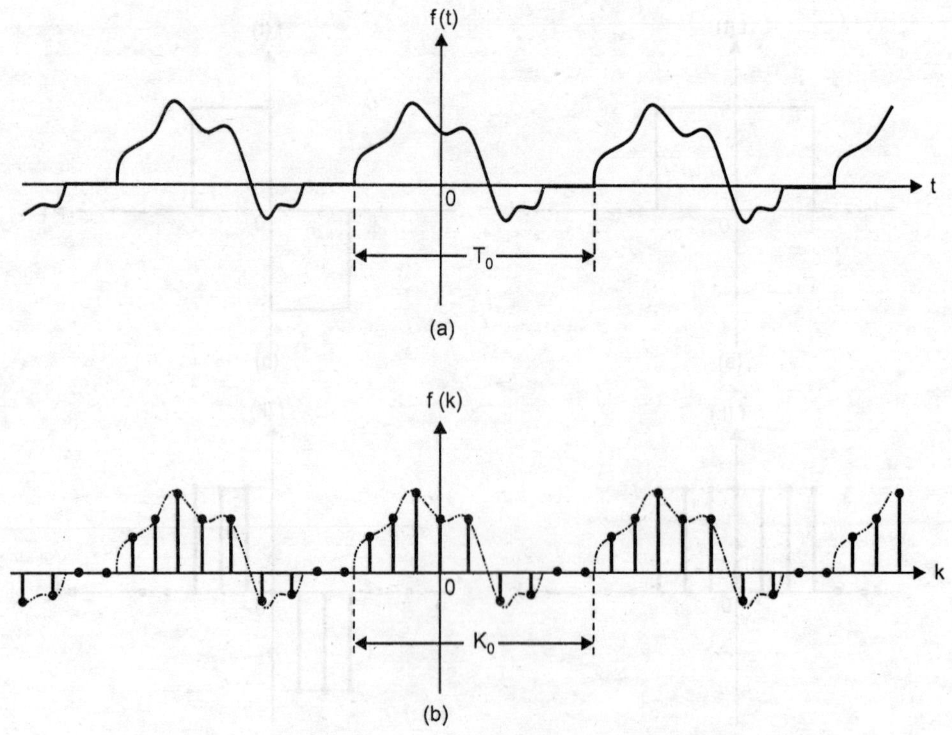

FIGURE 1.6 *(a) A continuous-time periodic signal $f(t)$ with fundamental period T_0*
(b) A discrete-time periodic signal $f(k)$ with fundamental period K_0.

The **reciprocal** of the **fundamental period** T_0 is called the **fundamental frequency** of the periodic signal $f(t)$ which is measured in hertz (Hz) or cycles per second and describes **how frequently** the periodic signal $f(t)$ **repeats** itself. Thus, we have

$$f_0 = \frac{1}{T_0} \qquad \text{...(1.6 A)}$$

Since there are 2π radians in one complete cycle, we can define the **angular frequency** which is measured in radians per second as

$$\omega_0 = \frac{2\pi}{T_0} \qquad \text{...(1.6 B)}$$

To simplify the terminology, ω_0 is often referred to simply as **frequency.**

By definition, a periodic signal $f(t)$ **remains unchanged** when time is **shifted** by one period. This means that a periodic signal must start at $t = -\infty$ because if it starts at some finite instant, say $t = 0$, the time-shifted signal $f(t + T_0)$ will start at $t = -T_0$ and $f(t + T_0)$ would not be the same as $f(t)$. Therefore, a periodic signal, by definition, **must start** at $t = -\infty$ and **continue forever**, as shown in Figure 1.6 *(a)*.

The second important property of a periodic signal $f(t)$ is that $f(t)$ can be generated by **periodic extension** of any segment of $f(t)$ of duration equal to **fundamental period** T_0. This means that we can generate $f(t)$ from any segment of $f(t)$ with duration of one period by placing this segment and the reproduction thereof **end to end repeatedly** on either side as illustrated in Figure 1.7.

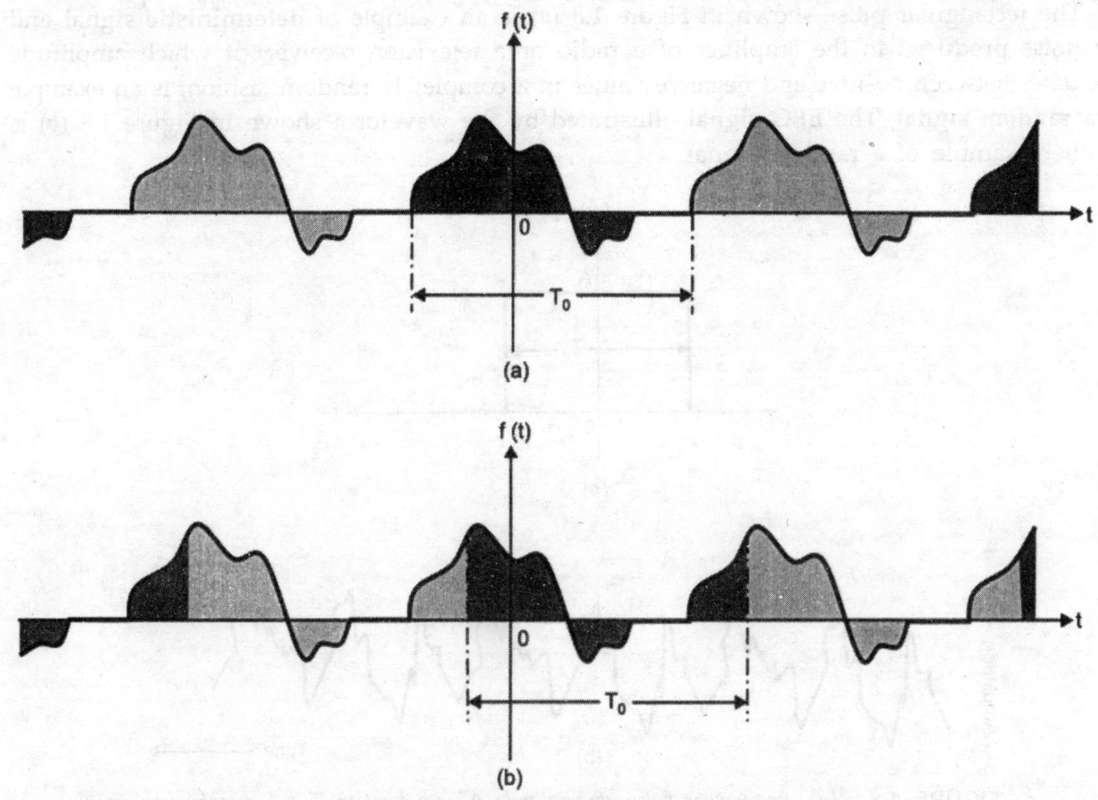

FIGURE 1.7 *Generation of a continuous-time periodic signal f(t) by periodic extension of any segment of duration T_0.*

In an analogous manner, a discrete-time signal $f(k)$ is said to be **periodic** if for some positive integer K_0

$$f(k) = f(k + K_0) \quad \text{for all } k. \qquad \text{...(1.7)}$$

If, however, no value of K_0 exists to satisfy the condition above, the signal $f(k)$ is called a **non-periodic signal**. Figure 1.6 (*b*) shows an example of a **periodic discrete-time signal** with period K_0.

The smallest value of K_0 that satisfies periodicity condition (1.7) is called the **fundamental period** of $f(k)$. Accordingly, the fundamental period K_0 defines the **duration** of one complete cycle of $f(k)$.

1.2.6 Deterministic and Random Signals

A signal having **no uncertainty** about it with respect to its value at any time is referred to as **deterministic signal**. On the other hand, a signal about which there is **uncertainty** before its actual occurrence is referred to as **random signal**. Such a signal may be viewed as belonging to a group of signals, with each signal in the group having a **different waveform**.

A deterministic signal may be modelled as completely specified function of time, whereas a random signal may be represented by its **probabilistic model** as each signal within the group has a certain **probability of occurrence**.

The rectangular pulse shown in Figure 1.8 (*a*) is an example of **deterministic signal** and the noise produced in the amplifier of a radio or a television receiver, of which amplitude fluctuates between positive and negative values in a completely random fashion, is an example of a **random signal**. The **EEG signal**, illustrated by the waveform shown in Figure 1.8 (*b*) is another example of a random signal.

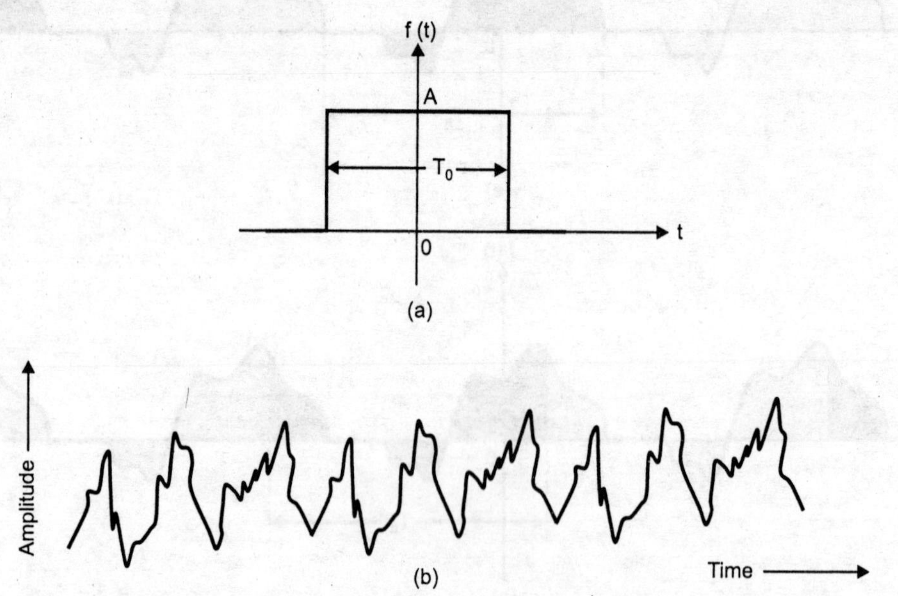

FIGURE 1.8 (*a*) *Rectangular pulse of amplitude A and duration* T_0*: A deterministic signal* (*b*) *Trace of EEG signal : A random signal.*

1.2.7 Energy and Power Signals

In many applications, the signals interacting in a physical system are directly associated to physical quantities confining **power** and **energy**. The **total energy** of the **continuous-time signal** $f(t)$ over the time interval $t_1 \le t \le t_2$ is defined as

$$E \overset{\Delta}{=} \int_{t_1}^{t_2} |f(t)|^2 \, dt \quad \text{...(1.8 A)}$$

and the **time-averaged power** of the same over the time interval $t_1 \le t \le t_2$ is defined as

$$P \overset{\Delta}{=} \frac{1}{t_2 - t_1} \int_{t_1}^{t_2} |f(t)|^2 \, dt \quad \text{...(1.8 B)}$$

In an analogous fashion, the **total energy** of a **discrete-time signal** $f(k)$ over the time interval $k_1 \le k \le k_2$ is defined as

$$E \overset{\Delta}{=} \sum_{k=k_1}^{k_2} |f(k)|^2 \quad \text{...(1.9 A)}$$

and the **time-averaged power** of the same over the time interval $k_1 \le k \le k_2$ is defined as

$$P \stackrel{\Delta}{=} \frac{1}{k_2 - k_1 + 1} \sum_{k=k_1}^{k_2} |f(k)|^2 \qquad \text{...(1.9 B)}$$

In many physical systems, however, we are interested in examining power and energy of continuous-time signal $f(t)$ or discrete-time signal $f(k)$ over an **infinite time interval** $-\infty < t < +\infty$ or $-\infty < k < +\infty$ respectively. In these cases, we define the total energy and time-averaged power as **limits** of equation (1.8) and (1.9) when the time interval increases **without bound**. Thus, we define the **total energy** of **continuous-time signal** $f(t)$ over an **infinite time interval** $-\infty < t < +\infty$ as

$$E_\infty \stackrel{\Delta}{=} \lim_{T_0 \to \infty} \left[\int_{-T_0}^{T_0} |f(t)|^2 \, dt \right]$$

or

$$E_\infty = \int_{-\infty}^{\infty} |f(t)|^2 \, dt \qquad \text{...(1.10 A)}$$

and the **time-averaged power** of the same over an **infinite time interval** $-\infty < t < +\infty$ as

$$P_\infty \stackrel{\Delta}{=} \lim_{T_0 \to \infty} \left[\frac{1}{2T_0} \int_{-T_0}^{T_0} |f(t)|^2 \, dt \right] \qquad \text{...(1.10 B)}$$

In an analogous fashion, we can define the **total energy** of a **discrete-time signal** $f(k)$ over an **infinite time interval** $-\infty < k < +\infty$ as

$$E_\infty \stackrel{\Delta}{=} \lim_{K_0 \to \infty} \left[\sum_{k=-K_0}^{K_0} |f(k)|^2 \right]$$

or

$$E_\infty = \sum_{k=-\infty}^{\infty} |f(k)|^2 \qquad \text{...(1.11 A)}$$

and the **time-averaged power** of the same over an **infinite time interval** $-\infty < k < +\infty$ as

$$P_\infty \stackrel{\Delta}{=} \lim_{K_0 \to \infty} \left[\frac{1}{2K_0 + 1} \sum_{k=-K_0}^{K_0} |f(k)|^2 \right] \qquad \text{...(1.11 B)}$$

Note that for some signals the integral in equation (1.10) or the sum in equation (1.11) **might not converge.** Such signals have **infinite energy**, while signals with $E_\infty < \infty$ have **finite energy**.

A signal is referred to as an **energy signal**, if and only if the total energy of the signal satisfies the condition

$$0 < E_\infty < \infty.$$

On the other hand, a signal is referred to as a **power signal**, if and only if the time-averaged power of the signal satisfies the condition

$$0 < P_\infty < \infty.$$

The energy and power classification of signals are **mutually exclusive.** Therefore, a signal cannot **simultaneously** be energy and power signal. In particular, an **energy signal** must have **zero average power,** since in the continuous-time case, for example, we see from equation (1.10 B) that

$$P_{\infty} = \lim_{T_0 \to \infty} \left[\frac{E_{\infty}}{2T_0} \right] = 0.$$

On the other hand, a **power signal** has **infinite energy.**

A continuous-time signal,

$$f(t) = \begin{cases} 1 \ ; & 0 \le t \le 1 \\ 0 \ ; & \text{Otherwise} \end{cases}$$

is an example of a **finite energy signal** since for the case $E_{\infty} = 1$ and $P_{\infty} = 0$.

Again, a discrete-time signal defined as

$$f(k) = 4 \qquad \text{for all } k$$

is an example of a **power signal** since it has **infinite energy but finite average power** $P_{\infty} = 16$.

There are also signals for which neither P_{∞} nor E_{∞} are finite, *e.g.*, a ramp signal defined by $f(t) = t$, is **neither an energy signal nor a power signals.**

It is also of interest to note that signals that are both **deterministic** and **non-periodic** are **energy signals,** whereas because of periodic repetition, periodic signals for which the area under $|f(t)|^2$ or $|f(k)|^2$ over one period is finite are usually viewed as **power signals;** however, **not all power signals are periodic.**

1.3 SIGNAL OPERATIONS

1.3.1 Reflection

A continuous-time signal $y(t)$ obtained by **replacing** time t with $-t$, such that

$$y(t) = f(-t) \qquad \text{...(1.12 A)}$$

represents a **reflected version** of $f(t)$ about the amplitude axis.

Similarly, a discrete-time signal $y(k)$ obtained by **replacing** time k with $-k$, such that

$$y(k) = f(-k) \qquad \text{...(1.12 B)}$$

represents a **reflected version** of $f(k)$ about the amplitude axis.

It is of special interest to note that an **even signal** is the **same** as its **reflected version.** On the other hand, **odd signal** is the **negative** of its **reflected version.**

1.3.2 Addition

A continuous-time signal $y(t)$ resulting from the **addition** of a pair of continuous-time signals $f_1(t)$ and $f_2(t)$ is defined as

$$y(t) = f_1(t) + f_2(t) \qquad \text{...(1.13 A)}$$

e.g., An **audio mixer** that combines music and voice signals.

Similarly, a discrete-time signal $y(k)$ resulting from the **addition** of a pair of discrete-time signals $f_1(k)$ and $f_2(k)$ is defined as

$$y(k) = f_1(k) + f_2(k) \qquad \text{...(1.13 B)}$$

1.3.3 Multiplication

A continuous-time signal $y(t)$ resulting from the **multiplication** of a pair of continuous-time signals $f_1(t)$ and $f_2(t)$ is defined as

$$y(t) = f_1(t) \cdot f_2(t) \qquad \text{...(1.14 A)}$$

e.g., An **amplitude modulated radio signal** which is obtained by **multiplication** of an **audio signal** plus a **d.c. component** and a high frequency sinusoidal signal called a **carrier wave.**

Similarly, a discrete-time signal $y(k)$ resulting from the **multiplication** of a pair of discrete-time signals $f_1(k)$ and $f_2(k)$ is defined as

$$y(k) = f_1(k) \cdot f_2(k) \qquad \text{...(1.14 B)}$$

1.3.4 Amplitude Scaling

A continuous-time signal $y(t)$ resulting from the **amplitude scaling** by a factor α applied to continuous-time signal $f(t)$ is defined as

$$y(t) = \alpha\, f(t) \qquad \text{...(1.15 A)}$$

e.g., An **electronic amplifier** that amplifies the amplitude of the input signals.

Similarly, a discrete-time signal $y(k)$ resulting from the **amplitude scaling** by a factor α applied to discrete-time signal $f(k)$ is defined as

$$y(k) = \alpha\, f(k) \qquad \text{...(1.15 B)}$$

1.3.5 Time Scaling

A continuous-time signal $y(t)$ resulting from the **time scaling** by a factor β applied to continuous-time signal $f(t)$ is defined as

$$y(t) = f(\beta t) \qquad \text{...(1.16 A)}$$

If $\beta > 1$, the signal $y(t)$ is referred to as a **compressed version** of the signal $f(t)$. On the other hand, when $0 < \beta < 1$, the signal $y(t)$ is referred to as an **expanded** or **stretched version** of the signal $f(t)$. Figure 1.9 well illustrates these operations.

In an analogous fashion, we can define the **time scaled version** of a **discrete-time signal** $f(k)$ as follows

$$y(k) = f(mk)\ ; \qquad m > 0 \qquad \text{...(1.16 B)}$$

which is defined only for **integer** values of m. If $m > 1$, then some values of the discrete-time signal $y(k)$ are **lost**.

1.3.6 Time Shifting

A continuous-time signal $y(t)$ resulting from the **shifting in time** by τ applied to continuous-time signal $f(t)$ is defined as

$$y(t) = f(t - \tau) \qquad \text{...(1.17 A)}$$

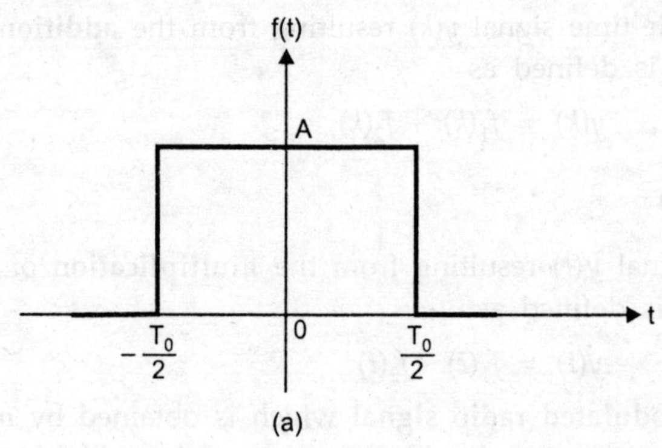

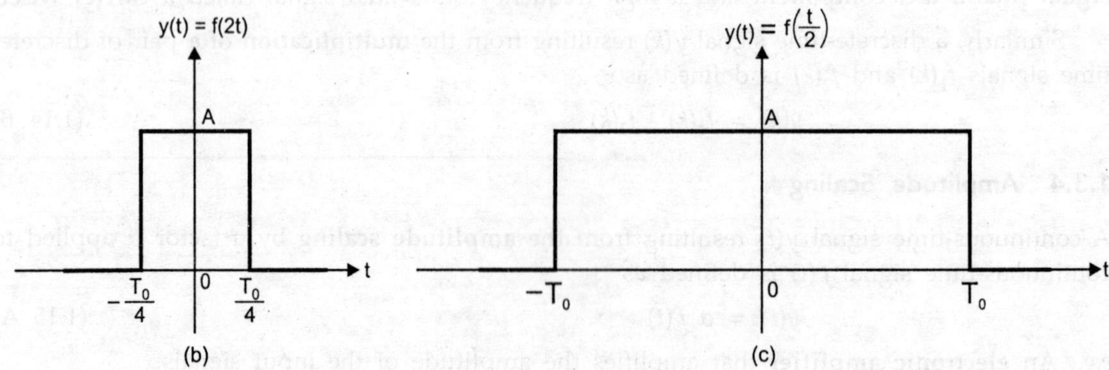

FIGURE 1.9 *Illustration of the time scaling operation :*
(a) Continuous-time signal f(t)
(b) Compressed version of f(t) by a factor of 2
(c) Stretched version of f(t) by a factor of 2.

If $\tau > 0$, the waveform representing $f(t)$ is shifted to the **right** along the time axis. On the other hand, when $\tau < 0$, the waveform representing $f(t)$ is shifted to the **left** along the time axis. Figure 1.10 well illustrates the fact.

In an analogous fashion, we can define the **time shifted version** of a **discrete-time signal** $f(k)$ as follows

$$y(k) = f(k - n) \qquad \text{...(1.17 B)}$$

which is defined only for **integer (either positive or negative)** values of n.

1.3.7 Joint Operation of Time Shifting and Time Scaling

A continuous-time signal $y(t)$ resulting from a **joint operation** of **time scaling** by a factor β and **shifting in time** by τ applied to continuous-time signal $f(t)$ is defined as

$$y(t) = f(\beta t - \tau) \qquad \text{...(1.18 A)}$$

In order to correctly obtain $y(t)$ from $f(t)$, the **time shifting** and **time scaling** operations must be performed in a **proper order**. It is suggested that the **time shifting operation should be performed first** on $f(t)$. This replaces t by $t - \tau$ resulting in an **intermediate signal** $\tilde{f}(t)$ defined by

$$\tilde{f}(t) = f(t - \tau)$$

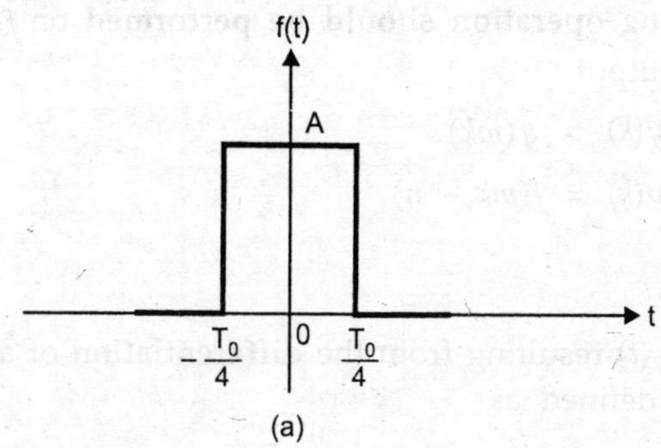

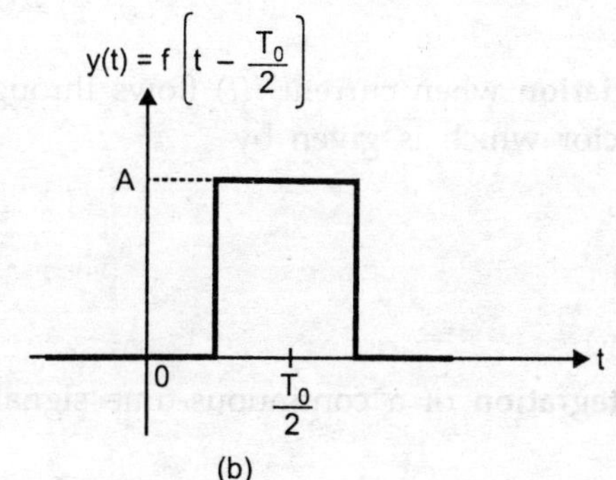

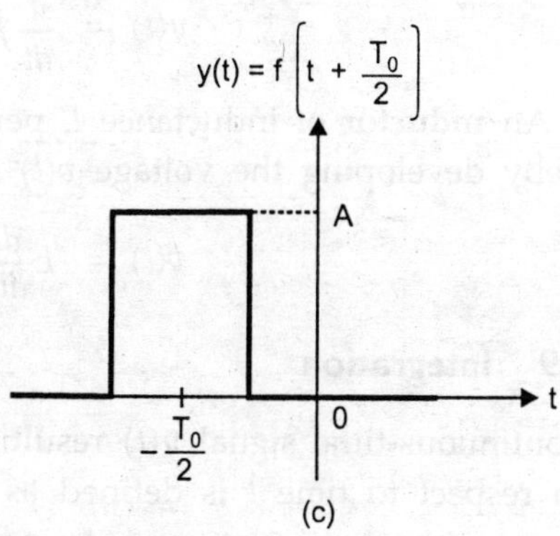

FIGURE 1.10 *Illustration of time shifting operation :*
(a) Continuous-time signal f(t)
(b) Time shifted version of f(t) by $\tau = T_0/2$
(c) Time shifted version of f(t) by $\tau = -T_0/2$.

Afterward, the time scaling operation should be performed on $\tilde{f}(t)$. This replaces t by βt, resulting in the desired output

$$y(t) = \tilde{f}(\beta t)$$

or,

$$y(t) = f(\beta t - \tau)$$

In an analogous fashion, we can define the **joint operation** of **time scaling** and **time shifting** applied to **discrete-time signal** $f(k)$ as follows

$$y(k) = f(mk - n) \quad ...(1.18\text{ B})$$

which is defined only for **integer (only positive)** values of m and **integer (either positive or negative)** values of n.

In order to correctly obtain $y(k)$ from $f(k)$, the **time shifting** and **time scaling** operations must be performed in a proper order. It is suggested that the **time shifting operation should be performed first** on $f(k)$. This replaces k by $k - n$ resulting in an **intermediate signal** $\tilde{f}(k)$ defined by

$$\tilde{f}(k) = f(k - n)$$

Afterward, the time scaling operation should be performed on $\tilde{f}(k)$. This replaces k by mk, resulting in the desired output

$$y(k) = \tilde{f}(mk)$$

or,

$$y(k) = f(mk - n)$$

1.3.8 Differentiation

A continuous-time signal $y(t)$ resulting from the **differentiation** of a continuous-time signal $f(t)$ with respect to time t is defined as

$$y(t) = \frac{d}{dt} f(t) \qquad ...(1.19)$$

e.g., An **inductor** of inductance L performs **differentiation** when current $i(t)$ flows through it thereby developing the voltage $v(t)$ across the inductor which is given by

$$v(t) = L\frac{d}{dt} i(t)$$

1.3.9 Integration

A continuous-time signal $y(t)$ resulting from the **integration** of a continuous-time signal $f(t)$ with respect to time t is defined as

$$y(t) = \int_{-\infty}^{t} f(\tau)\, d\tau \qquad ...(1.20)$$

e.g., A **capacitor** of capacitance C performs **integration** when current $i(t)$ flows through it thereby developing the voltage $v(t)$ across the capacitor which is given by

$$v(t) = \frac{1}{C} \int_{-\infty}^{t} i(\tau)\, d\tau$$

1.4 BASIC SIGNALS

There are several **basic signals** such as step signals, impulse signals, ramp signals, exponential and sinusoidal signals, *etc.*, that feature significantly in the study of signals and systems. Not only do these signals occur frequently, but they also serve as **basic building blocks** for the construction of more complex signals. In following sections, we shall describe the above mentioned basic signals, one by one.

1.4.1 Unit-Step Signal

The **continuous-time unit-step signal** is defined as

$$u(t) = \begin{cases} 1 \ ; & t \geq 0 \\ 0 \ ; & t < 0 \end{cases} \qquad ...(1.21)$$

It is said to have **discontinuity** at $t = 0$, since the value of $u(t)$ changes **instantaneously** from 0 to 1. A continuous-time unit-step signal is shown in Figure 1.11 (*a*).

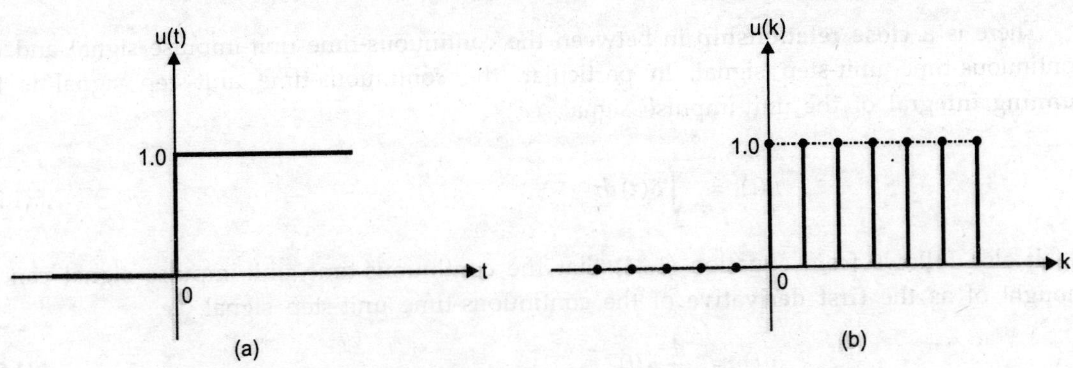

FIGURE 1.11 *Unit-step signals :*
(a) Continuous-time unit-step signal u(t)
(b) Discrete-time unit-step signal u(k).

In an analogous manner, we can define the **discrete-time unit-step signal** as

$$u(k) = \begin{cases} 1 & ; \; k \geq 0 \\ 0 & ; \; k < 0 \end{cases} \qquad \text{...(1.22)}$$

A discrete-time unit-step signal is shown in Figure 1.11 (*b*).

1.4.2 Unit-Impulse Signal

The **continuous-time unit-impulse signal** (also called **Dirac-Delta signal**, named after **P.A.M. Dirac**) is one of the simplest continuous-time signal and is defined by a **pair** of equations:

$$\delta(t) = 0 \qquad ; \; t \neq 0 \qquad \text{...(1.23 A)}$$

and

$$\int_{-\infty}^{\infty} \delta(t)\, dt = 1 \qquad \text{...(1.23 B)}$$

A continuous-time unit-impulse signal is shown in Figure 1.12 (*a*).

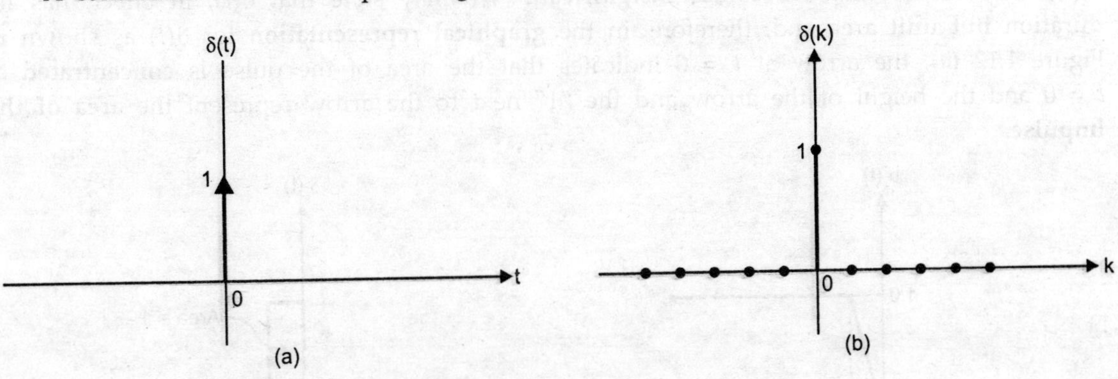

FIGURE 1.12 *Unit-impulse signals :*
(a) Continuous-time unit-impulse signal δ(t)
(b) Discrete-time unit-impulse signal δ(k).

There is a **close relationship** in between the continuous-time unit-impulse signal and the continuous-time unit-step signal. In particular, the continuous-time unit-step signal is the **running integral** of the unit-impulse signal, *i.e.*,

$$u(t) = \int_{-\infty}^{t} \delta(\tau)\, d\tau \qquad \text{...(1.24)}$$

It also follows from equation (1.24) that the continuous-time unit-impulse signal can be thought of as the **first derivative** of the continuous-time unit-step signal, *i.e.*,

$$\delta(t) = \frac{d}{dt} u(t) \qquad \text{...(1.25)}$$

Since the value of $u(t)$ changes **instantaneously** from 0 to 1, it is said to have **discontinuity** at $t = 0$ and, therefore, there is some **formal difficulty** with equation (1.25) as a representation of the unit-impulse signal. We, however, can interpret equation (1.25) by considering an approximated unit-step signal denoted by $u_\varepsilon(t)$ such that it rises from 0 to 1 in an **infinitesimal** time interval of length ε as illustrated in Figure 1.13 (*a*). It, therefore, follows that unit-step signal $u(t)$ is the **limit** of $u_\varepsilon(t)$ as $\varepsilon \to 0$, *i.e.*,

$$u(t) = \lim_{\varepsilon \to 0} u_\varepsilon(t) \qquad \text{...(1.26)}$$

Now, the derivative of $u_\varepsilon(t)$ is given by

$$\delta_\varepsilon(t) = \frac{d}{dt} u_\varepsilon(t) \qquad \text{...(1.27)}$$

where, $\delta_\varepsilon(t)$ is a short pulse of duration ε and with **unit area,** as shown in Figure 1.13 (*b*).

Now, one can visualize $\delta(t)$ in the **limiting form** as

$$\delta(t) = \lim_{\varepsilon \to 0} \delta_\varepsilon(t)$$

Thus, the unit-impulse signal $\delta(t)$ can be thought of as an **idealization** of the short pulse $\delta_\varepsilon(t)$ in the sense that ε becomes **insignificant**. We may state that $\delta(t)$, in effect, has **no duration but unit area** and, therefore, in the **graphical representation** for $\delta(t)$ as shown in Figure 1.12 (*a*), the arrow at $t = 0$ indicates that the area of the pulse is **concentrated** at $t = 0$ and the height of the arrow and the "1" next to the arrow represent the **area of the impulse**.

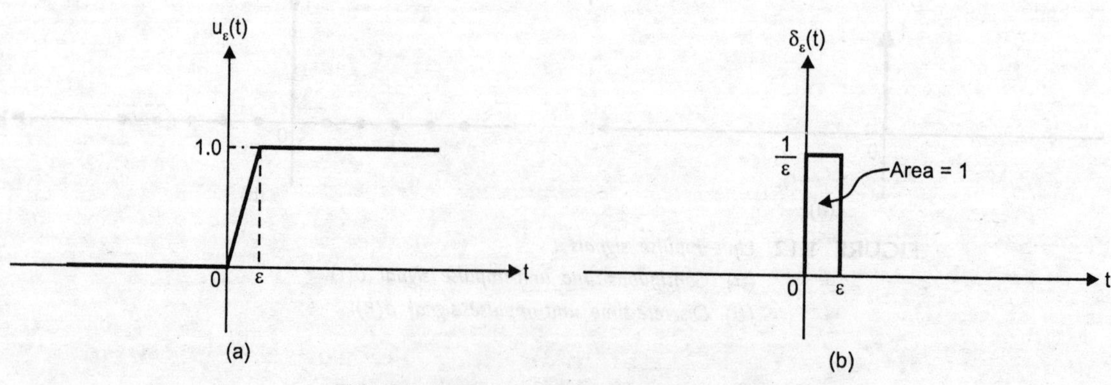

FIGURE 1.13 (*a*) *Approximated unit-step signal* $u_\varepsilon(t)$
(*b*) *Derivative of* $u_\varepsilon(t)$.

It is also of special interest to consider what happens when any continuous-time signal $f(t)$ is **multiplied** by the continuous-time unit-impulse signal $\delta(t)$. Since the impulse exists only at $t = 0$ and the value of $f(t)$ at $t = 0$ is $f(0)$, we may write

$$f(t)\ \delta(t) = f(0)\ \delta(t) \qquad ...(1.28\ A)$$

Similarly, if any continuous-time signal $f(t)$ is **multiplied** by an impulse $\delta(t - T_0)$ which is located at $t = T_0$, then we have

$$f(t)\ \delta(t - T_0) = f(T_0)\ \delta(t - T_0) \qquad ...(1.28\ B)$$

Now, we define the **discrete-time unit-impulse signal** as

$$\delta(k) = \begin{cases} 1 \ ; & k = 0 \\ 0 \ ; & k \neq 0 \end{cases} \qquad ...(1.29)$$

A discrete-time unit-impulse signal is shown in Figure 1.12 (*b*).

There is also a **close relationship** in between the discrete-time unit-impulse signal and the discrete-time unit-step signal. In particular, the discrete-time unit impulse signal is the **first difference** of the discrete-time step signal, *i.e.,*

$$\delta(k) = u(k) - u(k - 1) \qquad ...(1.30)$$

Further, since the impulse exists only at $k = 0$ and the value of any discrete-time signal $f(k)$ at $k = 0$ is $f(0)$, it follows that

$$f(k)\ \delta(k) = f(0)\ \delta(k) \qquad (1.31\ A)$$

Similarly, if any discrete-time signal $f(k)$ is **multiplied** by an impulse $\delta(k - K_0)$ which is located at $k = K_0$, then we have

$$f(k)\ \delta(k - K_0) = f(K_0)\ \delta(k - K_0) \qquad ...(1.31\ B)$$

1.4.3 Unit-Ramp Signal

The **continuous-time unit-ramp signal** is defined as

$$r(t) = \begin{cases} t \ ; & t \geq 0 \\ 0 \ ; & t < 0 \end{cases} \qquad ...(1.32)$$

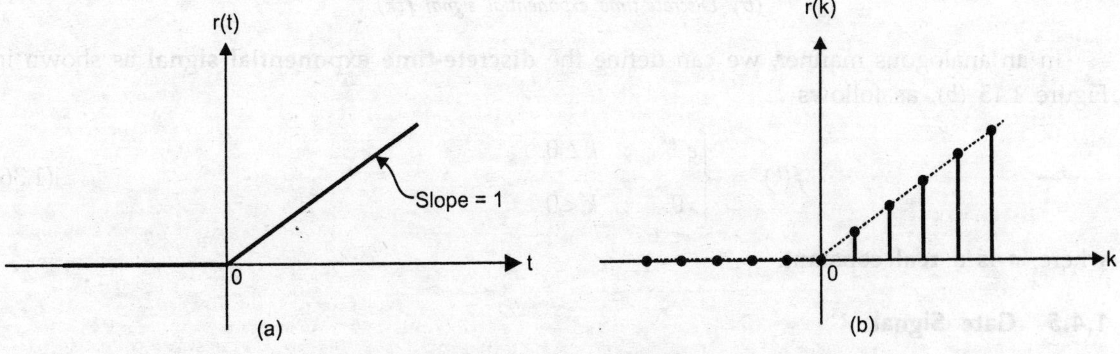

FIGURE 1.14 *Unit-ramp signals :*
(*a*) *Continuous-time unit-ramp signal r(t)*
(*b*) *Discrete-time unit-ramp signal r(k).*

Equivalently, we may write

$$r(t) = t\,u(t) \qquad ...(1.33)$$

A continuous-time unit-ramp signal is shown in Figure 1.14 (*a*).

In an analogous manner, we can define the **discrete-time unit-ramp signal** as

$$r(k) = \begin{cases} k & ; \quad k \geq 0 \\ 0 & ; \quad k < 0 \end{cases} \qquad ...(1.34)$$

Equivalently, we may write

$$r(k) = k\,u(k) \qquad ...(1.35)$$

A discrete-time unit-ramp signal is shown in Figure 1.14 (*b*).

1.4.4 Exponential Signal

The **continuous-time exponential signal** as shown in Figure 1.15 (*a*) is defined as

$$f(t) = \begin{cases} e^{-at} & ; \quad t \geq 0 \\ 0 & ; \quad t < 0 \end{cases}$$

where, a is a **real constant.**

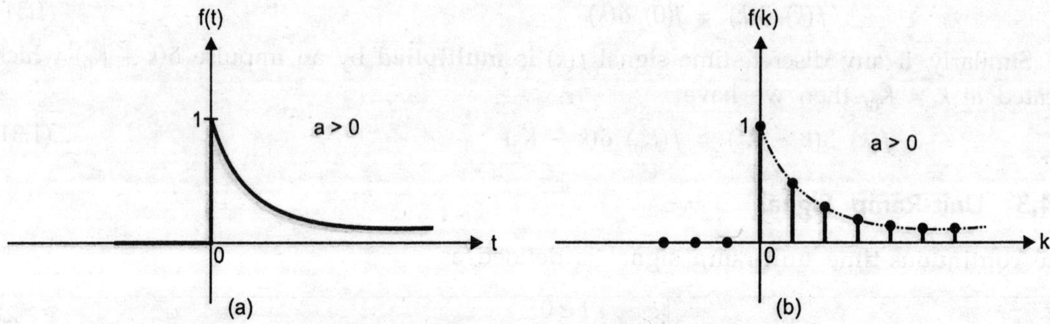

FIGURE 1.15 *Exponential signals*
(*a*) *Continuous-time exponential signal f(t)*
(*b*) *Discrete-time exponential signal f(k).*

In an analogous manner, we can define the **discrete-time exponential signal** as shown in Figure 1.15 (*b*), as follows

$$f(k) = \begin{cases} e^{-ak} & ; \quad k \geq 0 \\ 0 & ; \quad k < 0 \end{cases} \qquad ...(1.36)$$

where, a is a **real constant.**

1.4.5 Gate Signal

The **gate signal** is defined as

$$G_{T_1, T_2}(t) = u(t - T_1) - u(t - T_2) \qquad ...(1.37)$$

and is illustrated in Figure 1.16. This signal represents a **rectangular pulse of unit height** starting at $t = T_1$ and ends at $t = T_2$.

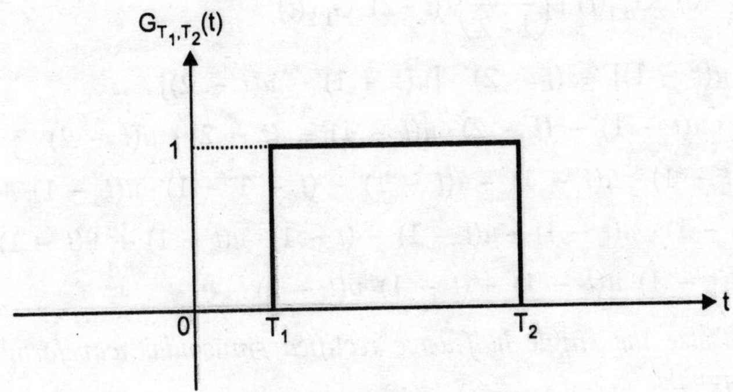

FIGURE 1.16 *Gate signal.*

The value of the signal within the gate remains unaffected. Thus, the gate signal provides a **gateway** through which only those values of the signal are **permitted to pass** that exist for the range $T_1 \leq t \leq T_2$.

The usefulness of the gate lies in the ease with which the unwanted portion of a continuous-time signal can be **omitted** without tedious graphical composition. Such signal is very useful in formation of complex waveforms.

Example 1.1. *Synthesize the triangular waveform as shown in Figure 1.17 in terms of basic signals.*

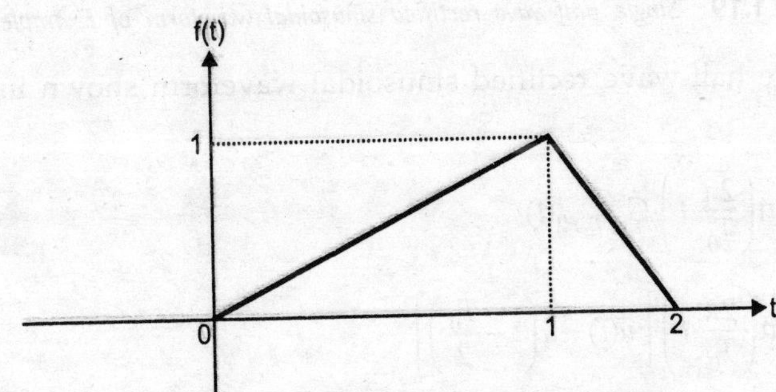

FIGURE 1.17 *Triangular waveform of Example 1.1.*

Solution. Before writing the synthesis equation, we first redraw the triangular waveform indicating the coordinates of every point as shown in Figure 1.18.

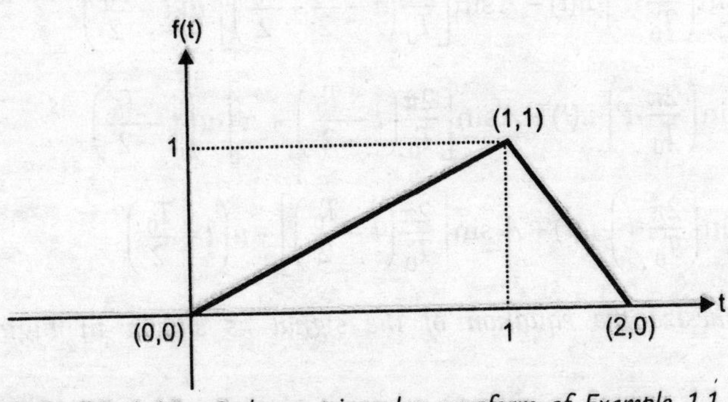

FIGURE 1.18 *Redrawn triangular waveform of Example 1.1.*

The triangular waveform shown in Figure 1.18 can now be expressed as follows:

$$f(t) = \left(\frac{1-0}{1-0}\right)\cdot(t-0)\cdot G_{0,1}(t) + \left(\frac{1-0}{1-2}\right)\cdot(t-2)\cdot G_{1,2}(t)$$

$$= t\cdot[u(t) - u(t-1)] - (t-2)\cdot[u(t-1) - u(t-2)]$$

$$= t\cdot u(t) - t\cdot u(t-1) - (t-2)\cdot u(t-1) + (t-2)\cdot u(t-2)$$

$$= t\cdot u(t) - (t-1)\cdot u(t-1) - u(t-1) - (t-1-1)\cdot u(t-1) + (t-2)\cdot u(t-2)$$

$$= t\cdot u(t) - (t-1)\cdot u(t-1) - u(t-1) - (t-1)\cdot u(t-1) + u(t-1) + (t-2)\cdot u(t-2)$$

$$= t\cdot u(t) - 2(t-1)\cdot u(t-1) + (t-1)\cdot u(t-2)$$

Example 1.2. *Synthesize the single half-wave rectified sinusoidal waveform as shown in Figure 1.19 in terms of basic signals.*

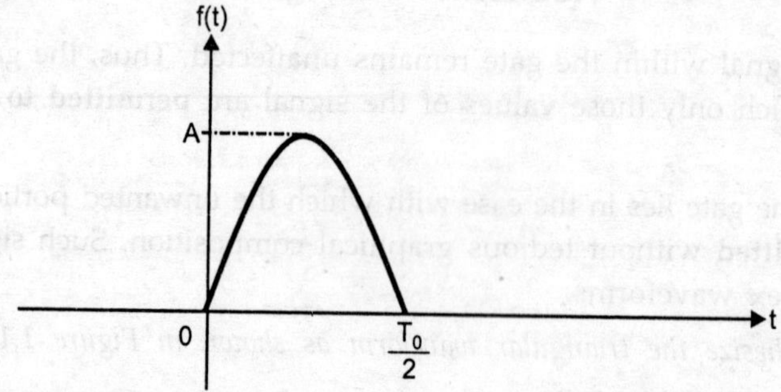

FIGURE 1.19 *Single half-wave rectified sinusoidal waveform of Example 1.2.*

Solution. The single half-wave rectified sinusoidal waveform shown in Figure 1.19 can be expressed as follows:

$$f(t) = A\sin\left(\frac{2\pi}{T_0}\cdot t\right)\cdot G_{0,T_0/2}(t)$$

$$= A\sin\left(\frac{2\pi}{T_0}\cdot t\right)\cdot\left[u(t) - u\left(t-\frac{T_0}{2}\right)\right]$$

$$= A\sin\left(\frac{2\pi}{T_0}\cdot t\right)\cdot u(t) - A\sin\left(\frac{2\pi}{T_0}\cdot t\right)\cdot u\left(t-\frac{T_0}{2}\right)$$

$$= A\sin\left(\frac{2\pi}{T_0}\cdot t\right)\cdot u(t) - A\sin\left[\frac{2\pi}{T_0}\left(t-\frac{T_0}{2}+\frac{T_0}{2}\right)\right]\cdot u\left(t-\frac{T_0}{2}\right)$$

$$= A\sin\left(\frac{2\pi}{T_0}\cdot t\right)\cdot u(t) - A\sin\left[\frac{2\pi}{T_0}\left(t-\frac{T_0}{2}\right)+\pi\right]\cdot u\left(t-\frac{T_0}{2}\right)$$

$$= A\sin\left(\frac{2\pi}{T_0}\cdot t\right)\cdot u(t) + A\sin\left[\frac{2\pi}{T_0}\left(t-\frac{T_0}{2}\right)\right]\cdot u\left(t-\frac{T_0}{2}\right)$$

Example 1.3. *Synthesize the equation of the signal as shown in Figure 1.20 in terms of ramp signal.*

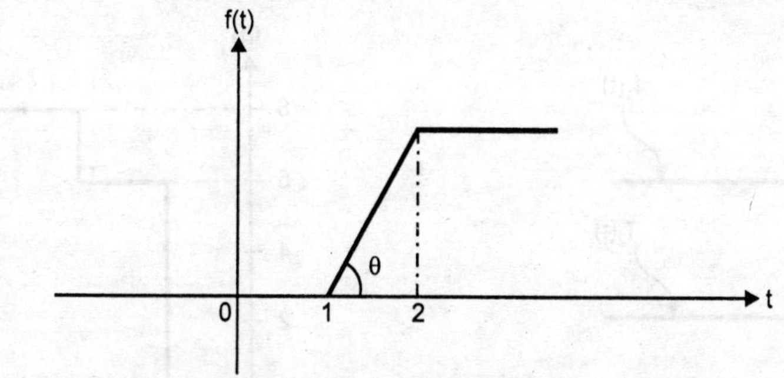

FIGURE 1.20 *Signal of Example 1.3.*

Solution. The signal of Figure 1.20 can be graphically separated as shown in Figure 1.21 so that

$$f(t) = f_1(t) + f_2(t)$$

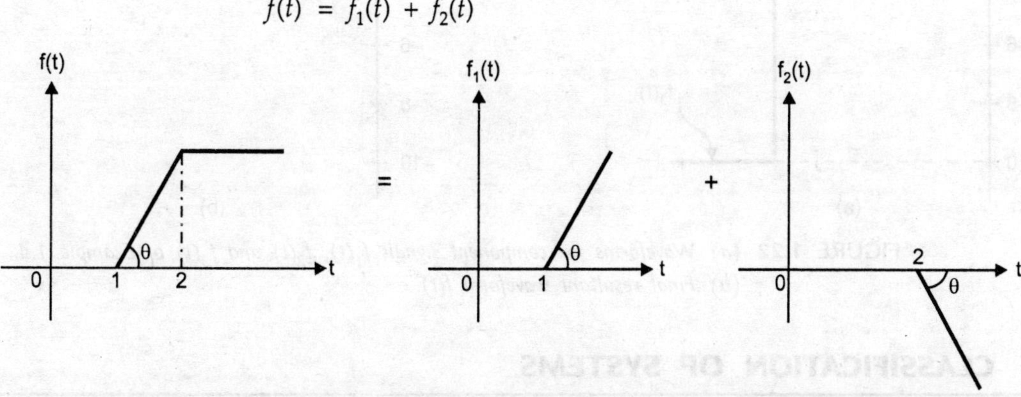

FIGURE 1.21 *Signal of Example 1.3 represented as the sum of two signals.*

Now, the equation of the signal can be expressed in terms of ramp signal as follows :

$$\begin{aligned} f(t) &= r(t-1)\cdot\tan\theta + r(t-2)\cdot\tan(-\theta) \\ &= r(t-1)\cdot\tan\theta - r(t-2)\cdot\tan\theta \\ &= [r(t-1) - r(t-2)]\cdot\tan\theta \end{aligned}$$

Example 1.4. *A wave equation is given by*

$$f(t) = 6u(t-1) + 2u(t-2) - 10u(t-3)$$

Draw the waveform.

Solution. Let us assume that

$$f_1(t) = 6u(t-1)$$

$$f_2(t) = 2u(t-2)$$

and $$f_3(t) = -10u(t-3)$$

so that $$f(t) = f_1(t) + f_2(t) + f_3(t)$$

The waveforms for component signals $f_1(t)$, $f_2(t)$ and $f_3(t)$ are drawn in Figure 1.22 (*a*) from which we obtain the final resultant waveform as shown in Figure 1.22 (*b*).

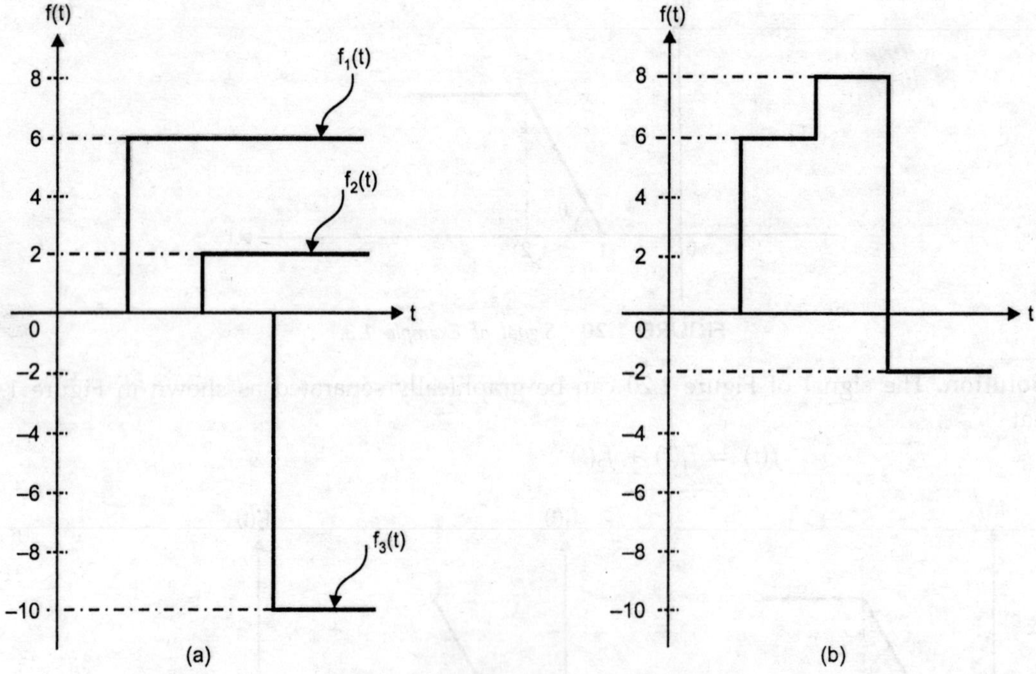

FIGURE 1.22 *(a) Waveforms for component signals $f_1(t)$, $f_2(t)$ and $f_3(t)$ of Example 1.4.*
(b) Final resultant waveform f(t).

1.5 CLASSIFICATION OF SYSTEMS

1.5.1 Continuous-Time and Discrete-Time Systems

A system in which **continuous-time input signals** are applied and results in **continuous-time output signals** is referred to as a **continuous-time system**. On the other hand, a system in which **discrete-time input signals** are applied and results in **discrete-time output signals** is referred to as a **discrete-time system**. Block-diagrams of continuous-time and discrete-time systems are illustrated in Figure 1.23.

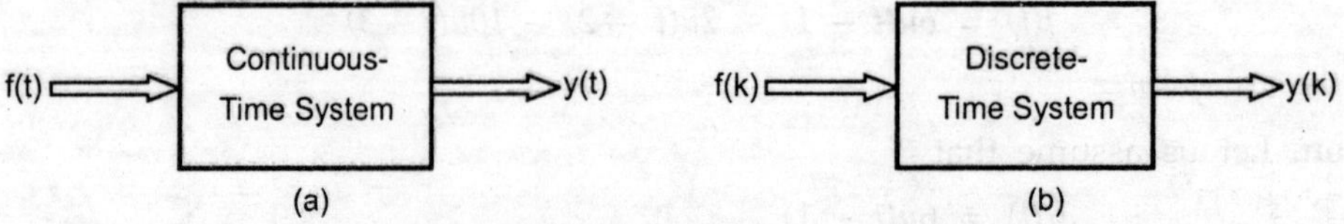

FIGURE 1.23 *(a) Block-diagram representation of a continuous-time system*
(b) Block-diagram representation of a discrete-time system.

1.5.2 Time-Invariant and Time-Varying Systems

Conceptually, a system is called **time-invariant system** if the **input-output relationship** of the system **does not change with time**. In other words, if the **response** of a system to a

delayed input and **delayed response** of the same are **equal**, the system is called **time-invariant system**, *i.e.*,

If $$y(t, \tau) = y(t - \tau) \qquad ...(1.38\ A)$$

where, $y(t, \tau)$ is the **response** of a continuous-time system to a **delayed input** and $y(t - \tau)$ is the **delayed response** of the same.

Then, the system is called **time-invariant continuous-time system.**

In an analogous manner, for discrete-time case, we state that

If $$y(k, n) = y(k - n) \qquad ...(1.38\ B)$$

where, $y(k, n)$ is the **response** of a discrete-time system to a **delayed input** and $y(k - n)$ is the **delayed response** of the same.

Then, the system is called **time-invariant discrete-time system.**

Conversely, a system is called **time-varying system** if the **input-output relationship** of the system **changes with time.** In other words, if the **response** of a system to a **delayed input** and **delayed response** of the same are **not equal,** the system is referred to as **time-varying system.**

1.5.3 Linear and Non-Linear Systems

The validity of the **principle of superposition** is the most fundamental property of the linear systems. In contrast to the linear systems, the principle of superposition does not hold good for non-linear systems. The concept of superposition implies that the **response** of a linear system to a **weighted sum of input signals** is **equal** to the **same weighted sum of output signals;** each output signal being **associated** with particular input signal acting on the system **independently** of all the other input signals.

In particular, suppose that $y_i(t)$; $i = 1, 2, 3, \ldots$ be the responses of a continuous-time system to the inputs $f_i(t)$; $i = 1, 2, 3, \ldots$ respectively. Then the system is **linear** if it satisfies the following properties :

1. The **additivity property,** *i.e.*, the response to $\sum_i f_i(t)$ is $\sum_i y_i(t)$, and
2. The **homogeneity property,** *i.e.*, the response to $a_i\, f_i(t)$ is $a_i\, y_i(t)$.

In an analogous manner, for discrete-time case, suppose that $y_i(k)$; $i = 1, 2, 3, \ldots$ be the responses of a discrete-time system to the inputs $f_i(k)$; $i = 1, 2, 3, \ldots$respectively. Then the system is **linear** if it satisfies the following properties:

1. The **additivity property,** *i.e.*, the response to $\sum_i f_i(k)$ is $\sum_i y_i(k)$, and
2. The **homogeneity property,** *i.e.*, the response to $a_i\, f_i(k)$ is $a_i\, y_i(k)$.

The two properties defining a linear system can be combined into a single statement recognized as the **superposition property:**

For Continuous-Time Case :

The response to $\sum_i a_i\, f_i(t)$ is $\sum_i a_i\, y_i(t)$.

For Discrete-Time Case :

The response to $\sum_i a_i f_i(k)$ is $\sum_i a_i y_i(k)$.

Here, a_i are **complex constants.**

The linear systems are important for the reason that their exact behaviour can be analyzed by standard techniques. However, systems in practice are **non-linear** in nature and, therefore, the analyses of non-linear systems is also of great importance.

Non-linearities such as saturation, dead zone, backlash, *etc.* are present **inherently** in the system because of the **non-ideality** of conventional control components and, therefore, are known as **incidental non-linearities.** Sometimes, however, non-linearities such as relay, *etc.* are purposely introduced in the system to improve systems characteristics. Such non-linearities are termed as **intentional non-linearities**. In the following sections we shall now discuss the basic features of the more **common physical non-linearities.**

1.5.3.1 Saturation

This is the most common phenomenon to many electrical, mechanical, magnetic and electromechanical control components. Ideally, the saturation phenomenon can be thought of as consisting of two distinct modes of operation. In the **low amplitude excitation mode,** the component operates **linearly**. However, in the **high amplitude excitation mode,** the output becomes **constant** and, thus, **independent** of the input magnitude.

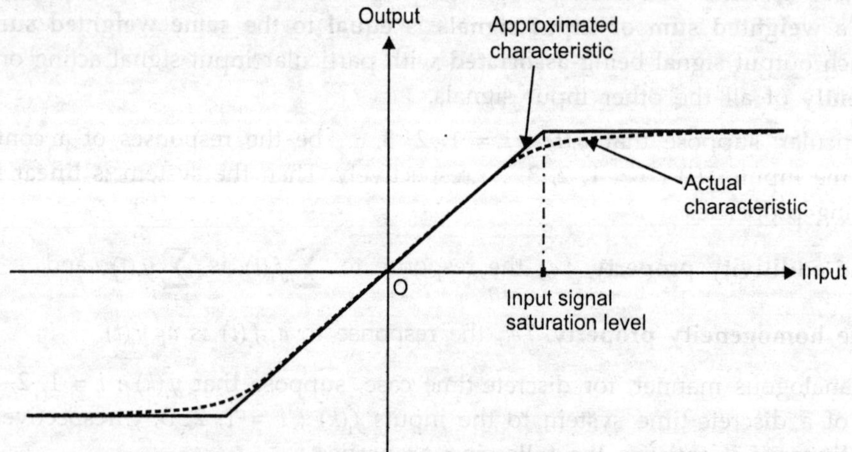

FIGURE 1.24 *Saturation non-linearity.*

All practical systems when excited by sufficiently large signals, exhibits the phenomenon of **saturation** due to limitations of physical capabilities of their components. One practical example is **amplifier** where the output varies **linearly** with the input for a limited range of operation and beyond that the output gets saturated, *i.e.*, the output tends to become **nearly constant** as illustrated in Figure 1.24. The point of discontinuity occurs at the input signal saturation level.

1.5.3.2 Dead Zone (or Dead Spot)

The phenomenon of **dead zone** or **dead spot** occurs because of **insensitivity** of certain components to sufficiently small input signals and can be cited in motors, actuators and DC servomotors. The input-output relationship for dead zone is illustrated in Figure 1.25.

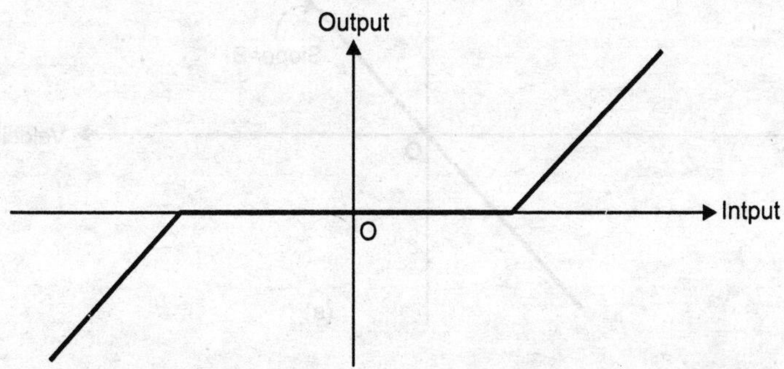

FIGURE 1.25 *Dead Zone (or dead spot) non-linearity.*

1.5.3.3 Hysteresis

The hysteresis sort of non-linearity occurs when increasing and decreasing portions of the input-output relationship follow different paths. For example, the **magnetization curve** of iron exhibits hysteresis and is illustrated in Figure 1.26.

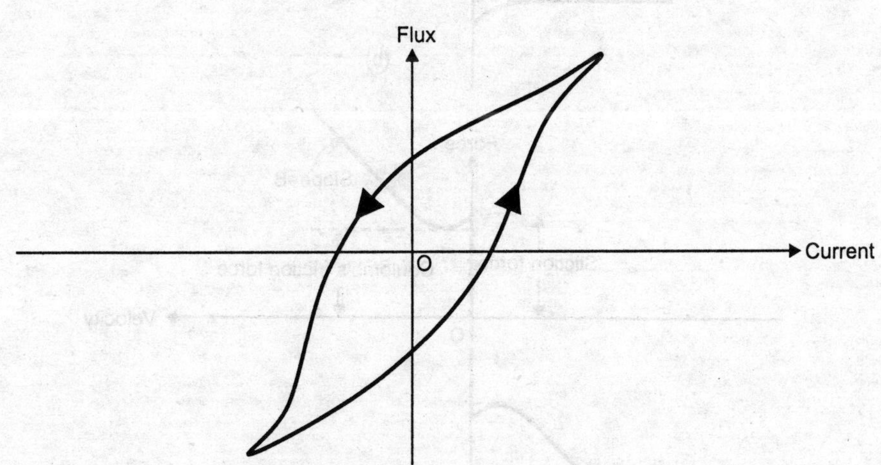

FIGURE 1.26 *Hysteresis non-linearity.*

1.5.3.4 Friction

Frictional forces **always oppose** the motion of a body whether it slides over any surface or rolls over the same. In mechanical systems, **viscous friction** is the predominant frictional force and is **proportional** to the relative velocity v between the moving surfaces, *i.e.*, for viscous friction the **force-velocity relation** is given by

$$F = Bv \qquad ...(1.39)$$

where, B is the coefficient of viscous friction. The force-velocity relationship of viscous friction is, thus, **linear** in nature and is shown in Figure 1.27 (*a*).

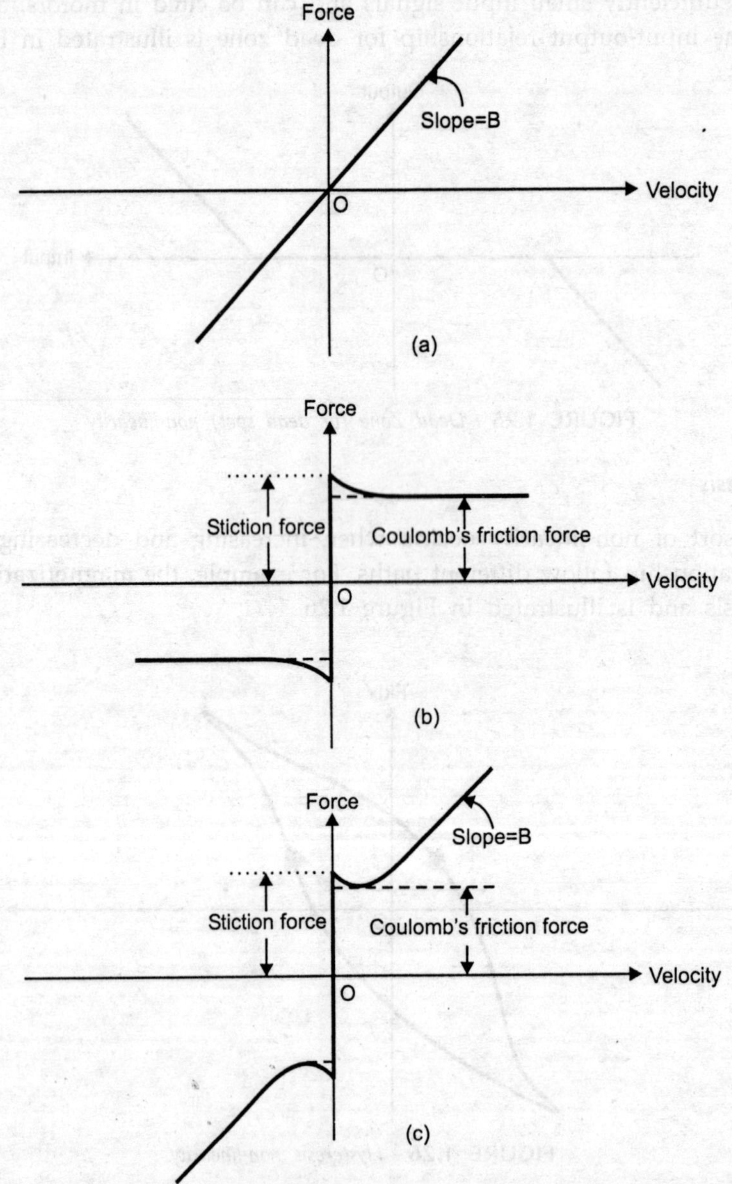

FIGURE 1.27 (*a*) *Viscous friction*
(*b*) *Stiction and Coulomb's friction*
(*c*) *Stiction, Coulomb's friction and Viscous friction.*

In addition to viscous friction, there exist two **non-linear frictional forces** also namely **Coulomb's friction** and **stiction**. Coulomb's friction is a **constant retarding force** and always opposes the relative motion between moving objects. The force of stiction is required to **initiate motion** of the body and is **always greater** than that of Coulomb's friction. This is due to the fact that more force is required to move a body from rest than to maintain it in motion. In real sense, as depicted in Figure 1.27 (*b*), the stiction force **gradually decreases** with velocity and changes over to Coulomb's frictional force. The **composite friction characteristic** is illustrated in Figure 1.27 (*c*).

1.5.3.5 Relay

A relay is a **non-linear power amplifier** that can provide large power amplification **inexpensively** and is, therefore, **intentionally** introduced in control systems. An ideal relay characteristic is shown in Figure 1.28 (*a*).

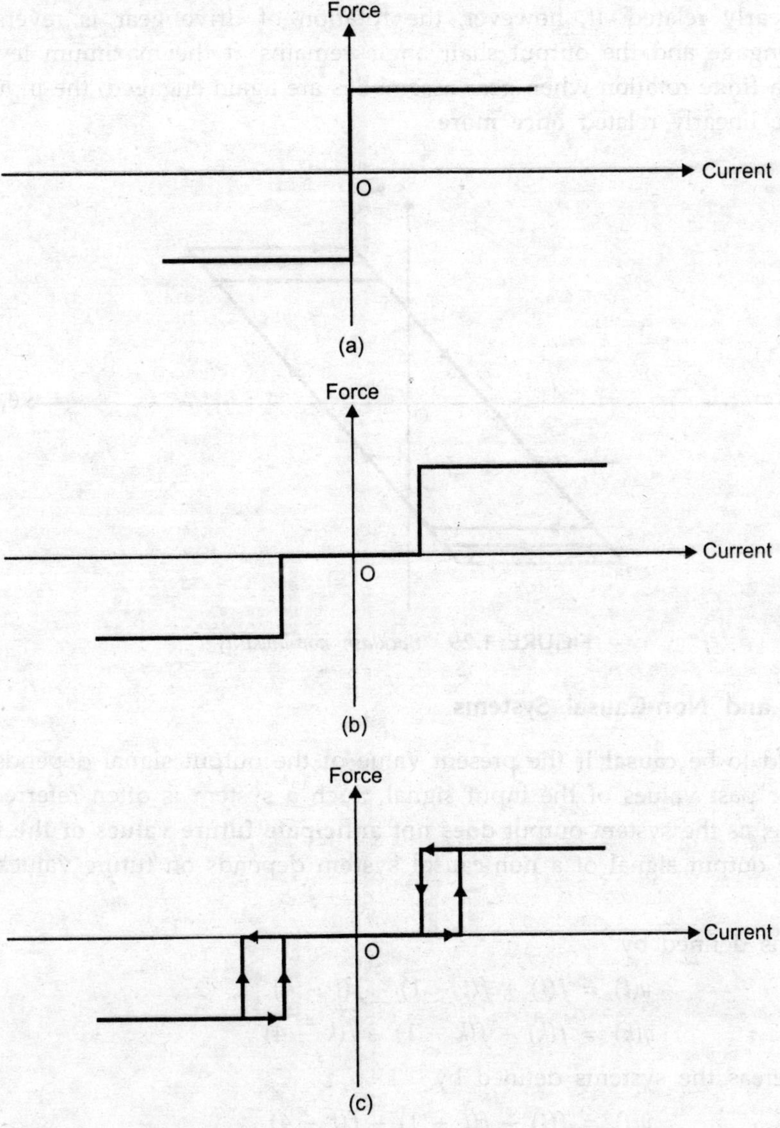

FIGURE 1.28 (*a*) *Ideal relay*
(*b*) *Relay with dead zone*
(*c*) *Relay with dead zone and hysteresis.*

In actual practice, a **dead zone** also is caused in the relay characteristic as seen in Figure 1.28 (*b*). This is due to the fact that a sufficient amount of current is required to actuate relay coils. Moreover, the relay characteristic always exhibits **hysteresis** since a large amount of current is needed to close the relay than the current at which the relay drops out. A relay characteristic with dead zone and hysteresis is illustrated in Figure 1.28 (*c*).

1.5.3.6 Backlash

Backlash results from the **freedom** existing in the **typical gear train** by which the **drive gear** may move through a finite rotation before engaging the **driven gear**. As long as the teeth of drive gear and those of driven gear are in contact, the input and output variables (shaft angles) are linearly related. If, however, the rotation of drive gear is reversed, the gear assemblies disengage and the output shaft angle remains at the maximum level previously attained. After a finite rotation when gear assemblies are again engaged, the input and output shaft angles are linearly related **once more**.

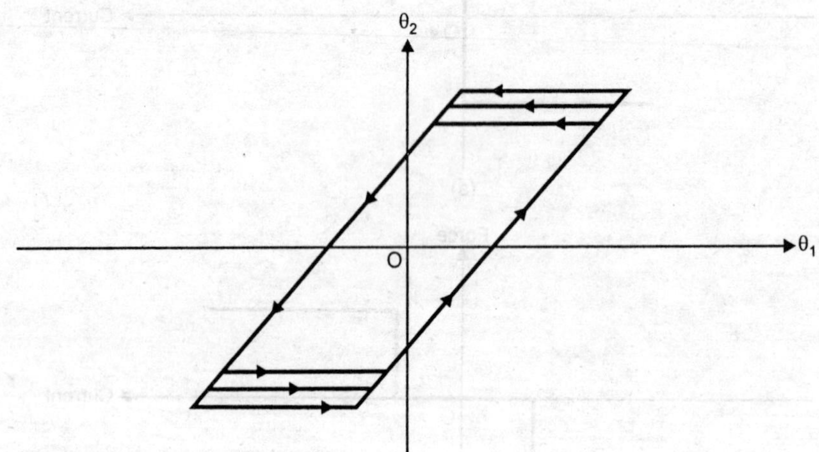

FIGURE 1.29 *Backlash non-linearity.*

1.5.4 Causal and Non-Causal Systems

A system is said to be **causal** if the present value of the output signal depends only on the present and / or past values of the input signal. Such a system is often referred to as being **non-anticipative**, as the system output **does not anticipate future values of the input**. On the other hand, the output signal of a **non-causal system** depends on future values of the input signal.

e.g., The systems defined by

$$y(t) = f(t) + f(t-1) - f(t-4)$$

and

$$y(k) = f(k) + f(k-1) - f(k-4)$$

are **causal**, whereas the systems defined by

$$y(t) = f(t) + f(t+1) - f(t-4)$$

and

$$y(k) = f(k) + f(k+1) - f(k-4)$$

are **non-causal**.

1.5.5 Static and Dynamic Systems

A system with an output signal that at any specific time depends on the value of the input signal only at that time is referred to as a **static system**. Static systems are also termed as **memoryless or instantaneous systems**. Physically, static systems contain **no energy-storage elements**. Mathematically, the **input-output relationship** for a static system must be of the form

$$y(t) = \psi(f(t), t) \qquad ...(1.40)$$

where, $\psi(\cdot)$ is a function, **possibly time dependent**, which depends on $f(t)$ only at the present time t.

In contrast, a **dynamic system** is one whose output depends on past or future values of the input in addition to the present time. If the system is also **causal**, the output of a dynamic system depends only on present and past values of the input. The temporal extent of past values on which the output depends defines how far the memory of the systems extends into past.

1.5.6 Stable and Unstable Systems

The **stability** of a system implies that the small changes in the system input (either in **system parameters** or in **initial conditions** of the system), do not result in large changes in the systems output. For any physical system, stability is one of the most important characteristics to be determined. Although there are various definitions of stability in common usage, the one we use is referred to as **bounded-input-bounded-output (BIBO) stability**. A system is said to be bounded-input-bounded-output (BIBO) stable if and only if every bounded input results in a bounded output. The **output** of such a system **does not diverge** if the **input does not diverge**. Thus, a system is BIBO stable if the output signal $y(t)$ satisfies the condition

$$|y(t)| \le M_y < \infty \qquad \text{for all } t$$

whenever, the input signal $f(t)$ satisfies the condition

$$|f(t)| \le M_f < \infty \qquad \text{for all } t$$

Both M_y and M_f represent some **finite positive numbers**. The condition for the BIBO stability of a discrete-time system can be described in a similar manner.

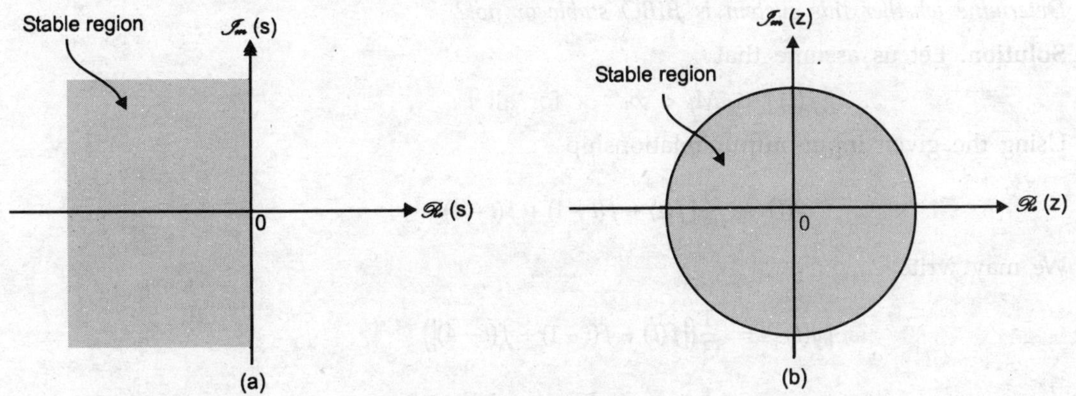

FIGURE 1.30 *(a) Region of stability in s-plane*
(b) Region of stability in z-plane.

A necessary condition for the **continuous-time system** to be **stable** is that the **roots** of the characteristic equation have **negative real parts** (or lie in the **left-half** of the **s-plane**), while for the **discrete-time system**, the **roots** of the characteristic equation are **inside** the **unit-circle** as shown in Figure 1.30 (*a*) and Figure 1.30 (*b*) respectively.

Example 1.5. *A system is described by the following input-output relationship*

$$y(t) = t\ f(t)$$

Determine whether this system is linear or not?

Solution. Let the input signal $f(t)$ be expressed as the weighted sum

$$f(t) = \sum_i a_i f_i(t)$$

We may then express the response of the system as

$$\begin{aligned} y(t) &= t\ f(t) \\ &= t \sum_i a_i f_i(t) \\ &= \sum_i a_i t\ f_i(t) \\ &= \sum_i a_i y_i(t) \end{aligned}$$

where, $y_i(t) = t\ f_i(t)$ is the output due to each input acting independently. The response to $\sum_i a_i f_i(t)$ is $\sum_i a_i f_i(t)$.

We thus see that the system under consideration satisfies the **superposition principle** and is, therefore, **linear.**

Example 1.6. *A system is described by the following input-output relationship*

$$y(t) = \frac{1}{3}[f(t) + f(t-1) + f(t-4)]$$

Determine whether this system is BIBO stable or not?

Solution. Let us assume that

$$|f(t)| \le M_f < \infty \qquad \text{for all } t$$

Using the given input-output relationship

$$y(t) = \frac{1}{3}[f(t) + f(t-1) + f(t-4)]$$

We may write

$$\begin{aligned} |y(t)| &= \frac{1}{3}(|f(t) + f(t-1) + f(t-4)|) \\ &\le \frac{1}{3}(|f(t)|+|f(t-1)|+|f(t-4)|) \end{aligned}$$

$$\leq \frac{1}{3}(M_f + M_f + M_f)$$

$$\leq M_f < \infty \qquad \text{for all } t$$

Hence the absolute value of the output signal $y(t)$ is **always less than the maximum absolute value** of the input signal $f(t)$ for all t, which shows that the system under consideration is **stable.**

1.6 ANALOGOUS SYSTEMS

Consider a **spring-mass-damper system** as shown in Figure 1.31. A **force** $F(t)$ is being applied on **mass** M which in turn results in a **displacement** $x(t)$ of the mass. There are three types of forces which resist the motion. These are

1. **Inertia Force :** It is defined as

$$F_M(t) = M\frac{d^2}{dt^2}x(t) \qquad \text{...(1.41 A)}$$

where, $F_M(t)$ = Inertia Force (N)

M = Mass of the system (Kg)

$x(t)$ = Displacement (m)

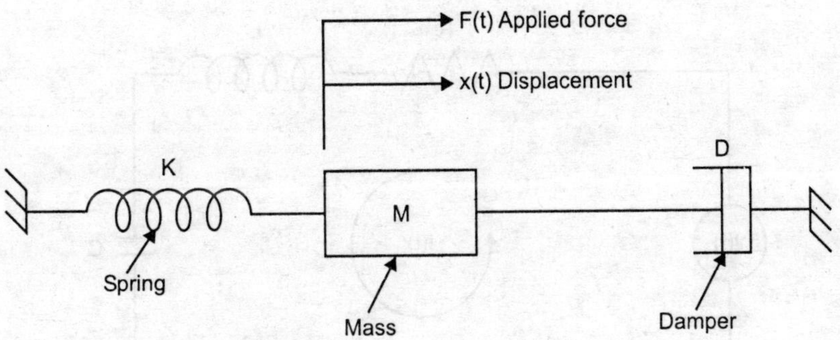

FIGURE 1.31 *Spring-mass-damper system.*

2. **Damping Force :** It is defined as

$$F_D(t) = D\frac{d}{dt}x(t) \qquad \text{...(1.41 B)}$$

where, $F_D(t)$ = Damping Force (N)

D = Coefficient of Viscous Damping (N/m/s)

$x(t)$ = Displacement (m)

3. **Spring Restoring Force :** It is defined as

$$F_K(t) = Kx(t) \qquad \text{...(1.41 C)}$$

where, $F_K(t)$ = Spring Force (N)

K = Spring Constant (N/m)

$x(t)$ = Displacement (m)

The **static equilibrium** of a **dynamic system** subjected to **externally driven forces** obeys the **D' Alembert's Principle** which can formally be stated as follows:

"For any body, the ***algebraic sum*** *of* ***externally applied forces*** *and the* ***forces resisting motion*** *in* ***any given direction*** *is zero".*

Thus, we can write

$$F(t) = F_M(t) + F_D(t) + F_K(t)$$

or
$$F(t) = M\frac{d^2}{dt^2}x(t) + D\frac{d}{dt}x(t) + Kx(t) \quad ...(1.42)$$

1.6.1 Force-Voltage Analogy

Now consider a **series RLC circuit** as shown in Figure 1.32. When applying **Kirchhoff's voltage law,** we may write

$$v(t) = Ri(t) + L\frac{d}{dt}i(t) + \frac{1}{C}\int_{-\infty}^{t} i(\tau)\,d\tau \quad ...(1.43)$$

We know that the **current** $i(t)$ is the **rate of flow of electric charge** $q(t)$, *i.e.,*

$$i(t) = \frac{d}{dt}q(t)$$

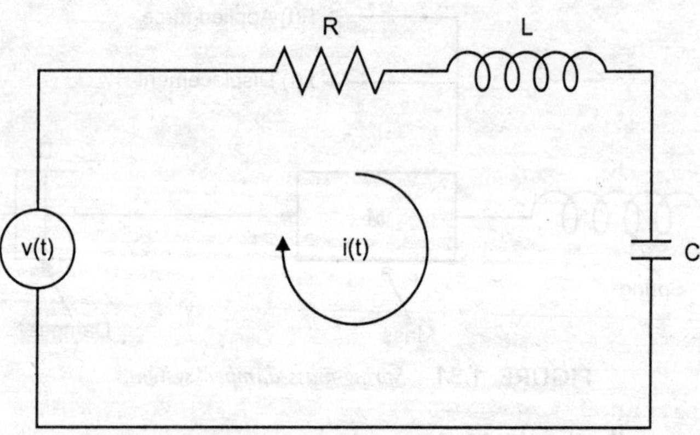

FIGURE 1.32 *Series RLC circuit.*

Thus, we may rewrite equation (1.43) as

$$v(t) = R\frac{d}{dt}q(t) + L\frac{d^2}{dt^2}q(t) + \frac{1}{C}q(t)$$

or,
$$v(t) = L\frac{d^2}{dt^2}q(t) + R\frac{d}{dt}q(t) + \frac{1}{C}q(t) \quad ...(1.44)$$

It is observed that equation (1.44) is **analogous** to equation (1.42) and following analogies are established:

TABLE 1.1 Force-Voltage Analogy

Mechanical System	*Electrical System*
Force $F(t)$	Voltage $v(t)$
Mass M	Inductance L
Coefficient of Viscous Damping D	Resistance R
Spring Constant K	Reciprocal of Capacitance $1/C$
Displacement $x(t)$	Electric Charge $q(t)$

As each individual term in equation (1.42) represents a **force** and each individual term in equation (1.44) represents a **voltage**, therefore, the analogy aforesaid is termed as **Force-Voltage Analogy**.

1.6.2 Force-Current Analogy

Now consider a **parallel RLC circuit** as shown in Figure 1.33. When applying **Kirchhoff's current law**, we may write

$$i(t) = \frac{1}{R}v(t) + \frac{1}{L}\int_{-\infty}^{t} v(\tau)d\tau + C\frac{d}{dt}v(t) \qquad ...(1.45)$$

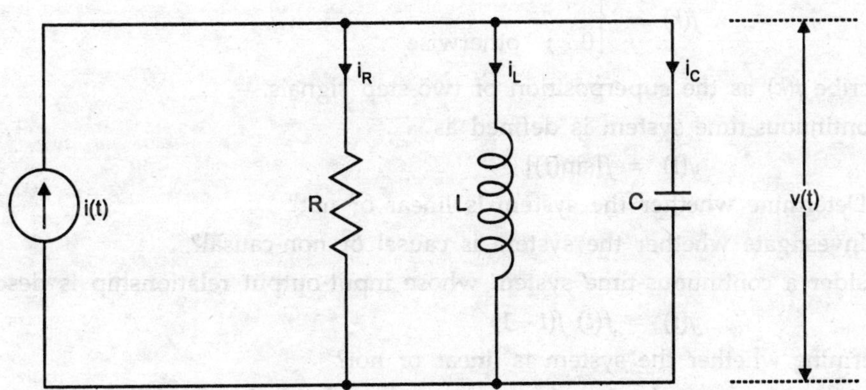

FIGURE 1.33 *Parallel RLC circuit.*

The **voltage** $v(t)$ is related to the **flux linkages** $\phi(t)$ **associated with the inductance** L as follows:

$$v(t) = \frac{d}{dt}\phi(t)$$

Thus, we may rewrite equation (1.45) as

$$i(t) = \frac{1}{R}\cdot\frac{d}{dt}\phi(t) + \frac{1}{L}\phi(t) + C\frac{d^2}{dt^2}\phi(t)$$

or,

$$i(t) = C\frac{d^2}{dt^2}\phi(t) + \frac{1}{R}\cdot\frac{d}{dt}\phi(t) + \frac{1}{L}\phi(t) \qquad ...(1.46)$$

It is observed that equation (1.46) is **analogous** to equation (1.42) and following analogies are established :

TABLE 1.2 Force-Current Analogy

Mechanical System	*Electrical System*
Force $F(t)$	Current $i(t)$
Mass M	Capacitance C
Coefficient of Viscous Damping D	Reciprocal of Resistance $1/R$
Spring Constant K	Reciprocal of Inductance $1/L$
Displacement $x(t)$	Flux Linkages $\phi(t)$

As each individual term in equation (1.42) represents a **force** and each individual term in equation (1.46) represents a **current**, therefore, the analogy aforesaid is termed as **Force-Current Analogy**.

PROBLEMS

1. A discrete-time signal $f(k)$ is defined by

$$f(k) = \begin{cases} 1 \;; & 0 \le k \le 9 \\ 0 \;; & \text{otherwise} \end{cases}$$

Describe $f(k)$ as the superposition of two-step signals.

2. A continuous-time system is defined as

$$y(t) = f[\sin(t)]$$

(*a*) Determine whether the system is linear or not?

(*b*) Investigate whether the system is causal or non-causal?

3. Consider a continuous-time system whose input-output relationship is described as

$$y(t) = f(t)\, f(t-1)$$

Determine whether the system is linear or not?

4. Consider the system having the input-output relationship as follows

$$y(t) = f^2(t)$$

Determine whether the system is linear or not?

5. Consider a discrete-time system whose input-output relationship is described as

$$y(k) = a^k f(k)$$

where $a > 1$. Determine whether the system is BIBO stable or not?

6. Consider the system defined as

$$y(t) = f(2t)$$

Determine whether the system is time-invariant or not?

7. A continuous-time system is defined as

$$y(t) = \sin[f(t)]$$

Determine whether the system is time-invariant or not?

8. Consider the continuous-time signal

$$f(t) = \delta(t + 2) - \delta(t - 2)$$

Calculate the value of E_∞ for the signal

$$y(t) = \int_{-\infty}^{t} f(\tau)d\tau$$

9. Synthesize the saw-tooth waveform as shown in Figure 1.34 in terms of basic signals.

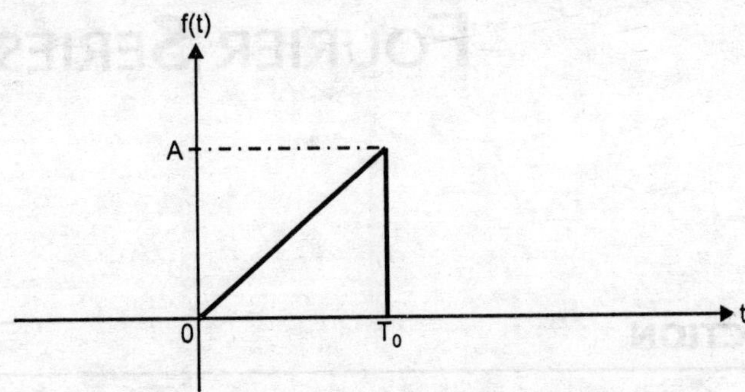

FIGURE 1.34 *Saw-tooth waveform of Problem (9).*

10. Synthesize the square waveform as shown in Figure 1.35 in terms of basic signals.

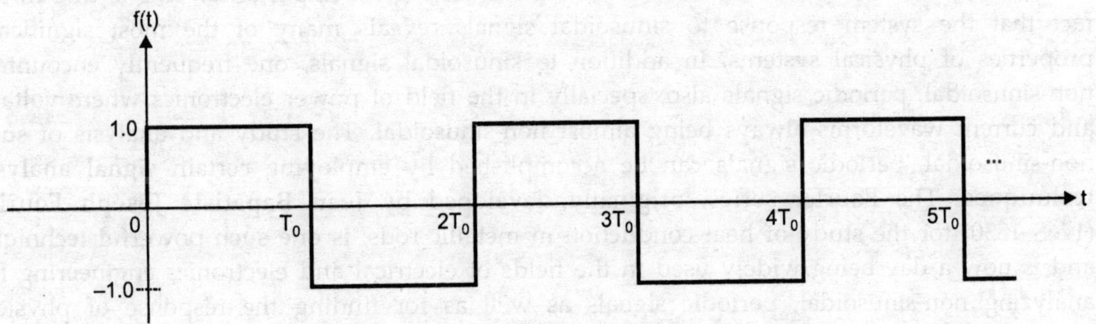

FIGURE 1.35 *Square waveform of Problem (10).*

11. Synthesize the square waveform as shown in Figure 1.36 in terms of basic signals. Assume $T_0 = 1$ Sec.

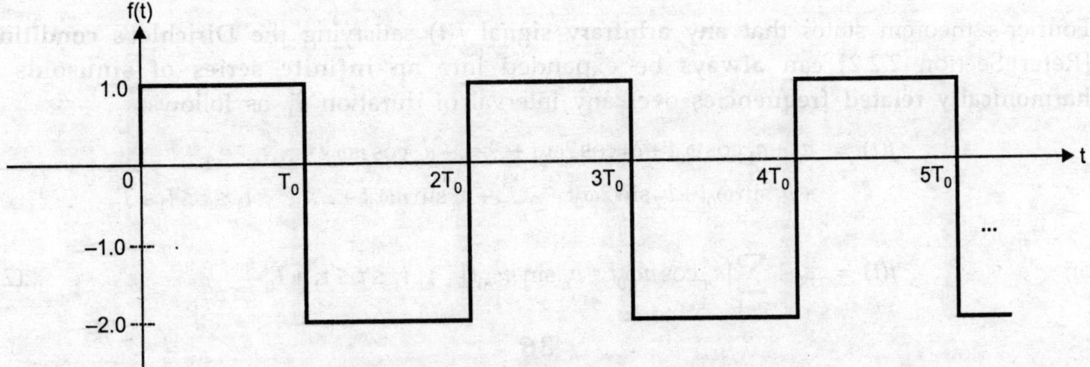

FIGURE 1.36 *Square waveform of Problem (11).*

Chapter 2

FOURIER SERIES ANALYSIS

2.1 INTRODUCTION

Sinusoidal signals are of common occurrence in many physical and engineering problems. In areas like control systems engineering, communication systems engineering, power electronics *etc.*, the **response of systems to sinusoidal signals** is of great importance. This is due to the fact that the system response to sinusoidal signals reveals many of the most significant properties of physical systems. In addition to sinusoidal signals, one frequently encounters non-sinusoidal, periodic signals also; specially in the field of power electronics where voltage and current waveforms always being almost non-sinusoidal. The study and analysis of such non-sinusoidal, periodic signals can be accomplished by employing certain **signal analysis techniques**. The Fourier series, originally developed by **Jean Bapatiste Joseph Fourier** (1768–1830) for the study of heat conduction in metallic rods, is one such powerful technique and is now a day being widely used in the fields of electrical and electronics engineering for analyzing non-sinusoidal, periodic signals as well as for finding the response of physical systems to such signals.

2.2 FOURIER SERIES REPRESENTATIONS

Fourier's theorem states that any **arbitrary signal** $f(t)$ satisfying the **Dirichlet's conditions** [Refer Section 2.2.2] can **always** be expanded into an **infinite series of sinusoids of harmonically related frequencies** over any interval of duration T_0 as follows:

$$\begin{aligned} f(t) &= a_0 + a_1 \cos \omega_0 t + a_2 \cos 2\omega_0 t + \ldots\ldots + a_n \cos n\omega_0 t + \ldots\ldots \\ &\quad + b_1 \sin \omega_0 t + b_2 \sin 2\omega_0 t + \ldots\ldots + b_n \sin n\omega_0 t + \ldots\ldots \;;\; t_1 \le t \le t_1 + T_0 \end{aligned}$$

or,

$$f(t) = a_0 + \sum_{n=1}^{\infty} [a_n \cos n\omega_0 t + b_n \sin n\omega_0 t] \;;\; t_1 \le t \le t_1 + T_0 \qquad \ldots(2.1)$$

where, $$\omega_0 = \frac{2\pi}{T_0}$$

The representation of the signal $f(t)$ in the form of equation (2.1) over a finite interval $[t_1, t_1 + T_0]$ is referred to as the **trigonometric form** of the **Fourier series.** The **Fourier coefficient** a_0 is a **constant** (zero frequency component) and is, therefore, referred to as the **d.c. component.** One can determine a_0 from

$$a_0 = \frac{1}{T_0}\int_{t_1}^{t_1+T_0} f(t)dt \quad \text{...(2.2 A)}$$

which is the **average value** of the signal $f(t)$ over a finite interval $[t_1, t_1 + T_0]$. The **Fourier coefficients** a_n and b_n are determined from

$$a_n = \frac{2}{T_0}\int_{t_1}^{t_1+T_0} f(t)\cos n\omega_0 t\, dt \;;\; n = 1, 2, 3, \ldots\ldots \quad \text{...(2.2 B)}$$

and $$b_n = \frac{2}{T_0}\int_{t_1}^{t_1+T_0} f(t)\sin n\omega_0 t\, dt \;;\; n = 1, 2, 3, \ldots\ldots \quad \text{...(2.2 C)}$$

In equation (2.1) the component having **fundamental frequency** equal to ω_0 is called the **fundamental component** or the **first harmonic component**. The component with **twice** the frequency of fundamental component is referred to as the **second harmonic component.** More generally, the component having frequency $n\omega_0$ is referred to as the nth **harmonic component.** Each of these harmonic components can completely be specified by three parameters *viz.* **amplitude, phase** and **frequency**. We have observed that in Fourier series representation of equation (2.1), the third parameter, *i.e.*, **frequency** of each harmonic component is **automatically** fixed by the fundamental frequency ω_0. However, the **amplitude** and **phase information** of each harmonic component can be determined by means of the **Fourier coefficients** a_n and b_n. This fact is more clearly brought out by combining the cosine and sine terms in equation (2.1) into a single cosine term of the same frequency using the trigonometric identity

$$a_n \cos n\omega_0 t + b_n \sin n\omega_0 t = C_n \cos(n\omega_0 t + \theta_n) \quad \text{...(2.3)}$$

where, $$C_n = \sqrt{(a_n^2 + b_n^2)} \quad \text{...(2.4 A)}$$

and $$\theta_n = -\tan^{-1}\left(\frac{b_n}{a_n}\right) \quad \text{...(2.4 B)}$$

For consistency writing $C_0 = a_0$, we can represent the trigonometric form of the Fourier series (2.1) in a **compact trigonometric form** as

$$f(t) = C_0 + \sum_{n=1}^{\infty} C_n \cos(n\omega_0 t + \theta_n) \;;\; t_1 \le t \le t_1 + T_0 \quad \text{...(2.5)}$$

The coefficient C_n is the **amplitude** of the nth harmonic component whereas the **phase information** is given by θ_n which represents the **relative phase angle (delay angle)** of the nth harmonic component from the fundamental one. C_n and θ_n can be obtained from equations (2.4).

A more convenient and elegant representation of the Fourier series can be obtained merely by expressing the sine and cosine terms in equation (2.1) as **exponential function** with **complex constants**. We have the following relations

$$\cos n\omega_0 t = \frac{1}{2}(e^{jn\omega_0 t} + e^{-jn\omega_0 t}) \qquad \text{...(2.6 A)}$$

and

$$\sin n\omega_0 t = \frac{1}{2j}(e^{jn\omega_0 t} - e^{-jn\omega_0 t}) \qquad \text{...(2.6 B)}$$

In terms of the exponential expressions (2.6), the trigonometric series (2.1) becomes

$$f(t) = a_0 + \sum_{n=1}^{\infty}\left[a_n\left(\frac{e^{jn\omega_0 t} + e^{-jn\omega_0 t}}{2}\right) + b_n\left(\frac{e^{jn\omega_0 t} - e^{-jn\omega_0 t}}{2j}\right)\right]$$

or,

$$f(t) = a_0 + \sum_{n=1}^{\infty}\left[\left(\frac{a_n - jb_n}{2}\right)e^{jn\omega_0 t} + \left(\frac{a_n + jb_n}{2}\right)e^{-jn\omega_0 t}\right] \qquad \text{...(2.7)}$$

The coefficients of the exponential terms in equation (2.7) are **complex conjugates.** Defining them as

$$F_n = \left(\frac{a_n - jb_n}{2}\right) \qquad \text{...(2.8 A)}$$

and

$$F_{-n} = \left(\frac{a_n + jb_n}{2}\right) \qquad \text{...(2.8 B)}$$

and writing $F_0 = a_0$, equation (2.7) modifies to

$$f(t) = F_0 + \sum_{n=1}^{\infty}\left[F_n e^{jn\omega_0 t} + F_{-n} e^{-jn\omega_0 t}\right]$$

or,

$$f(t) = \left[F_0 + \sum_{n=1}^{\infty} F_n e^{jn\omega_0 t}\right] + \sum_{n=1}^{\infty} F_{-n} e^{-jn\omega_0 t}$$

or,

$$f(t) = \sum_{n=0}^{\infty} F_n e^{jn\omega_0 t} + \sum_{n=1}^{-\infty} F_n e^{jn\omega_0 t}$$

or,

$$f(t) = \sum_{n=-\infty}^{\infty} F_n e^{jn\omega_0 t} \;;\; t_1 \le t \le t_1 + T_0 \qquad \text{...(2.9)}$$

The representation of the signal $f(t)$ in the form of equation (2.9) over a finite interval $[t_1, t_1 + T_0]$ is referred to as the **exponential** or **complex form** of the **Fourier series.** The coefficient F_n can be determined merely by substituting the values of a_n and b_n from equations (2.2) in equation (2.8 A). We have

$$F_n = \frac{1}{2}(a_n - jb_n)$$

or,

$$F_n = \frac{1}{2}\left[\frac{2}{T_0}\int_{t_1}^{t_1+T_0} f(t)\cos n\omega_0 t \, dt\right] - \frac{j}{2}\left[\frac{2}{T_0}\int_{t_1}^{t_1+T_0} f(t)\sin n\omega_0 t \, dt\right]$$

or, $$F_n = \frac{1}{T_0}\int_{t_1}^{t_1+T_0} f(t)[\cos n\omega_0 t - j\sin n\omega_0 t]\,dt$$

or, $$F_n = \frac{1}{T_0}\int_{t_1}^{t_1+T_0} f(t)\,e^{-jn\omega_0 t}\,dt \;;\; n = 0, \pm 1, \pm 2, \pm 3, \ldots\ldots, \pm\infty \quad \ldots(2.10)$$

Equations (2.9) and (2.10) are required for representation of the signal $f(t)$ in the **exponential** or **complex form** of the **Fourier series** over a finite interval $[t_1, t_1 + T_0]$.

In equation (2.9), the component for $n = 0$ is a **constant.** The components for $n = \pm 1$ both have **fundamental frequency** equal to ω_0 and are collectively referred to as the **fundamental components** or the **first harmonic components**. The two components for $n = \pm 2$ having **twice** the frequency of the fundamental components and are collectively referred to as the **second harmonic components.** More generally, the components having frequencies $\pm n\omega_0$ are collectively referred to as the nth **harmonic components.**

The amplitude and phase information of each harmonic component can be determined by means of the complex coefficient F_n. This fact is more clearly brought out by an easy mathematical manipulation in equation (2.9) as follows:

Writing the **complex coefficient** F_n in **polar form** as

$$F_n = |F_n|\,e^{j\varphi_n} \quad \ldots(2.11)$$

Substituting above polar form of the complex coefficient F_n into equation (2.9), we obtain

$$f(t) = \sum_{n=-\infty}^{\infty} |F_n|\,e^{j\varphi_n} e^{jn\omega_0 t}$$

or, $$f(t) = \sum_{n=-\infty}^{-1} |F_n|\,e^{j(n\omega_0 t+\varphi_n)} + F_0 + \sum_{n=1}^{\infty} |F_n|\,e^{j(n\omega_0 t+\varphi_n)}$$

or, $$f(t) = F_0 + \sum_{n=1}^{\infty}\left[\,|F_n|\,e^{j(n\omega_0 t+\varphi_n)} + |F_{-n}|e^{j(-n\omega_0 t+\varphi_{-n})}\right]$$

or, $$f(t) = F_0 + \sum_{n=1}^{\infty}\left[\,|F_n|\,e^{j(n\omega_0 t+\varphi_n)} + |F_n|e^{-j(n\omega_0 t+\varphi_n)}\right]$$ [Refer Section 2.3]

or, $$f(t) = F_0 + \sum_{n=1}^{\infty}\left[2\,|F_n|\left(\frac{e^{j(n\omega_0 t+\varphi_n)} + e^{-j(n\omega_0 t+\varphi_n)}}{2}\right)\right]$$

or, $$f(t) = F_0 + \sum_{n=1}^{\infty} 2\,|F_n|\,\cos(n\omega_0 t + \varphi_n) \;;\; t_1 \le t \le t_1 + T_0 \quad \ldots(2.12)$$

The magnitude $|F_n|$ is the **amplitude** of the nth harmonic component in the exponential form of the Fourier series representation (2.9) whereas the **phase information** is given by φ_n which represents the **relative phase angle (delay angle)** of the nth harmonic component from the fundamental one.

Comparing equations (2.5) and (2.12), we obtain some important results for different forms of the Fourier series as follows:

$$C_0 = F_0 = a_0 \quad \text{...(2.13 A)}$$

$$C_n = 2|F_n| = \sqrt{(a_n^2 + b_n^2)} \quad \text{...(2.13 B)}$$

and

$$\theta_n = \varphi_n = -\tan^{-1}\left(\frac{b_n}{a_n}\right) \quad \text{...(2.13 C)}$$

The **exponential** or **complex form** of the Fourier series has a number of **advantages** over the **trigonometric form**. Only **one integral** has to be calculated in the former as **against three integrals** in the later. Moreover, for mathematical manipulations the exponential form (2.9) of the Fourier series is often more convenient than the trigonometric form (2.1) of the same.

2.2.1 Concept of Negative Frequency

The exponential form (2.9) of the Fourier series represents an **arbitrary signal** $f(t)$ over a **finite interval of duration** T_0 by a **linear combination** of **complex-valued exponential functions** of **frequencies** 0, $\pm\omega_0$, $\pm 2\omega_0$, $\pm 3\omega_0$, ... and so on. The appearance of negative frequencies in the exponential form of the Fourier series representation is a **result** of mathematical manipulations that convert sinusoidal functions into a pair of exponential functions. Actually, negative frequency signals are not physical signals, rather these are **purely mathematical concepts** required to give a **compact mathematical model** of a **real signal** (*viz.* cos $n\omega_0 t$ or sin $n\omega_0 t$) by a combination of **complex exponentials of positive and negative frequencies** (*viz.* $e^{jn\omega_0 t}$ and $e^{-jn\omega_0 t}$).

2.2.2 Existence of the Fourier Series: Dirichlet's Conditions

The Fourier series expansion of a signal $f(t)$ contains an **infinite** number of terms. The sum of these terms at any time **must equal** the value of the signal at that time, *i.e.*, the Fourier series must **converge** at every point of $f(t)$. The **sufficient conditions** for the **uniform convergence** of a Fourier series are called **Dirichlet's conditions.** The Dirichlet conditions are as follows:

Condition (1): *The signal $f(t)$ must be **absolutely integrable** over any period, i.e.,*

$$\int_{T_0} |f(t)|\, dt < \infty$$

For the Fourier series to exist, the Fourier coefficients a_0, a_n and b_n in equations (2.2) must be **finite**. If a signal $f(t)$ satisfies the condition aforesaid, the **existence** of a Fourier series is **guaranteed**, but the series **may not converge** at every point.

A signal $f(t) = \frac{1}{t}$ for $0 < t \le 1$, which is periodic with $T_0 = 1$, **violates the first Dirichlet condition** stated above. This signal is illustrated in Figure 2.1.

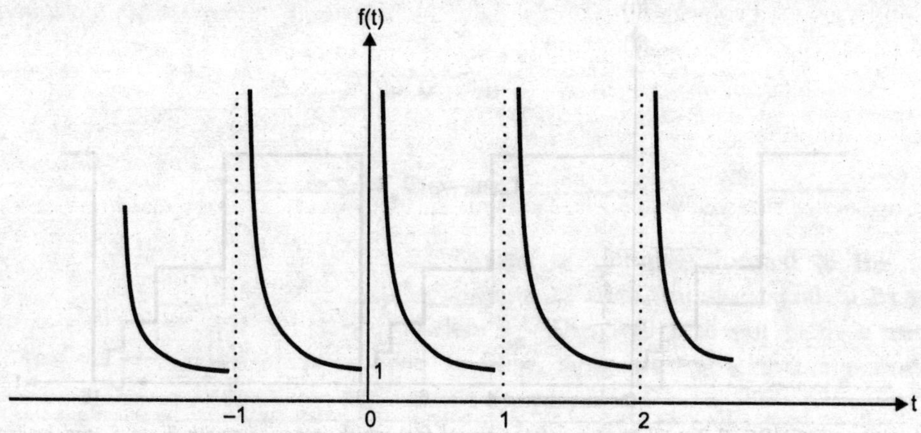

FIGURE 2.1 *Illustration of the signal* $f(t) = \frac{1}{t}$ *for* $0 < t \leq 1$ *which is periodic with* $T_0 = 1$ *and violating first Dirichlet condition.*

Condition (2): *The signal f(t) can have only a* ***finite number of maxima and minima*** *in any finite interval of time.*

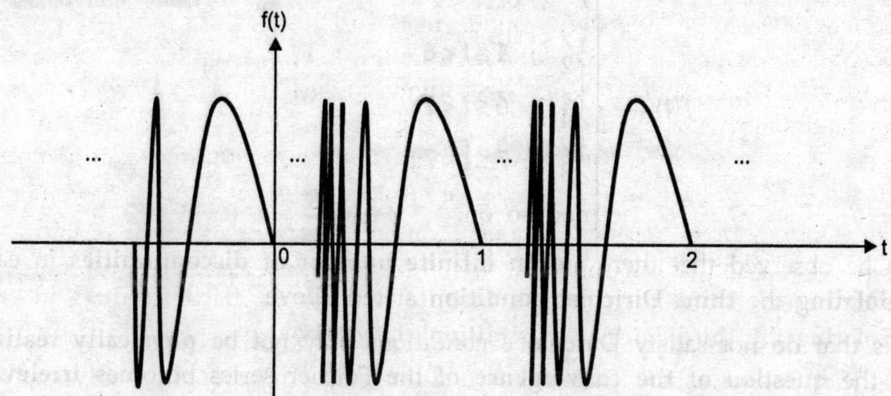

FIGURE 2.2 *Illustration of the signal* $f(t) = \sin\left(\frac{2\pi}{t}\right)$ *for* $0 < t \leq 1$ *which is periodic with* $T_0 = 1$ *and violating second Dirichlet condition.*

A signal $f(t) = \sin\left(\frac{2\pi}{t}\right)$ for $0 < t \leq 1$, which is periodic with $T_0 = 1$, **violates** the **second Dirichlet condition** stated above as the signal *f*(*t*) has an **infinite number of maxima and minima** in the interval (0, 1). However, this signal meets the **first Dirichlet condition** as

$$\int_0^1 |f(t)|\, dt < 1$$

This signal is illustrated in Figure 2.2.

Condition (3): *The signal f(t) can have only a* ***finite number of discontinuities*** *in any finite interval of time.*

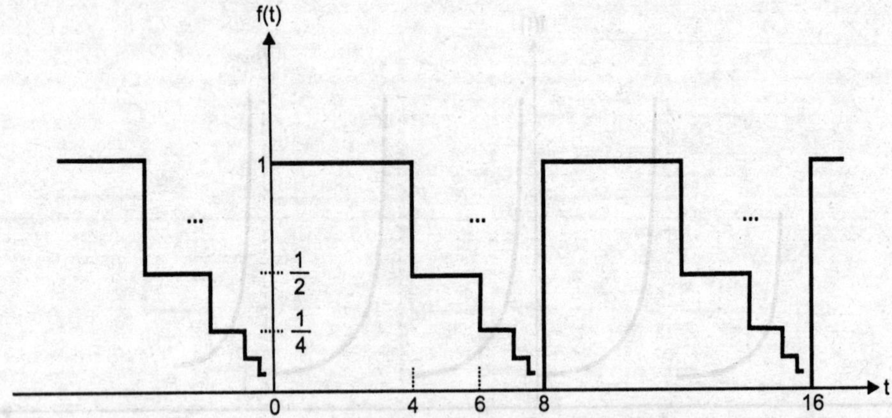

FIGURE 2.3 *Illustration of the signal f(t) which is periodic with T_0 = 8 and having an infinite number of discontinuities in each period thereby violating third Dirichlet condition.*

A signal $f(t)$, as illustrated in Figure 2.3, is periodic with $T_0 = 8$ and is composed of an **infinite number of sections** ; each of which is **half the width** and **half the height** of the previous section, *i.e.*,

$$f(t) = \begin{cases} 1 & ; \ 0 \le t < 4 \\ \frac{1}{2} & ; \ 4 \le t < 6 \\ \frac{1}{4} & ; \ 6 \le t < 7 \\ \frac{1}{8} & ; \ 7 \le t < 7.5 \\ \text{and so on.} \end{cases}$$

It can be observed that there are an **infinite number of discontinuities** in each period thereby **violating** the **third Dirichlet condition** stated above.

Signals that do not satisfy Dirichlet's conditions may not be **physically realizable** and, therefore, the **question of the convergence** of the Fourier series becomes **irrelevant** in the analysis of physical systems.

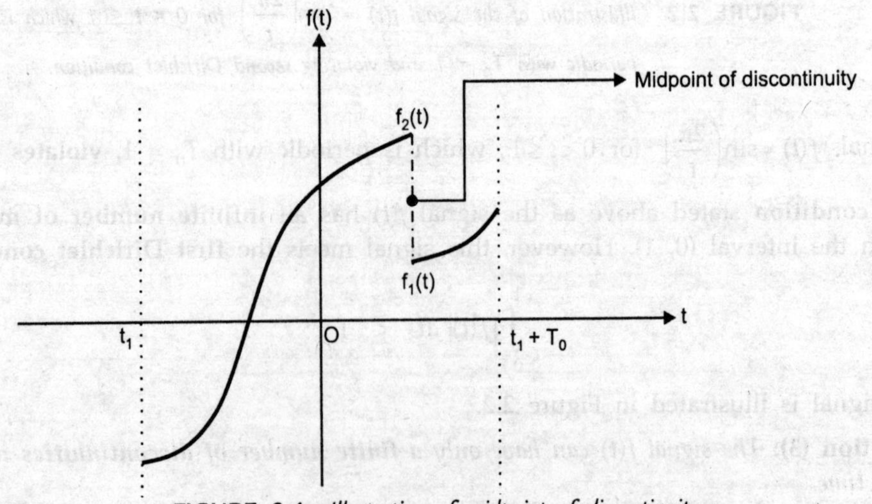

FIGURE 2.4 *Illustration of midpoint of discontinuity.*

For a **continuous periodic signal,** the Fourier series **converges** and **equals** the **original signal** for all t. However, for a periodic signal having a finite number of discontinuities in each period, the Fourier series **converges** and **equals** the **original signal** at every **continuous point** except at the **points of discontinuity** at which the series **converges** to a value equal to that of the **midpoint of discontinuity,** *i.e.,* at a point of discontinuity such as that of Figure 2.4, the Fourier series converges **towards** the value $\frac{1}{2}[f_1(t)+f_2(t)]$.

In like manner, if the signal **assumes different values** at the **boundaries** of the **interval of definition,** the Fourier series **converges** at these points **towards** the value $\frac{1}{2}[f(t_1)+f(t_1+T_0)]$.

2.2.3 Fourier Series Representation Over The Entire Interval ($-\infty < t < +\infty$)

So far, we have considered the Fourier series representation of an arbitrary signal $f(t)$ over a **finite interval** $[t_1, t_1 + T_0]$. If we denote the right hand side of equation (2.5) by $\psi(t)$, *i.e.,*

$$\psi(t) = C_0 + \sum_{n=1}^{\infty} C_n \cos[n\omega_0 t + \theta_n]$$

Then,

$$\psi(t + T_0) = C_0 + \sum_{n=1}^{\infty} C_n \cos[n\omega_0 (t + T_0) + \theta_n]$$

$$= C_0 + \sum_{n=1}^{\infty} C_n \cos[n\omega_0 t + 2n\pi + \theta_n] \qquad \left(\text{since } \omega_0 = \frac{2\pi}{T_0}\right)$$

$$= C_0 + \sum_{n=1}^{\infty} C_n \cos[n\omega_0 t + \theta_n]$$

$$= \psi(t) \qquad \text{for all } t$$

This shows that the **Fourier series** $\psi(t)$ of a signal $f(t)$ is a **periodic function** in which the **segment** of **signal** $f(t)$ over the interval $[t_1, t_1 + T_0]$ **repeats periodically** with period T_0. Thus, when we represent a signal $f(t)$ by the Fourier series $\psi(t)$ over a finite interval $[t_1, t_1 + T_0]$, the Fourier series $\psi(t)$ is **equal** to the signal $f(t)$ over **this interval alone.** Outside this interval the signal $f(t)$ and the corresponding Fourier series $\psi(t)$ **need not be equal** [Refer Figure 2.5]. The Fourier series $\psi(t)$, however, repeats periodically with period T_0 outside the interval $[t_1, t_1 + T_0]$.

If the signal $f(t)$ happens itself to be periodic with period T_0, the Fourier series $\psi(t)$ representing this periodic signal $f(t)$ over a finite interval $[t_1, t_1 + T_0]$ will also represent $f(t)$ for the **entire interval** $(-\infty, +\infty)$. Thus, the trigonometric Fourier series representing a periodic signal $f(t)$ over the entire interval $(-\infty, +\infty)$ is given by

$$f(t) = a_0 + \sum_{n=1}^{\infty} [a_n \cos n\omega_0 t + b_n \sin n\omega_0 t] \quad ; \quad -\infty < t < +\infty \qquad \text{...(2.14 A)}$$

where, the Fourier coefficients are expressed as

$$a_0 = \frac{1}{T_0} \int_{T_0} f(t)dt \qquad \text{...(2.14 B)}$$

$$a_n = \frac{2}{T_0}\int_{T_0} f(t)\cos n\omega_0 t\, dt \ ; \ n = 1, 2, 3, \ldots\ldots \quad \text{...(2.14 C)}$$

and

$$b_n = \frac{2}{T_0}\int_{T_0} f(t)\sin n\omega_0 t\, dt \ ; \ n = 1, 2, 3, \ldots\ldots \quad \text{...(2.14 D)}$$

Furthermore, the exponential Fourier series representing a periodic signal $f(t)$ over the entire interval $(-\infty, +\infty)$ is given by

$$f(t) = \sum_{n=-\infty}^{\infty} F_n\, e^{jn\omega_0 t} \ ; \quad -\infty < t < +\infty \quad \text{...(2.15 A)}$$

where,

$$F_n = \frac{1}{T_0}\int_{T_0} f(t)\, e^{-jn\omega_0 t}\, dt \ ; \ n = 0, \pm 1, \pm 2, \pm 3, \ldots\ldots, \pm\infty \quad \text{...(2.15 B)}$$

Note that, the choice of t_1 in equations (2.2) and (2.10) is **immaterial** in case of a **periodic signal**, *i.e.*, the **Fourier series expansion** of a **periodic signal** is **unique**, irrespective of the location of t_1 in the signal waveform. We, therefore, may choose any value of t_1 in equations (2.2) and (2.10). In other words, we may perform the integration over any interval of T_0. Here, $\int_{T_0} (\cdot)\, dt$ denotes that the integration is performed over any interval of T_0.

Henceforth, unless otherwise specified, we shall consider the case of **periodic signals only** in our discussions.

2.3 FOURIER SPECTRA

A periodic signal can be characterized **graphically** in the **frequency domain** by making two plots. The first being a plot of the **amplitude of harmonic components versus frequency,** known as the **amplitude spectrum** and the second being a plot of the **relative phase angle (delay angle) versus frequency,** called the **phase spectrum.** These two plots together are referred to as the **Fourier spectra** or the **frequency spectra.** However, the frequency components being **discrete,** the spectra consist of a set of lines equally spaced at interval of ω_0 and are referred to as **line spectra.**

If the spectra are obtained from the complex or exponential form of the Fourier series, lines are present at both **positive and negative frequencies**. Such spectra are referred to as **double-sided line spectra.** However, in **single-sided line spectra,** as the name implies, lines are present only at **positive frequencies.** Such spectra are obtained **either** from the **compact trigonometric form** (2.5) of the Fourier series **or** by writing the **exponential form** (2.9) of the Fourier series as a **cosine series** (2.12) by representing the **complex coefficient** F_n in **polar form** (2.11). The coefficients F_n and F_{-n} given by equations (2.8) are **complex conjugates** of each other, *i.e.*,

$$F_{-n} = F_n^*$$

If F_n is a **general complex number,** then

$$F_n = |F_n|\, e^{j\varphi_n} \quad \text{and} \quad F_{-n} = |F_n|\, e^{-j\varphi_n}$$

$$\therefore \qquad |F_{-n}| = |F_n|$$

It, therefore, follows that the **amplitude spectrum** is **symmetrical** about the vertical axis passing through the origin and is an **even function** of the frequency $n\omega_0$. Moreover, the phase of F_n is φ_n and the phase of F_{-n} is $-\varphi_n$. It is obvious that the **phase spectrum** is **anti-symmetrical** about the vertical axis passing through the origin and is an **odd function** of the frequency $n\omega_0$.

Furthermore, equations (2.13) reveal that the **height of lines** in exponential amplitude spectrum is **half** of the same in trigonometric amplitude spectrum for $n \geq 1$ and the exponential phase spectrum is **identical** to the trigonometric phase spectrum for $n \geq 0$. The height of lines rapidly decreases for a **fast convergent series**. However, for signals having **discontinuities** such as the triangular, saw-tooth *etc.*, the line height decreases slowly. Therefore, line spectra also give an **approximate indication** of the **number of terms (harmonics)** required in the **synthesis (generation)** of a periodic signal.

Example 2.1. *A signal, as shown in Figure 2.5 (a), is given by $f(t) = e^{-t/2}$; $0 \leq t \leq \pi$. Find the Fourier series representation of the signal in (a) trigonometric form (b) exponential form. Also sketch the corresponding spectra.*

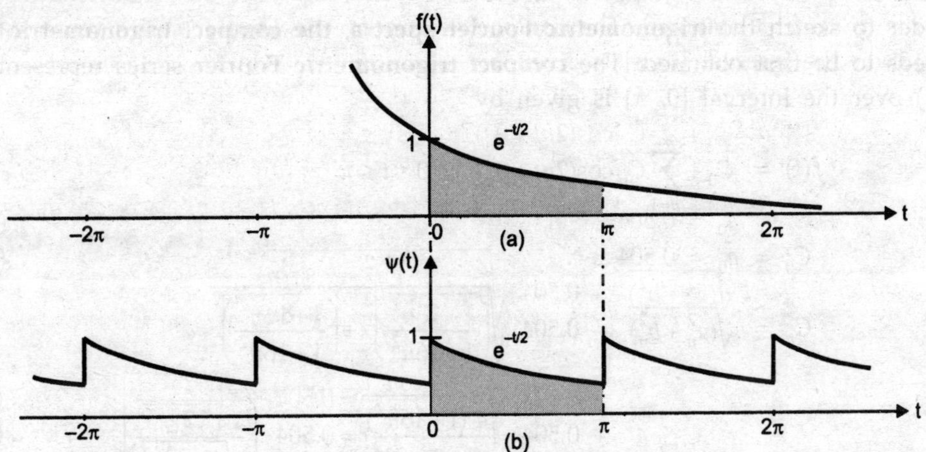

FIGURE 2.5 *(a) Signal f(t)*
(b) Its Fourier series ψ(t) repeating periodically with period π.

Solution. The **trigonometric Fourier series** representing the signal $f(t)$ over the interval $[0, \pi]$ is given by

$$f(t) = a_0 + \sum_{n=1}^{\infty} [a_n \cos n\omega_0 t + b_n \sin n\omega_0 t] \quad ; \quad 0 \leq t \leq \pi \qquad \text{...(2.16)}$$

where, $$\omega_0 = \frac{2\pi}{T_0} = 2 \text{ rad/sec.}$$

The Fourier coefficients are evaluated as follows:

$$a_0 = \frac{1}{\pi}\int_0^{\pi} e^{-t/2} dt = 0.504$$

$$a_n = \frac{2}{\pi}\int_0^{\pi} e^{-t/2}\cos 2nt\, dt = 0.504\left(\frac{2}{1+16n^2}\right)$$

$$b_n = \frac{2}{\pi}\int_0^{\pi} e^{-t/2}\sin 2nt\, dt = 0.504\left(\frac{8n}{1+16n^2}\right)$$

Putting all these values in equation (2.16), we have

$$f(t) = 0.504 + \sum_{n=1}^{\infty}\left[0.504\left(\frac{2}{1+16n^2}\right)\cos 2nt + 0.504\left(\frac{8n}{1+16n^2}\right)\sin 2nt\right] \quad ; \quad 0 \le t \le \pi$$

or,

$$f(t) = 0.504\left[1 + \sum_{n=1}^{\infty}\left\{\frac{2}{1+16n^2}(\cos\ 2nt + 4n\sin 2nt)\right\}\right] \quad ; \quad 0 \le t \le \pi \qquad ...(2.17)$$

Equation (2.17) gives the **trigonometric Fourier series** representing the signal $f(t)$ over the interval $[0,\ \pi]$.

***Note that** in Figure 2.5, the Fourier series $\psi(t)$, which denotes the right-hand side of equation (2.17), is a **periodic function** in which the **segment of signal** $f(t)$ over the interval $[0,\ \pi]$ **repeats periodically** with period π. [Refer Section 2.2.3].*

In order to sketch the **trigonometric Fourier spectra**, the **compact trigonometric Fourier series** needs to be first obtained. The **compact trigonometric Fourier series** representing the signal $f(t)$ over the interval $[0,\ \pi]$ is given by

$$f(t) = C_0 + \sum_{n=1}^{\infty} C_n\cos(2nt + \theta_n) \quad ; \quad 0 \le t \le \pi \qquad ...(2.18)$$

where,

$$C_0 = a_0 = 0.504 \qquad ...(2.19\ A)$$

$$C_n = \sqrt{(a_n^2 + b_n^2)} = 0.504\sqrt{\left(\frac{2}{1+16n^2}\right)^2 + \left(\frac{8n}{1+16n^2}\right)^2}$$

$$= 0.504\sqrt{\frac{4\,(1+16n^2)}{(1+16n^2)^2}} = 0.504\left(\frac{2}{\sqrt{1+16n^2}}\right) \qquad ...(2.19\ B)$$

and

$$\theta_n = -\tan^{-1}\left(\frac{b_n}{a_n}\right) = -\tan^{-1}\left[\frac{\left(\frac{8n}{1+16n^2}\right)}{\left(\frac{2}{1+16n^2}\right)}\right] = -\tan^{-1}\ [4n] \qquad ...(2.19\ C)$$

Putting all these values in equation (2.18), we have

$$f(t) = 0.504 + 0.504\sum_{n=1}^{\infty}\left(\frac{2}{\sqrt{1+16n^2}}\right)\cos\ (2nt - \tan^{-1}4n\ ) \ ; \ 0 \le t \le \pi \qquad (2.20)$$

Equation (2.20) gives the **compact trigonometric Fourier series** representing the signal $f(t)$ over the interval $[0,\ \pi]$. The **amplitudes** and **relative phase angles** are computed as follows:

$C_0 = 0.504$ and $\theta_0 = 0°$

$C_1 = 0.244$ and $\theta_1 = -75.96°$

$$C_2 = 0.125 \quad \text{and} \quad \theta_2 = -82.87°$$
$$C_3 = 0.084 \quad \text{and} \quad \theta_3 = -85.24°$$
$$C_4 = 0.063 \quad \text{and} \quad \theta_4 = -86.42°$$
$$C_5 = 0.051 \quad \text{and} \quad \theta_5 = -87.14°$$
$$C_6 = 0.042 \quad \text{and} \quad \theta_6 = -87.61°$$
$$C_7 = 0.036 \quad \text{and} \quad \theta_7 = -87.95°$$
$$\vdots \qquad\qquad\qquad \vdots$$

and so on.

The **trigonometric (or single-sided) Fourier spectra** are shown in Figure 2.6.

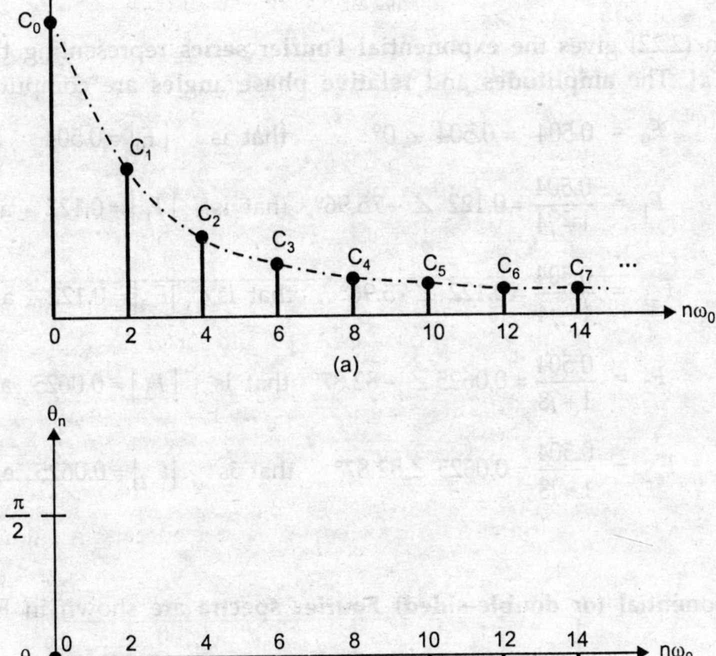

FIGURE 2.6 *Trigonometric (or single-sided) Fourier spectra of Example 2.1*
(a) Amplitude spectrum
(b) Phase spectrum.

Now, the **exponential Fourier series** representing the signal $f(t)$ over the interval $[0, \pi]$ is given by

$$f(t) = \sum_{n=-\infty}^{\infty} F_n \, e^{j2nt} \; ; \quad 0 \le t \le \pi \qquad ...(2.21)$$

where, $$F_n = \frac{1}{\pi}\int_0^{\pi} e^{-t/2}\, e^{-j2nt}\, dt = \frac{1}{\pi}\int_0^{\pi} e^{-\left(\frac{1}{2}+j2n\right)t}\, dt$$

$$= \frac{-1}{\pi\left(\frac{1}{2}+j2n\right)} \cdot e^{-\left(\frac{1}{2}+j2n\right)t}\Big|_0^{\pi} = \frac{0.504}{1+j4n}$$

Putting F_n in equation (2.21), we have

$$f(t) = \sum_{n=-\infty}^{\infty}\left[\frac{0.504}{1+j4n} \cdot e^{j2nt}\right]; \quad 0 \le t \le \pi$$

or, $$f(t) = 0.504\sum_{n=-\infty}^{\infty}\left[\frac{1}{1+j4n} \cdot e^{j2nt}\right]; \quad 0 \le t \le \pi \qquad \text{...(2.22)}$$

Equation (2.22) gives the **exponential Fourier series** representing the signal $f(t)$ over the interval $[0, \pi]$. The **amplitudes** and **relative phase angles** are computed as follows:

$$F_0 = 0.504 = 0.504 \angle 0° \quad \text{that is} \quad |F_0| = 0.504 \quad \text{and} \quad \varphi_0 = 0°$$

$$F_1 = \frac{0.504}{1+j4} = 0.122 \angle -75.96° \quad \text{that is} \quad |F_1| = 0.122 \quad \text{and} \quad \varphi_1 = -75.96°$$

$$F_{-1} = \frac{0.504}{1-j4} = 0.122 \angle 75.96° \quad \text{that is} \quad |F_{-1}| = 0.122 \quad \text{and} \quad \varphi_{-1} = 75.96°$$

$$F_2 = \frac{0.504}{1+j8} = 0.0625 \angle -82.87° \quad \text{that is} \quad |F_2| = 0.0625 \quad \text{and} \quad \varphi_2 = -82.87°$$

$$F_{-2} = \frac{0.504}{1-j8} = 0.0625 \angle 82.87° \quad \text{that is} \quad |F_{-2}| = 0.0625 \quad \text{and} \quad \varphi_{-2} = 82.87°$$

$$\vdots \qquad\qquad \vdots \qquad\qquad \vdots \qquad\qquad \vdots$$

and so on.

The **exponential (or double-sided) Fourier spectra** are shown in Figure 2.7.

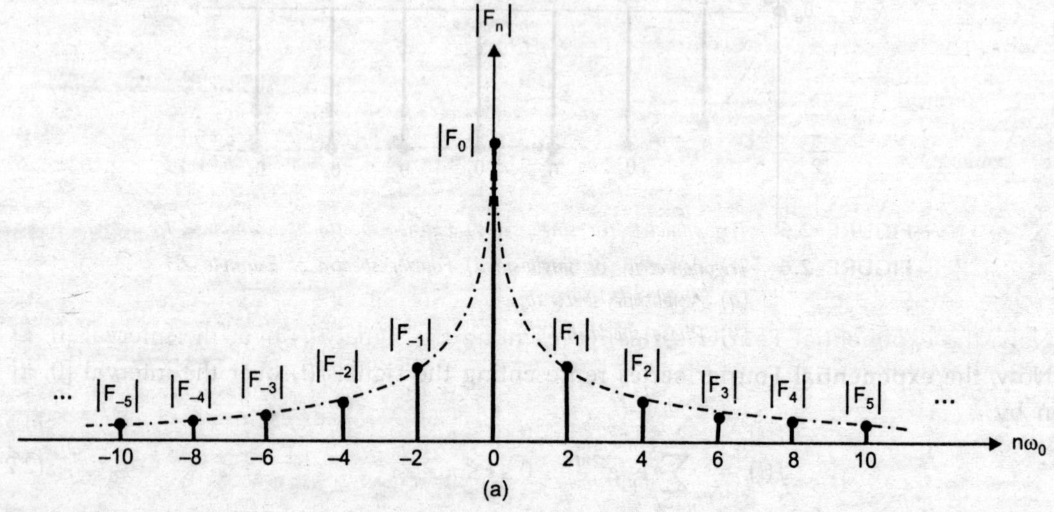

(a)

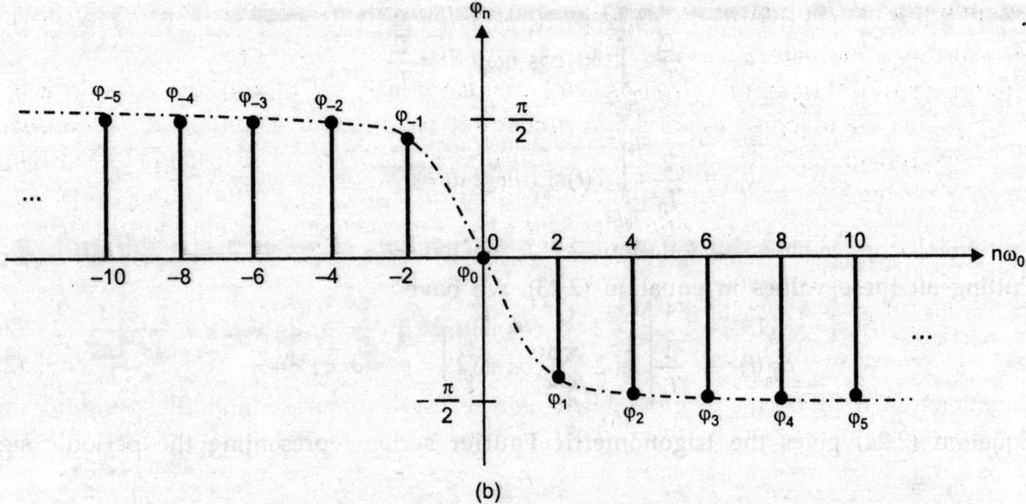

FIGURE 2.7 *Exponential (or double-sided) Fourier spectra of Example 2.1*
(a) Amplitude spectrum
(b) Phase spectrum.

Example 2.2. *Find (a) trigonometric Fourier series (b) exponential Fourier series for the periodic impulse train* $\delta_{T_0}(t)$ *as shown in Figure 2.8. Also sketch the corresponding spectra.*

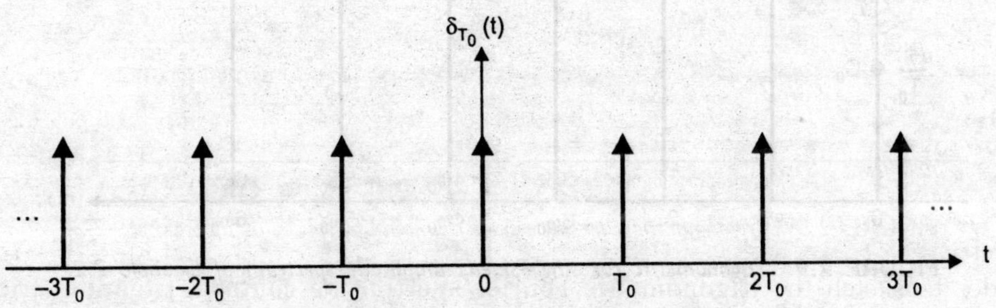

FIGURE 2.8 *Periodic impulse-train.*

Solution. The **trigonometric Fourier series** representing the periodic signal $\delta_{T_0}(t)$ is given by

$$\delta_{T_0}(t) = a_0 + \sum_{n=1}^{\infty}[a_n \cos n\omega_0 t + b_n \sin n\omega_0 t] \quad ; \; -\infty < t < +\infty \qquad ...(2.23)$$

where, $\omega_0 = \dfrac{2\pi}{T_0}$

The Fourier coefficients are evaluated by choosing the interval of integration $\left[-\dfrac{T_0}{2}, \dfrac{T_0}{2}\right]$ and recognizing that over this interval $\delta_{T_0}(t) = \delta(t)$ as follows:

$$a_0 = \frac{1}{T_0}\int_{-T_0/2}^{T_0/2} \delta(t)dt = \frac{1}{T_0}$$

$$a_n = \frac{2}{T_0} \int_{-T_0/2}^{T_0/2} \delta(t) \cos n\omega_0 t \, dt = \frac{2}{T_0}$$

$$b_n = \frac{2}{T_0} \int_{-T_0/2}^{T_0/2} \delta(t) \sin n\omega_0 t \, dt = 0$$

Putting all these values in equation (2.23), we have

$$\delta_{T_0}(t) = \frac{1}{T_0}\left[1 + 2\sum_{n=1}^{\infty} \cos n\omega_0 t\right] \quad ; \quad -\infty < t < +\infty \qquad \text{...(2.24)}$$

Equation (2.24) gives the **trigonometric Fourier series** representing the periodic signal $\delta_{T_0}(t)$.

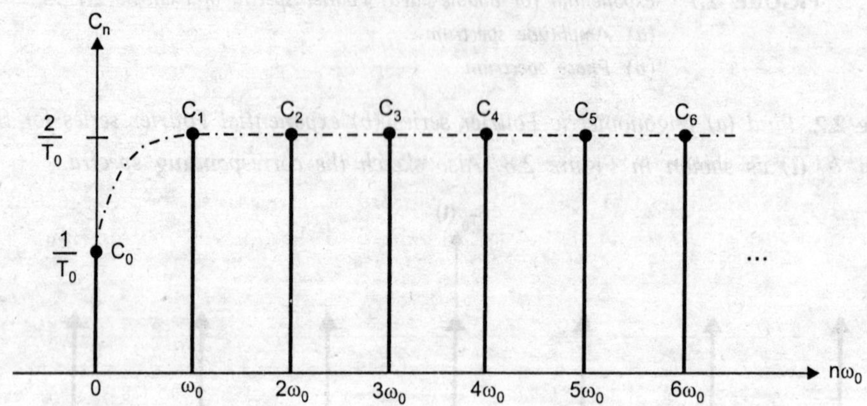

FIGURE 2.9 *Trigonometric (or single-sided) amplitude spectrum of Example 2.2.*

In order to sketch the **trigonometric Fourier spectra,** the **compact trigonometric Fourier series** needs to be first obtained. The **compact trigonometric Fourier series** representing the signal $\delta_{T_0}(t)$ is given by

$$\delta_{T_0}(t) = C_0 + \sum_{n=1}^{\infty} C_n \cos(n\omega_0 t + \theta_n) \quad ; \quad -\infty < t < +\infty \qquad \text{...(2.25)}$$

where,

$$C_0 = a_0 = \frac{1}{T_0} \qquad \text{...(2.26 A)}$$

$$C_n = \sqrt{(a_n^2 + b_n^2)} = \frac{2}{T_0} \qquad \text{...(2.26 B)}$$

and

$$\theta_n = -\tan^{-1}\left(\frac{b_n}{a_n}\right) = -\tan^{-1}[0] = 0° \qquad \text{...(2.26 C)}$$

Putting all these values in equation (2.25), we have

$$\delta_{T_0}(t) = \frac{1}{T_0} + \sum_{n=1}^{\infty} \frac{2}{T_0}\cos(n\omega_0 t) \; ; \quad -\infty < t < +\infty \qquad ...(2.27)$$

Equation (2.27) gives the **compact trigonometric Fourier series** representing the periodic signal $\delta_{T_0}(t)$. All **relative phase angles** are zero, therefore, no need of phase spectrum. The **trigonometric (or single-sided) amplitude spectrum** is shown in Figure 2.9.

Now, the **exponential Fourier series** representing the periodic signal $\delta_{T_0}(t)$ is given by

$$\delta_{T_0}(t) = \sum_{n=-\infty}^{\infty} F_n\, e^{jn\omega_0 t} \; ; \quad -\infty < t < +\infty \qquad ...(2.28)$$

where,

$$F_n = \frac{1}{T_0}\int_{T_0} \delta_{T_0}(t) e^{-jn\omega_0 t}\, dt$$

Choosing the interval of integration $\left[-\frac{T_0}{2}, \frac{T_0}{2}\right]$ and recognizing that over this interval $\delta_{T_0}(t) = \delta(t)$, we have

$$F_n = \frac{1}{T_0}\int_{-T_0/2}^{T_0/2} \delta(t) e^{-jn\omega_0 t}\, dt = \frac{1}{T_0}$$

Putting F_n in equation (2.28), we have

$$\delta_{T_0}(t) = \frac{1}{T_0}\sum_{n=-\infty}^{\infty} e^{jn\omega_0 t} \; ; \quad -\infty < t < +\infty \qquad ...(2.29)$$

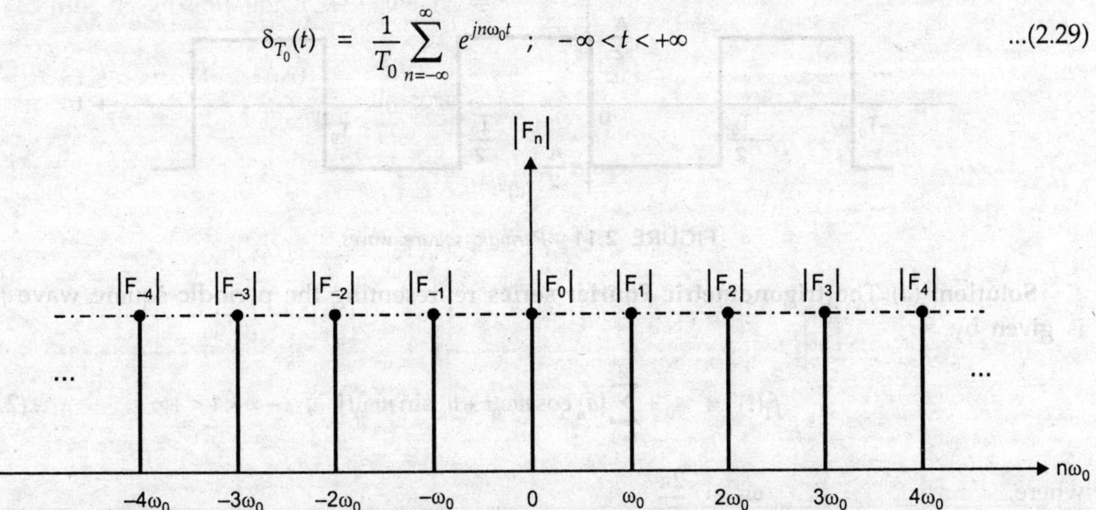

FIGURE 2.10 *Exponential (or double-sided) amplitude spectrum of Example 2.2.*

Equation (2.29) gives the **exponential Fourier series** representing the periodic signal $\delta_{T_0}(t)$. All **relative phase angles** are **zero**, therefore, no need of phase spectrum. The **exponential (or double-sided) amplitude spectrum** is shown in Figure 2.10.

Example 2.3. *Find the trigonometric Fourier series for the periodic square wave for different cases as depicted in Figure 2.11. Compare as well as comment on results obtained.*

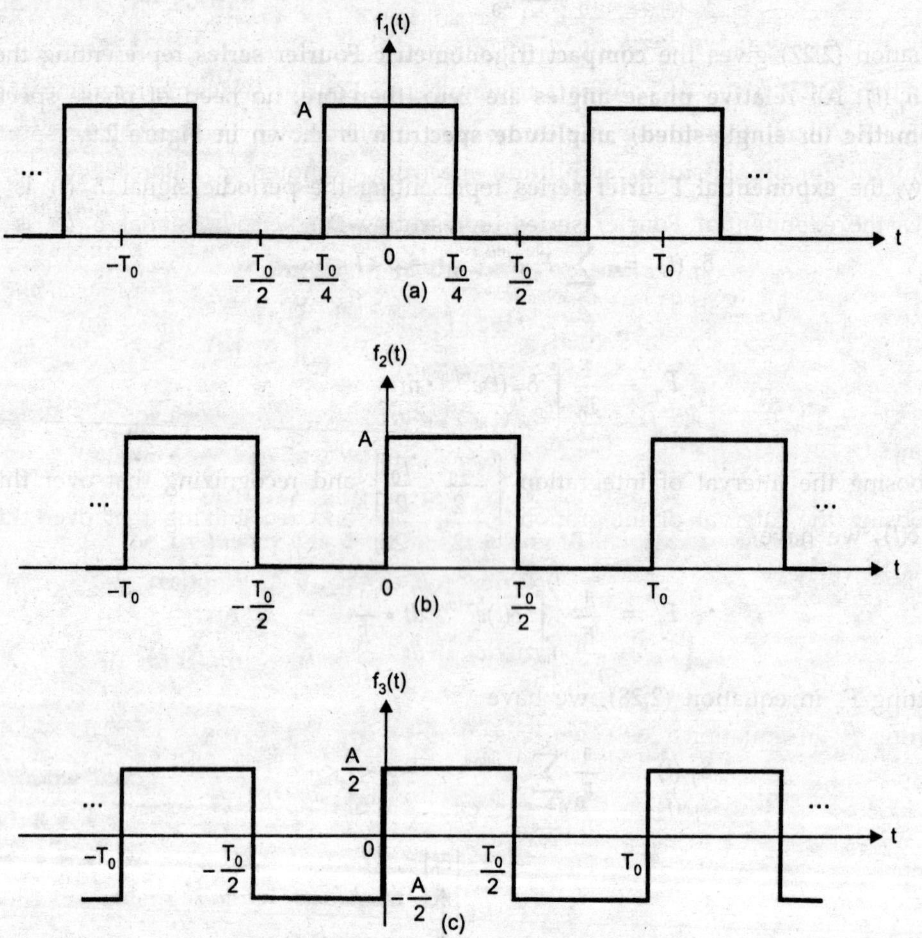

FIGURE 2.11 *Periodic square waves.*

Solution. (*a*) The **trigonometric Fourier series** representing the periodic square wave $f_1(t)$ is given by

$$f_1(t) = a_0 + \sum_{n=1}^{\infty}[a_n \cos n\omega_0 t + b_n \sin n\omega_0 t] \quad ; \quad -\infty < t < +\infty \qquad ...(2.30)$$

where,
$$\omega_0 = \frac{2\pi}{T_0}$$

Choosing the interval of integration $\left[-\frac{T_0}{2}, \frac{T_0}{2}\right]$ and recognizing that $\omega_0 T_0 = 2\pi$, the Fourier coefficients are evaluated as follows:

$$a_0 = \frac{1}{T_0}\int_{-T_0/2}^{T_0/2} f_1(t)dt = \frac{1}{T_0}\int_{-T_0/4}^{T_0/4} A\,dt = \frac{A}{2}$$

$$a_n = \frac{2}{T_0}\int_{-T_0/2}^{T_0/2} f_1(t)\cos n\omega_0 t\, dt$$

$$= \frac{2}{T_0}\int_{-T_0/2}^{-T_0/4} f_1(t)\cos n\omega_0 t\, dt + \frac{2}{T_0}\int_{-T_0/4}^{T_0/4} f_1(t)\cos n\omega_0 t\, dt + \frac{2}{T_0}\int_{T_0/4}^{T_0/2} f_1(t)\cos n\omega_0 t\, dt$$

$$= \frac{2}{T_0}\int_{-T_0/4}^{T_0/4} A\cos n\omega_0 t\, dt = \frac{2A}{T_0}\left[\frac{\sin n\omega_0 t}{n\omega_0}\right]_{-T_0/4}^{T_0/4}$$

$$= \frac{2A}{n\pi}\sin\left(\frac{n\pi}{2}\right)$$

$$b_n = \frac{2}{T_0}\int_{-T_0/2}^{T_0/2} f_1(t)\sin n\omega_0 t\, dt$$

$$= \frac{2}{T_0}\int_{-T_0/2}^{-T_0/4} f_1(t)\sin n\omega_0 t\, dt + \frac{2}{T_0}\int_{-T_0/4}^{T_0/4} f_1(t)\sin n\omega_0 t\, dt + \frac{2}{T_0}\int_{T_0/4}^{T_0/2} f_1(t)\sin n\omega_0 t\, dt$$

$$= \frac{2}{T_0}\int_{-T_0/4}^{T_0/4} A\sin n\omega_0 t\, dt = \frac{2A}{T_0}\left[-\frac{\cos n\omega_0 t}{n\omega_0}\right]_{-T_0/4}^{T_0/4}$$

$$= 0$$

Putting all these values in equation (2.30), we have

$$f_1(t) = \frac{A}{2} + \sum_{n=1}^{\infty}\left[\frac{2A}{n\pi}\sin\left(\frac{n\pi}{2}\right)\cos n\omega_0 t\right] \quad ; \quad -\infty < t < +\infty \qquad \text{...(2.31 A)}$$

Equation (2.31 A) gives the **trigonometric Fourier series** representing the periodic signal $f_1(t)$. In an expanded form, we have

$$f_1(t) = \frac{A}{2} + \frac{2A}{\pi}\cos\omega_0 t - \frac{2A}{3\pi}\cos 3\omega_0 t + \frac{2A}{5\pi}\cos 5\omega_0 t - \frac{2A}{7\pi}\cos 7\omega_0 t + \ldots\ldots \quad ; \quad -\infty < t < +\infty \qquad \text{...(2.31 B)}$$

(*b*) The **trigonometric Fourier series** representing the periodic square wave $f_2(t)$ is given by

$$f_2(t) = a_0 + \sum_{n=1}^{\infty}[a_n \cos n\omega_0 t + b_n \sin n\omega_0 t] \quad ; \quad -\infty < t < +\infty \qquad \text{...(2.32)}$$

where, $\omega_0 = \dfrac{2\pi}{T_0}$

Choosing the interval of integration $[0, T_0]$ and recognizing that $\omega_0 T_0 = 2\pi$, the Fourier coefficients are evaluated as follows:

$$a_0 = \frac{1}{T_0}\int_0^{T_0} f_2(t)dt$$

$$= \frac{1}{T_0}\int_0^{T_0/2} f_2(t)dt + \frac{1}{T_0}\int_{T_0/2}^{T_0} f_2(t)dt$$

$$= \frac{1}{T_0}\int_0^{T_0/2} A\,dt = \frac{A}{2}$$

$$a_n = \frac{2}{T_0}\int_0^{T_0} f_2(t)\cos n\omega_0 t\,dt$$

$$= \frac{2}{T_0}\int_0^{T_0/2} f_2(t)\cos n\omega_0 t\,dt + \frac{2}{T_0}\int_{T_0/2}^{T_0} f_2(t)\cos n\omega_0 t\,dt$$

$$= \frac{2}{T_0}\int_0^{T_0/2} A\cos n\omega_0 t\,dt = \frac{2A}{T_0}\left[\frac{\sin n\omega_0 t}{n\omega_0}\right]_0^{T_0/2}$$

$$= 0$$

$$b_n = \frac{2}{T_0}\int_0^{T_0} f_2(t)\sin n\omega_0 t\,dt$$

$$= \frac{2}{T_0}\int_0^{T_0/2} f_2(t)\sin n\omega_0 t\,dt + \frac{2}{T_0}\int_{T_0/2}^{T_0} f_2(t)\sin n\omega_0 t\,dt$$

$$= \frac{2}{T_0}\int_0^{T_0/2} A\sin n\omega_0 t\,dt = \frac{2A}{T_0}\left[-\frac{\cos n\omega_0 t}{n\omega_0}\right]_0^{T_0/2}$$

$$= \frac{A}{n\pi}(1-\cos n\pi)$$

Putting all these values in equation (2.32), we have

$$f_2(t) = \frac{A}{2} + \sum_{n=1}^{\infty}\left[\frac{A}{n\pi}(1-\cos n\pi)\cos n\omega_0 t\right] \quad ; \quad -\infty < t < +\infty \qquad \text{...(2.33 A)}$$

Equation (2.33 A) gives the **trigonometric Fourier series** representing the periodic signal $f_2(t)$. In an expanded form, we have

$$f_2(t) = \frac{A}{2} + \frac{2A}{\pi}\sin\omega_0 t + \frac{2A}{3\pi}\sin 3\omega_0 t + \frac{2A}{5\pi}\sin 5\omega_0 t + \frac{2A}{7\pi}\sin 7\omega_0 t + \ldots\ldots \quad ; \quad -\infty < t < +\infty \qquad \text{...(2.33 B)}$$

(*c*) The **trigonometric Fourier series** representing the periodic square wave $f_3(t)$ is given by

$$f_3(t) = a_0 + \sum_{n=1}^{\infty}[a_n\cos n\omega_0 t + b_n\sin n\omega_0 t] \quad ; \quad -\infty < t < +\infty \qquad \text{...(2.34)}$$

where, $$\omega_0 = \frac{2\pi}{T_0}$$

Choosing the interval of integration $[0, T_0]$ and recognizing that $\omega_0 T_0 = 2\pi$, the Fourier coefficients are evaluated as follows:

$$a_0 = \frac{1}{T_0}\int_0^{T_0} f_3(t)dt = \frac{1}{T_0}\int_0^{T_0/2} f_3(t)dt + \frac{1}{T_0}\int_{T_0/2}^{T_0} f_3(t)dt$$

$$= \frac{1}{T_0}\int_0^{T_0/2}\left(\frac{A}{2}\right)dt + \frac{1}{T_0}\int_{T_0/2}^{T_0}\left(-\frac{A}{2}\right)dt$$

$$= 0$$

$$a_n = \frac{2}{T_0}\int_0^{T_0} f_3(t)\cos n\omega_0 t\, dt$$

$$= \frac{2}{T_0}\int_0^{T_0/2} f_3(t)\cos n\omega_0 t\, dt + \frac{2}{T_0}\int_{T_0/2}^{T_0} f_3(t)\cos n\omega_0 t\, dt$$

$$= \frac{2}{T_0}\int_0^{T_0/2}\left(\frac{A}{2}\right)\cos n\omega_0 t\, dt + \frac{2}{T_0}\int_{T_0/2}^{T_0}\left(-\frac{A}{2}\right)\cos n\omega_0 t\, dt$$

$$= \frac{A}{T_0}\left[\frac{\sin n\omega_0 t}{n\omega_0}\right]_0^{T_0/2} - \frac{A}{T_0}\left[\frac{\sin n\omega_0 t}{n\omega_0}\right]_{T_0/2}^{T_0}$$

$$= 0$$

$$b_n = \frac{2}{T_0}\int_0^{T_0} f_3(t)\sin n\omega_0 t\, dt$$

$$= \frac{2}{T_0}\int_0^{T_0/2} f_3(t)\sin n\omega_0 t\, dt + \frac{2}{T_0}\int_{T_0/2}^{T_0} f_3(t)\sin n\omega_0 t\, dt$$

$$= \frac{2}{T_0}\int_0^{T_0/2}\left(\frac{A}{2}\right)\sin n\omega_0 t\, dt + \frac{2}{T_0}\int_{T_0/2}^{T_0}\left(-\frac{A}{2}\right)\sin n\omega_0 t\, dt$$

$$= \frac{2A}{T_0}\left[-\frac{\cos n\omega_0 t}{n\omega_0}\right]_0^{T_0/2} + \frac{2A}{T_0}\left[-\frac{\cos n\omega_0 t}{n\omega_0}\right]_{T_0/2}^{T_0}$$

$$= \frac{A}{n\pi}(1-\cos n\pi)$$

Putting all these values in equation (2.34), we have

$$f_3(t) = \frac{A}{2} + \sum_{n=1}^{\infty}\left[\frac{A}{n\pi}(1-\cos n\pi)\cos n\omega_0 t\right] \quad ; \quad -\infty < t < +\infty \qquad \text{...(2.35 A)}$$

Equation (2.35 A) gives the **trigonometric Fourier series** representing the periodic signal $f_3(t)$. In an expanded form, we have

$$f_3(t) = \frac{2A}{\pi}\sin\omega_0 t + \frac{2A}{3\pi}\sin 3\omega_0 t + \frac{2A}{5\pi}\sin 5\omega_0 t + \frac{2A}{7\pi}\sin 7\omega_0 t + \ldots\ldots \quad ; \quad -\infty < t < +\infty \qquad \text{...(2.35 B)}$$

Comparing equations (2.31 B), (2.33 B) and (2.35 B), we observe that the Fourier series representation of periodic square wave $f_1(t)$ contains **cosine terms only** and all sine terms are zero, whereas the same of periodic square wave $f_2(t)$ and $f_3(t)$ contain **sine terms only** and all cosine terms are zero. Further, all the Fourier series representations contain **only odd harmonics.**

As illustrated in Figure 2.11, the periodic square wave $f_2(t)$ is obtained by **shifting the vertical axis** of the periodic square wave $f_1(t)$. Taking into account the above observations, we may conclude that horizontal shifting of vertical axis **does not add any new frequencies**; it merely **alters** the **magnitudes of sine and cosine terms.**

Similarly, the periodic square wave $f_3(t)$ is obtained by **shifting the horizontal axis,** as illustrated in Figure 2.11, of the periodic square wave $f_2(t)$. Taking into account the above observations, we may conclude that vertical shifting of horizontal axis only **alters** the **average** or the **constant term**; it has **no effect on other components** of the Fourier series.

2.4 EFFECTS OF SYMMETRY

In equation (2.31 B), we have $b_n = 0$ for all n and $a_n = 0$ for all even values of n. Similarly, in equation (2.33 B) and (2.35 B), we have $a_n = 0$ for all n and $b_n = 0$ for all even values of n. This is not an accident rather is because of **certain types of symmetry** associated with the signal waveforms which result in some of the Fourier coefficients being absent in the Fourier series representation of the signal.

If we could predict right at the beginning which of the Fourier coefficients would be zero by recognizing such symmetries, a considerable amount of labor may be saved in the Fourier series analysis. It can be shown that the Fourier series of **any even periodic signal** consists of **cosine terms only** and the series of **any odd periodic signal** consists of **sine terms only.** Further, the Fourier series of **any periodic signal** in which every half-cycle is an **inverted version** of the adjacent one, consists of **odd harmonics** only; all even harmonic components are zero. Now, we shall **summarize** what we have just stated as follows:

(*a*) If $f(t) = f(-t)$; Even Symmetry

Then **all** the **sine terms** in the Fourier series **vanish,** *i.e.,* $b_n = 0$. This is verified by Example 2.3 (*a*).

(*b*) If $f(t) = -f(-t)$; Odd Symmetry

Then the **d.c.** and **all** the **cosine terms** in the Fourier series **vanish,** *i.e.,* $a_0 = a_n = 0$. This is verified by Example 2.3 (*c*).

(*c*) If $f\left(t \pm \frac{T_0}{2}\right) = -f(t)$; Half-wave (or Mirror or Rotation) Symmetry

Then all the coefficients of **even harmonics** in the Fourier series **vanish**, *i.e.*, the Fourier series contains **only odd harmonics**. This is verified by Example 2.3 where the Fourier series expansions (2.31 B), (2.33 B) and (2.35 B) of corresponding periodic square waves of Figure 2.11 contain only those frequency components which are **odd multiple** of ω_0.

Now, consider the Fourier series representation (2.33 B) for periodic square wave shown in Figure 2.11 (*b*), which also contains **sine terms** only and, therefore, it must possess the **odd symmetry**. But, this fact is **not apparent** from the waveform of Figure 2.11 (*b*). However, the odd symmetry of the waveform is more clearly brought out by **shifting the horizontal axis** so as to remove the average or d.c. value of the periodic square wave. It, therefore, implies that **the even or odd symmetry should be examined only after removing the average or d.c. value of the periodic signal by suitable shifting of the horizontal axis.**

It should be stressed in this connection that **even or odd symmetry** is **not a fundamental property** of the periodic signal rather depends upon the **location of the vertical axis**, which is **arbitrary** and the choice, in general, exists over the entire interval $(-\infty, +\infty)$. Thus, an **even** periodic signal can be made **odd** by simply **shifting the vertical axis**. However, the **shifting of either axis** does not modify the presence or absence of **even or odd harmonics**, which is an **intrinsic property** of the periodic signal. These can be verified by Figure 2.11, where periodic square waveform (*a*) possesses **even symmetry** and can be made to possess **odd symmetry** simply by shifting the vertical axis by $\frac{T_0}{4}$ resulting in periodic square waveform (*b*), but in either case the Fourier series representations contain **only odd harmonics.**

Also, one can easily verify that if **either** of the above symmetries is associated with the periodic signal, the **limit of integration** in the expression of a_n and b_n reduces to **half the period with twice the signal amplitude.**

2.5 THE GIBBS PHENOMENON

After expanding a periodic signal into a Fourier series, it is attempted to **reconstruct** it by summing a number of harmonic components. To get a feel of the idea of **approximating** an arbitrary signal by the **partial sum** of the harmonic components, let us consider the trigonometric Fourier series representation of the periodic square wave given by equation (2.31 B).

If we take partial sum of first two terms of the Fourier series in equation (2.31 B), *i.e.*, the d.c. component and the fundamental harmonic component for approximating the periodic square wave, we get the reconstructed waveform as shown in Figure 2.12 (*a*). If we include the third harmonic also in the partial sum, we get the reconstructed waveform as shown in Figure 2.12 (*b*). Figure 2.12 (*c*) shows the approximation of periodic square wave by the partial sum of terms up to the fifth harmonic.

It is found that the reconstructed curve approximates the original function **better and better** as the number of harmonics is increased in the partial sum, **except** in the vicinity of the discontinuities, where the partial sum converges to $\frac{A}{2}$, the **midpoint** of discontinuity, as shown in Figure 2.12 (*d*) [Refer Section 2.2.2].

Another interesting fact, as illustrated in Figure 2.12 (*d*), is that there occurs an **overshoot** on both sides of the discontinuity. One would think that the amplitude of this overshoot would tend to zero as the number of terms in the Fourier series is increased to infinity, but this is not the case.

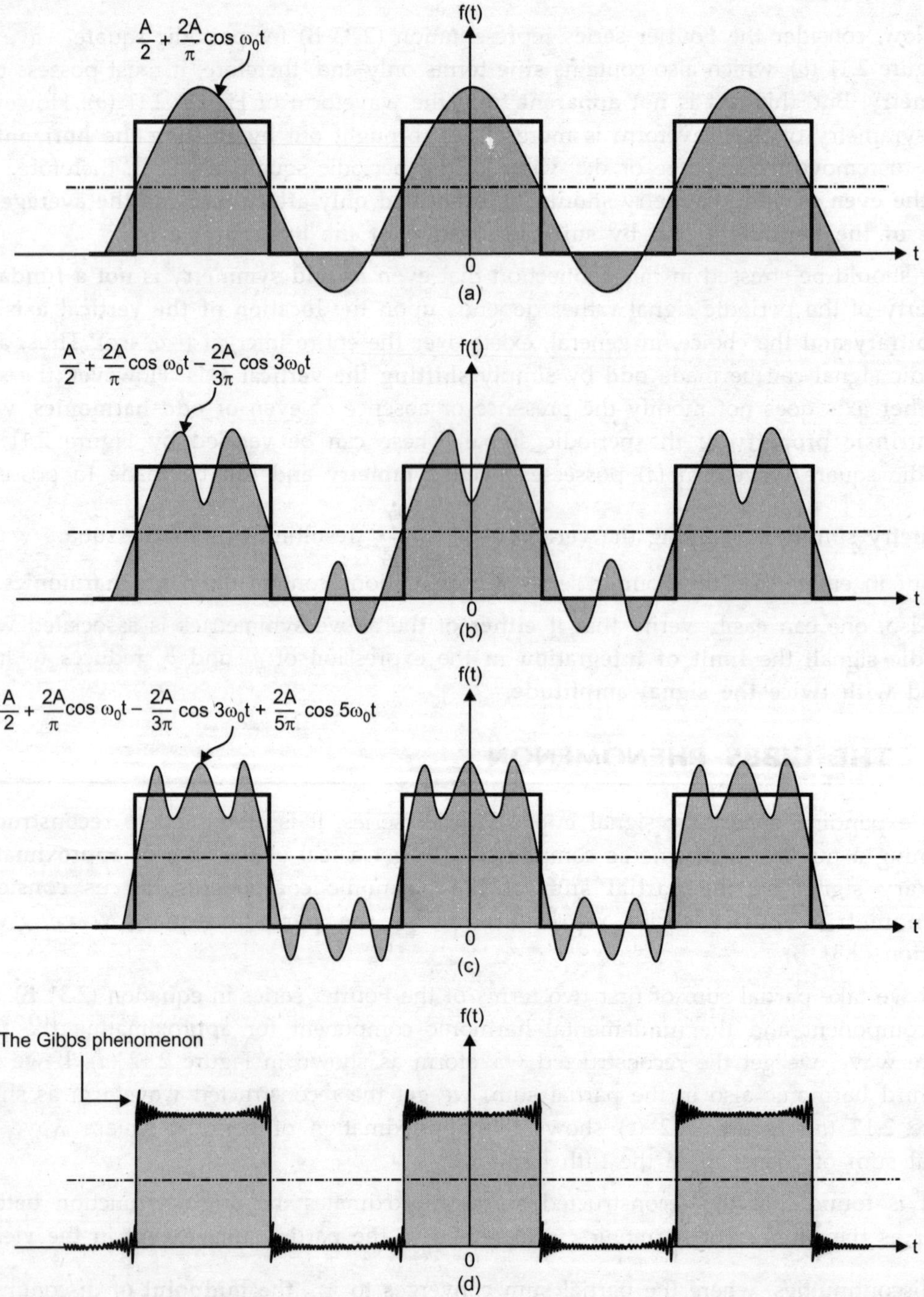

FIGURE 2.12 *Approximation of the periodic square wave by the partial-sum of harmonic components: An illustration of the Gibbs phenomenon.*

Josiah Gibbs, a famous mathematical physicist, showed that for a discontinuity of certain height, the partial sum exhibits an overshoot of 9% of the height of discontinuity; no matters how large number of harmonic components are included in the partial sum. This behaviour has come to be known as the **Gibbs phenomenon.** It reveals that at a point of discontinuity, the given function cannot be approximated to a tolerance of better than ±9%, even if a very large number of terms are used in the Fourier series representation. Of course, a **redeeming feature** is that the **time occupied by this overshoot does tend to zero** and, therefore, in practice the **overshoot may not have much effect.**

2.6 PARSEVAL'S RELATION

The **Parseval's relationships** state that the **time-averaged power** in a **periodic signal** $f(t)$ is equal to the **sum of the time-averaged powers** in all its **harmonic components.** The power of a continuous-time periodic signal $f(t)$ is given by

$$P = \frac{1}{T_0} \int_{-T_0}^{T_0} |f(t)|^2 \, dt$$

where it is assumed that $f(t)$ may be **complex-valued** in general. Recognizing that

$$|f(t)|^2 = f(t) f^*(t)$$

and that $f^*(t)$ is represented in the **Fourier series** as

$$f^*(t) = \sum_{n=-\infty}^{\infty} F_n^* \, e^{-jn\omega_0 t}$$

Thus, we have

$$P = \frac{1}{T_0} \int_{-T_0}^{T_0} f(t) \left[\sum_{n=-\infty}^{\infty} F_n^* \, e^{-jn\omega_0 t} \right] dt$$

Interchanging the order of integration and summation so as to obtain

$$P = \sum_{n=-\infty}^{\infty} F_n^* \left[\frac{1}{T_0} \int_{-T_0}^{T_0} f(t) \, e^{-jn\omega_0 t} \, dt \right]$$

or,

$$P = \sum_{n=-\infty}^{\infty} F_n^* \, F_n$$

Hence, we conclude that

$$\frac{1}{T_0} \int_{-T_0}^{T_0} |f(t)|^2 \, dt = \sum_{n=-\infty}^{\infty} |F_n|^2$$

The quantity $|F_n|^2$ is termed as the **average power** in the nth **harmonic component** of the periodic signal $f(t)$.

Example 2.4. *Find the trigonometric Fourier representation of the periodic saw-tooth waveform as shown in Figure 2.13.*

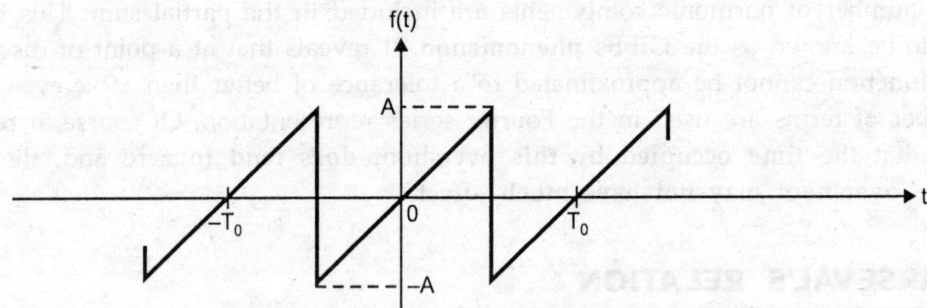

FIGURE 2.13 *Periodic saw-tooth waveform of Example 2.4.*

Solution. The **trigonometric Fourier series** representing the periodic saw-tooth waveform $f(t)$ is given by

$$f(t) = a_0 + \sum_{n=1}^{\infty}[a_n \cos n\omega_0 t + b_n \sin n\omega_0 t] \quad ; \quad -\infty < t < +\infty \qquad ...(2.36)$$

where, $$\omega_0 = \frac{2\pi}{T_0}$$

Before evaluating Fourier coefficients, we shall first examine the given saw-tooth waveform for **symmetry conditions**, and we observe that

$$f(t) = -f(-t) \quad ; \quad \text{Odd Symmetry}$$

Therefore, the **d.c.** and **all** the **cosine terms** in the Fourier series **vanish**, *i.e.*, $a_0 = a_n = 0$ and the Fourier series consists of **sine terms only.** We have

$$f(t) = \sum_{n=1}^{\infty} b_n \sin n\omega_0 t \quad ; \quad -\infty < t < +\infty \qquad ...(2.37)$$

The periodic waveform shown in Figure 2.13 can be described as

$$f(t) = \left(\frac{2A}{T_0}t\right) \quad ; \quad -\frac{T_0}{2} \le t \le \frac{T_0}{2}$$

Choosing the interval of integration $\left[-\frac{T_0}{2}, \frac{T_0}{2}\right]$ and recognizing that $\omega_0 T_0 = 2\pi$, the Fourier coefficient b_n is evaluated as follows:

$$b_n = \frac{2}{T_0}\int_{-T_0/2}^{T_0/2} f(t)\sin n\omega_0 t \, dt$$

$$= \frac{4}{T_0}\int_{0}^{T_0/2}\left(\frac{2A}{T_0}t\right)\sin n\omega_0 t \, dt \qquad \text{[Refer Section 2.4]}$$

$$= \frac{8A}{T_0^2}\int_{0}^{T_0/2} t \sin n\omega_0 t \, dt$$

$$= \frac{8A}{T_0^2}\left[\frac{1}{n^2\omega_0^2}\sin n\omega_0 t - \frac{t}{n\omega_0}\cos n\omega_0 t\right]_0^{T_0/2}$$

$$= \frac{8A}{T_0^2}\left[\frac{1}{n^2\omega_0^2}\sin n\pi - \frac{T_0}{2n\omega_0}\cos n\pi\right]$$

$$= \frac{4A}{T_0^2}\left[-\frac{T_0}{n\omega_0}\cos n\pi\right] = -\frac{2A}{n\pi}\cos n\pi$$

Putting above value in equation (2.37), we have

$$f(t) = \sum_{n=1}^{\infty}\left[\left(-\frac{2A}{n\pi}\cos n\pi\right)\sin n\omega_0 t\right] \quad ; \quad -\infty < t < +\infty \qquad \text{...(2.38 A)}$$

Equation (2.38 A) gives the **trigonometric Fourier series** representing the periodic saw-tooth waveform $f(t)$. In an expanded form, we have

$$f(t) = \frac{2A}{\pi}\sin\omega_0 t - \frac{2A}{2\pi}\sin 2\omega_0 t + \frac{2A}{3\pi}\sin 3\omega_0 t - \frac{2A}{4\pi}\sin 4\omega_0 t + \ldots\ldots \quad ; \quad -\infty < t < +\infty$$

...(2.38 B)

It is important to note that if we choose the interval of integration as $[0, T_0]$, the integration would be carried out in two parts, *i.e.*, from 0 to $\frac{T_0}{2}$ and again from $\frac{T_0}{2}$ to T_0, since the governing equations of the waveform are different for the two intervals. On the other hand, if we choose the interval of integration as $\left[-\frac{T_0}{2}, \frac{T_0}{2}\right]$, the **full period is covered** and the governing equation of the waveform is the same for the entire interval and, therefore, there is only one integration.

Example 2.5. *Obtain the trigonometric Fourier series for the signal as shown in Figure 2.14.*

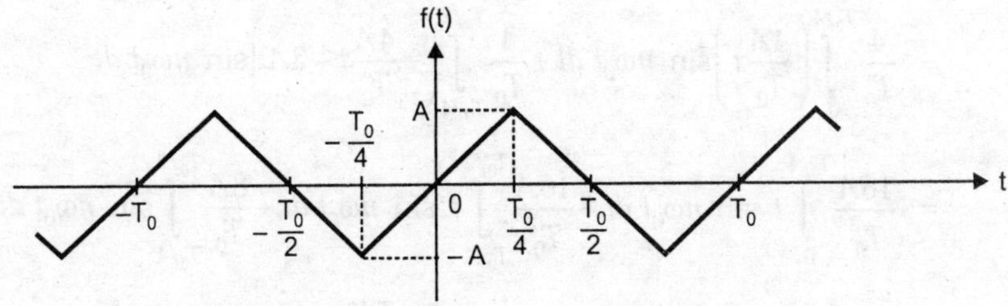

FIGURE 2.14 *Periodic signal of Example 2.5.*

Solution. The **trigonometric Fourier series** representing the periodic triangular waveform $f(t)$ is given by

$$f(t) = a_0 + \sum_{n=1}^{\infty}[a_n\cos n\omega_0 t + b_n\sin n\omega_0 t] \quad ; \quad -\infty < t < +\infty \qquad \text{...(2.39)}$$

where, $\omega_0 = \frac{2\pi}{T_0}$

Before evaluating Fourier coefficients, we shall first examine the given triangular waveform for **symmetry conditions**, and we observe that

$$f(t) = -f(-t) \quad ; \quad \text{Odd Symmetry}$$

Therefore, the **d.c.** and **all** the **cosine terms** in the Fourier series **vanish**, *i.e.*, $a_0 = a_n = 0$ and the Fourier series consists of **sine terms only.**

Also, $$f\left(t \pm \frac{T_0}{2}\right) = -f(t) \quad ; \quad \text{Half-wave (or Mirror or Rotation) Symmetry}$$

Therefore, all the coefficients of **even harmonics** in the Fourier series **vanish**, *i.e.*, the Fourier series contains **only odd harmonics**. We have

$$f(t) = \sum_{n=1}^{\infty} b_n \sin n\omega_0 t \quad ; \quad -\infty < t < +\infty \qquad \text{...(2.40)}$$

The periodic waveform shown in Figure 2.14 can be described as

$$f(t) = \begin{cases} \dfrac{4A}{T_0}t & ; \quad 0 \le t \le \dfrac{T_0}{4} \\ -\dfrac{4A}{T_0}t + 2A & ; \quad \dfrac{T_0}{4} \le t \le \dfrac{3T_0}{4} \\ \dfrac{4A}{T_0}t - 4A & ; \quad \dfrac{3T_0}{4} \le t \le T_0 \end{cases}$$

Choosing the interval of integration $[0, T_0]$ and recognizing that $\omega_0 T_0 = 2\pi$, the Fourier coefficient b_n is evaluated as follows:

$$b_n = \frac{2}{T_0}\int_0^{T_0} f(t)\sin n\omega_0 t\, dt$$

$$= \frac{4}{T_0}\int_0^{T_0/2} f(t)\sin n\omega_0 t\, dt \qquad \text{[Refer Section 2.4]}$$

$$= \frac{4}{T_0}\int_0^{T_0/4}\left(\frac{4A}{T_0}t\right)\sin n\omega_0 t\, dt + \frac{4}{T_0}\int_{T_0/4}^{T_0/2}\left(-\frac{4A}{T_0}t + 2A\right)\sin n\omega_0 t\, dt$$

$$= \frac{16A}{T_0^2}\int_0^{T_0/4} t\sin n\omega_0 t\, dt - \frac{16A}{T_0^2}\int_{T_0/4}^{T_0/2} t\sin n\omega_0 t\, dt + \frac{8A}{T_0}\int_{T_0/4}^{T_0/2}\sin n\omega_0 t\, dt$$

$$= \frac{16A}{T_0^2}\left[\frac{1}{n^2\omega_0^2}\sin n\omega_0 t - \frac{t}{n\omega_0}\cos n\omega_0 t\right]_0^{T_0/4}$$

$$- \frac{16A}{T_0^2}\left[\frac{1}{n^2\omega_0^2}\sin n\omega_0 t - \frac{t}{n\omega_0}\cos n\omega_0 t\right]_{T_0/4}^{T_0/2}$$

$$+ \frac{8A}{T_0}\left[-\frac{\cos n\omega_0 t}{n\omega_0}\right]_{T_0/4}^{T_0/2}$$

$$= \frac{16A}{T_0^2}\left[\frac{1}{n^2\omega_0^2}\sin\left(\frac{n\pi}{2}\right)-\frac{T_0}{4n\omega_0}\cos\left(\frac{n\pi}{2}\right)\right]$$

$$-\frac{16A}{T_0^2}\left[\frac{1}{n^2\omega_0^2}\sin n\pi-\frac{T_0}{2n\omega_0}\cos n\pi\right]$$

$$+\frac{16A}{T_0^2}\left[\frac{1}{n^2\omega_0^2}\sin\left(\frac{n\pi}{2}\right)-\frac{T_0}{4n\omega_0}\cos\left(\frac{n\pi}{2}\right)\right]$$

$$+\frac{8A}{n\omega_0 T_0}\left[-\cos n\pi+\cos\left(\frac{n\pi}{2}\right)\right]$$

$$= \frac{4A}{n^2\pi^2}\sin\left(\frac{n\pi}{2}\right)-\frac{2A}{n\pi}\cos\left(\frac{n\pi}{2}\right)-0+\frac{4A}{n\pi}\cos n\pi$$

$$+\frac{4A}{n^2\pi^2}\sin\left(\frac{n\pi}{2}\right)-\frac{2A}{n\pi}\cos\left(\frac{n\pi}{2}\right)$$

$$-\frac{4A}{n\pi}\cos n\pi+\frac{4A}{n\pi}\cos\left(\frac{n\pi}{2}\right)$$

$$= \frac{8A}{n^2\pi^2}\sin\left(\frac{n\pi}{2}\right)$$

Putting above value in equation (2.40), we have

$$f(t) = \sum_{n=1}^{\infty}\left[\frac{8A}{n^2\pi^2}\sin\left(\frac{n\pi}{2}\right)\sin n\omega_0 t\right] \quad ; \quad -\infty < t < +\infty \qquad \text{...(2.41 A)}$$

Equation (2.41 A) gives the **trigonometric Fourier series** representing the periodic triangular waveform $f(t)$. In an expanded form, we have

$$f(t) = \frac{8A}{\pi^2}\sin\omega_0 t-\frac{8A}{9\pi^2}\sin 3\omega_0 t+\frac{8A}{25\pi^2}\sin 5\omega_0 t-\frac{8A}{49\pi^2}\sin 7\omega_0 t+\ldots\ldots \quad ; \quad -\infty < t < +\infty \qquad \text{...(2.41 B)}$$

Example 2.6. *Obtain the trigonometric Fourier series for the signal as shown in Figure 2.15 (a).*

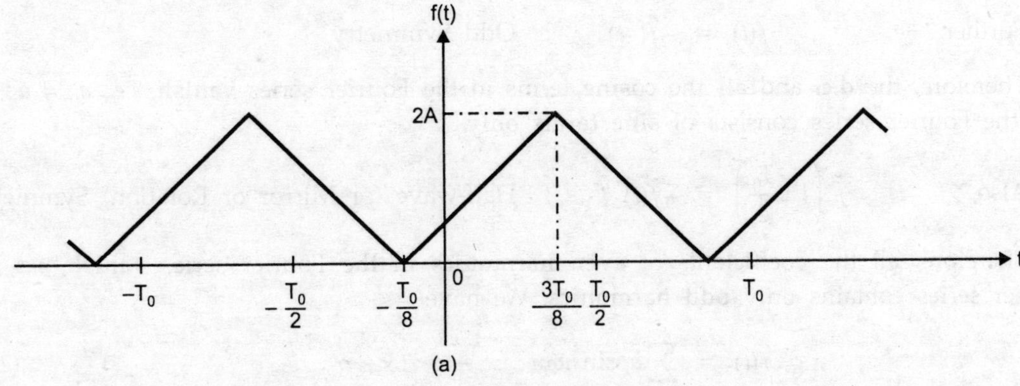

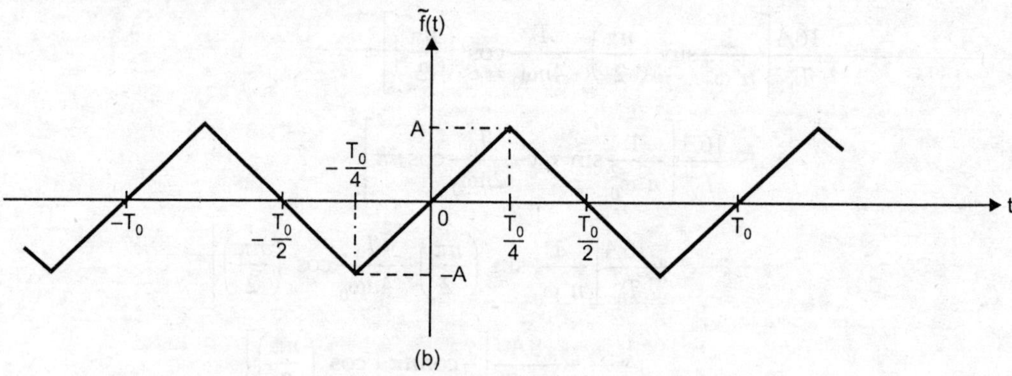

FIGURE 2.15 *(a) Periodic signal waveform of Example 2.6*
(b) Final shifted waveform.

Solution. The **trigonometric Fourier series** representing the periodic triangular waveform $f(t)$ is given by

$$f(t) = a_0 + \sum_{n=1}^{\infty}[a_n \cos n\omega_0 t + b_n \sin n\omega_0 t] \quad ; \quad -\infty < t < +\infty \qquad ...(2.42)$$

where, $$\omega_0 = \frac{2\pi}{T_0}$$

Before evaluating Fourier coefficients, we shall first examine the given triangular waveform for **symmetry conditions**, and we find that the symmetry is **not apparent** from the waveform of Figure 2.15 (*a*). However, the **odd symmetry** of the waveform is more clearly brought out by **shifting the horizontal axis** so as to remove the average or d.c. value of the periodic triangular wave **followed by shifting the vertical axis** by $\frac{T_0}{8}$ to the right resulting in periodic triangular waveform $\tilde{f}(t)$ as shown in Figure 2.15 (*b*). We have

$$\tilde{f}(t) = f\left(t + \frac{T_0}{8}\right) - A$$

or, $$f(t) = \tilde{f}\left(t - \frac{T_0}{8}\right) + A$$

Further, $$\tilde{f}(t) = -\tilde{f}(-t) \quad ; \quad \text{Odd Symmetry}$$

Therefore, the **d.c.** and **all** the **cosine terms** in the Fourier series **vanish,** *i.e.*, $a_0 = a_n = 0$ and the Fourier series consists of **sine terms only.**

Also, $$\tilde{f}\left(t \pm \frac{T_0}{2}\right) = -\tilde{f}(t) \quad ; \quad \text{Half-wave (or Mirror or Rotation) Symmetry}$$

Therefore, all the coefficients of **even harmonics** in the Fourier series **vanish,** *i.e.*, the Fourier series contains **only odd harmonics**. We have

$$\tilde{f}(t) = \sum_{n=1}^{\infty} b_n \sin n\omega_0 t \quad ; \quad -\infty < t < +\infty \qquad ...(2.43)$$

The periodic waveform shown in Figure 2.15 (*b*) can be described as

$$\tilde{f}(t) = \begin{cases} \dfrac{4A}{T_0}t & ; \quad 0 \le t \le \dfrac{T_0}{4} \\ -\dfrac{4A}{T_0}t + 2\text{A} & ; \quad \dfrac{T_0}{4} \le t \le \dfrac{3T_0}{4} \\ \dfrac{4A}{T_0}t - 4\text{A} & ; \quad \dfrac{3T_0}{4} \le t \le T_0 \end{cases}$$

Choosing the interval of integration $[0, T_0]$ and recognizing that $\omega_0 T_0 = 2\pi$, the Fourier coefficient b_n is evaluated as follows

$$b_n = \frac{2}{T_0}\int_0^{T_0} \tilde{f}(t)\sin n\omega_0 t\, dt$$

$$= \frac{4}{T_0}\int_0^{T_0/2} \tilde{f}(t)\sin n\omega_0 t\, dt \qquad \text{[Refer Section 2.4]}$$

$$= \frac{16A}{T_0^2}\int_0^{T_0/4} t\sin n\omega_0 t\, dt - \frac{16A}{T_0^2}\int_{T_0/4}^{T_0/2} t\sin n\omega_0 t\, dt + \frac{8A}{T_0}\int_{T_0/4}^{T_0/2} \sin n\omega_0 t\, dt$$

$$= \frac{16A}{T_0^2}\left[\frac{1}{n^2\omega_0^2}\sin n\omega_0 t - \frac{t}{n\omega_0}\cos n\omega_0 t\right]_0^{T_0/4}$$

$$-\frac{16A}{T_0^2}\left[\frac{1}{n^2\omega_0^2}\sin n\omega_0 t - \frac{t}{n\omega_0}\cos n\omega_0 t\right]_{T_0/4}^{T_0/2}$$

$$+\frac{8A}{T_0}\left[-\frac{\cos n\omega_0 t}{n\omega_0}\right]_{T_0/4}^{T_0/2}$$

$$= \frac{16A}{T_0^2}\left[\frac{1}{n^2\omega_0^2}\sin\left(\frac{n\pi}{2}\right) - \frac{T_0}{4n\omega_0}\cos\left(\frac{n\pi}{2}\right)\right]$$

$$-\frac{16A}{T_0^2}\left[\frac{1}{n^2\omega_0^2}\sin n\pi - \frac{T_0}{2n\omega_0}\cos n\pi\right]$$

$$+\frac{16A}{T_0^2}\left[\frac{1}{n^2\omega_0^2}\sin\left(\frac{n\pi}{2}\right) - \frac{T_0}{4n\omega_0}\cos\left(\frac{n\pi}{2}\right)\right]$$

$$+\frac{8A}{n\omega_0 T_0}\left[-\cos n\pi + \cos\left(\frac{n\pi}{2}\right)\right]$$

$$= \frac{4A}{n^2\pi^2}\sin\left(\frac{n\pi}{2}\right) - \frac{2A}{n\pi}\cos\left(\frac{n\pi}{2}\right) - 0 + \frac{4A}{n\pi}\cos n\pi$$

$$+ \frac{4A}{n^2\pi^2}\sin\left(\frac{n\pi}{2}\right) - \frac{2A}{n\pi}\cos\left(\frac{n\pi}{2}\right)$$

$$- \frac{4A}{n\pi}\cos n\pi + \frac{4A}{n\pi}\cos\left(\frac{n\pi}{2}\right)$$

$$= \frac{8A}{n^2\pi^2}\sin\left(\frac{n\pi}{2}\right)$$

Putting above value in equation (2.43), we have

$$\tilde{f}(t) = \sum_{n=1}^{\infty}\left[\frac{8A}{n^2\pi^2}\sin\left(\frac{n\pi}{2}\right)\sin n\omega_0 t\right] \quad ; \quad -\infty < t < +\infty \qquad \text{...(2.44 A)}$$

Equation (2.44 A) gives the **trigonometric Fourier series** representing the periodic triangular waveform $\tilde{f}(t)$. In an expanded form, we have

$$\tilde{f}(t) = \frac{8A}{\pi^2}\sin\omega_0 t - \frac{8A}{9\pi^2}\sin 3\omega_0 t + \frac{8A}{25\pi^2}\sin 5\omega_0 t - \frac{8A}{49\pi^2}\sin 7\omega_0 t + \ldots\ldots \quad ; \quad -\infty < t < +\infty$$

$$\text{...(2.44 B)}$$

However, $$f(t) = \tilde{f}\left(t - \frac{T_0}{8}\right) + A$$

$\therefore$
$$f(t) = A + \frac{8A}{\pi^2}\sin\omega_0\left(t - \frac{T_0}{8}\right) - \frac{8A}{9\pi^2}\sin 3\omega_0\left(t - \frac{T_0}{8}\right)$$

$$+ \frac{8A}{25\pi^2}\sin 5\omega_0\left(t - \frac{T_0}{8}\right) - \frac{8A}{49\pi^2}\sin 7\omega_0\left(t - \frac{T_0}{8}\right) + \ldots\ldots$$

or,
$$f(t) = A + \frac{8A}{\pi^2}\left[\sin\omega_0 t\cos\left(\frac{\omega_0 T_0}{8}\right) - \cos\omega_0 t\sin\left(\frac{\omega_0 T_0}{8}\right)\right]$$

$$- \frac{8A}{9\pi^2}\left[\sin 3\omega_0 t\cos\left(\frac{3\omega_0 T_0}{8}\right) - \cos 3\omega_0 t\sin\left(\frac{3\omega_0 T_0}{8}\right)\right] + \ldots\ldots$$

or,
$$f(t) = A + \frac{8A}{\pi^2}\left[\sin\omega_0 t\cos\left(\frac{\pi}{4}\right) - \cos\omega_0 t\sin\left(\frac{\pi}{4}\right)\right]$$

$$- \frac{8A}{9\pi^2}\left[\sin 3\omega_0 t\cos\left(\frac{3\pi}{4}\right) - \cos 3\omega_0 t\sin\left(\frac{3\pi}{4}\right)\right] + \ldots\ldots \quad ; \quad -\infty < t < +\infty$$

$$\text{...(2.45)}$$

Equation (2.45) gives the **trigonometric Fourier series** representing the periodic triangular waveform $f(t)$.

Example 2.7. *Obtain the trigonometric Fourier series for the signal as shown in Figure 2.16 (a).*

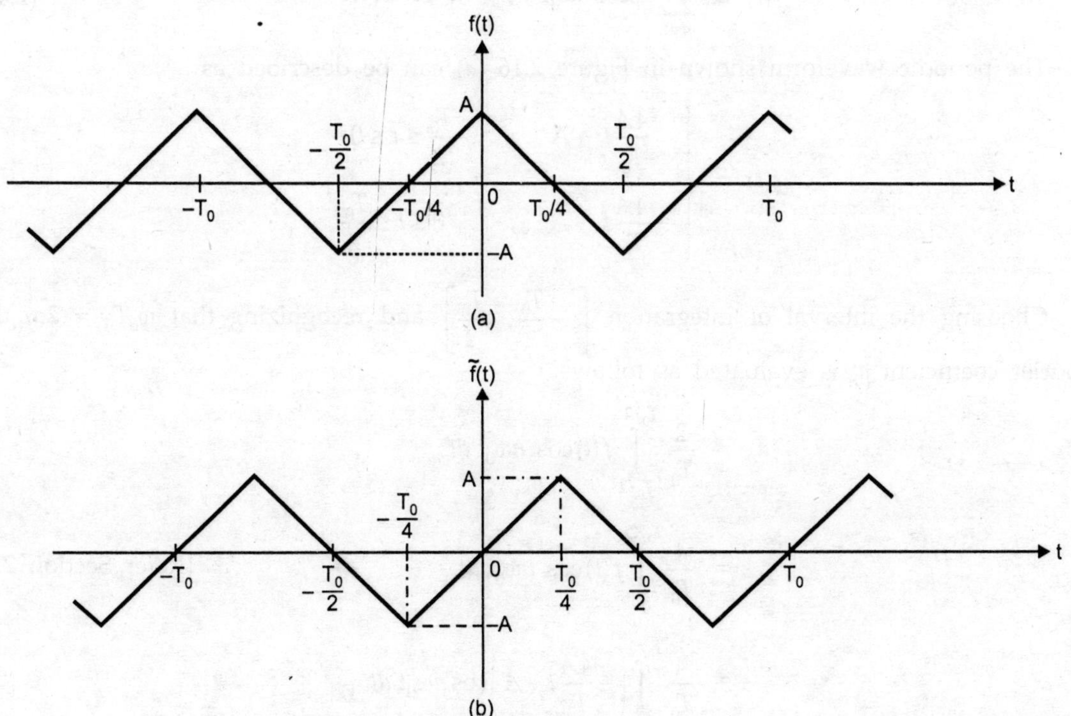

FIGURE 2.16 *(a) Periodic signal waveform of Example 2.7*
(b) Final shifted waveform.

Solution. The **trigonometric Fourier series** representing the periodic triangular waveform $f(t)$ is given by

$$f(t) = a_0 + \sum_{n=1}^{\infty}[a_n \cos n\omega_0 t + b_n \sin n\omega_0 t] \quad ; \quad -\infty < t < +\infty \qquad ...(2.46)$$

where, $$\omega_0 = \frac{2\pi}{T_0}$$

Before evaluating Fourier coefficients, we shall first examine the given triangular waveform for **symmetry conditions**, and we observe that

$$f(t) = f(-t) \quad ; \quad \text{Even Symmetry}$$

Therefore, **all** the **sine terms** in the Fourier series **vanish**, *i.e.*, $b_n = 0$ and the Fourier series consists of **cosine terms only.**

Also, $$f\left(t \pm \frac{T_0}{2}\right) = -f(t) \quad ; \quad \text{Half-wave (or Mirror or Rotation) Symmetry}$$

Therefore, all the coefficients of **even harmonics** in the Fourier series **vanish**, *i.e.*, the Fourier series contains **only odd harmonics.**

Moreover, the **average value** of the periodic signal also **vanishes**, *i.e.*, $a_0 = 0$.

We have
$$f(t) = \sum_{n=1}^{\infty} a_n \cos n\omega_0 t \quad ; \quad -\infty < t < +\infty \qquad ...(2.47)$$

The periodic waveform shown in Figure 2.16 (*a*) can be described as

$$f(t) = \begin{cases} \frac{4A}{T_0} t + \text{A} \quad ; & -\frac{T_0}{2} \le t \le 0 \\ -\frac{4A}{T_0} t + \text{A} \quad ; & 0 \le t \le \frac{T_0}{2} \end{cases}$$

Choosing the interval of integration $\left[-\frac{T_0}{2}, \frac{T_0}{2}\right]$ and recognizing that $\omega_0 T_0 = 2\pi$, the Fourier coefficient a_n is evaluated as follows:

$$a_n = \frac{2}{T_0} \int_{-T_0/2}^{T_0/2} f(t) \cos n\omega_0 t \, dt$$

$$= \frac{4}{T_0} \int_0^{T_0/2} f(t) \cos n\omega_0 t \, dt \qquad \text{[Refer Section 2.4]}$$

$$= \frac{4}{T_0} \int_0^{T_0/2} \left(-\frac{4A}{T_0} t + A \right) \cos n\omega_0 t \, dt$$

$$= -\frac{16A}{T_0^2} \int_0^{T_0/2} t \cos n\omega_0 t \, dt + \frac{4A}{T_0} \int_0^{T_0/2} \cos n\omega_0 t \, dt$$

$$= -\frac{16A}{T_0^2} \left[\frac{1}{n^2 \omega_0^2} \cos n\omega_0 t + \frac{t}{n\omega_0} \sin n\omega_0 t \right]_0^{T_0/2} + \frac{4A}{T_0} \left[\frac{\sin n\omega_0 t}{n\omega_0} \right]_0^{T_0/2}$$

$$= -\frac{16A}{T_0^2} \left[\frac{1}{n^2 \omega_0^2} \cos n\pi + \frac{T_0}{2n\omega_0} \sin n\pi \right] + \frac{16A}{T_0^2} \left[\frac{1}{n^2 \omega_0^2} \right] + \frac{4A}{n\omega_0 T_0} \sin n\pi$$

$$= -\frac{4A}{n^2 \pi^2} \cos n\pi + \frac{4A}{n^2 \pi^2} = \frac{4A}{n^2 \pi^2} (1 - \cos n\pi)$$

Putting above value in equation (2.47), we have

$$f(t) = \sum_{n=1}^{\infty} \left[\frac{4A}{n^2 \pi^2} (1 - \cos n\pi) \cos n\omega_0 t \right] \quad ; \quad -\infty < t < +\infty \qquad ...(2.48 \text{ A})$$

Equation (2.48 A) gives the **trigonometric Fourier series** representing the periodic triangular waveform $f(t)$. In an expanded form, we have

$$f(t) = \frac{8A}{\pi^2} \cos \omega_0 t + \frac{8A}{9\pi^2} \cos 3\omega_0 t + \frac{8A}{25\pi^2} \cos 5\omega_0 t + \frac{8A}{49\pi^2} \cos 7\omega_0 t + \ldots\ldots \quad ; \quad -\infty < t < +\infty$$

...(2.48 B)

Alternative Method

The **trigonometric Fourier series** representing the periodic triangular waveform $f(t)$ is given by

$$f(t) = a_0 + \sum_{n=1}^{\infty}[a_n \cos n\omega_0 t + b_n \sin n\omega_0 t] \quad ; \quad -\infty < t < +\infty \qquad ...(2.49)$$

where,
$$\omega_0 = \frac{2\pi}{T_0}$$

Before evaluating Fourier coefficients, we shall first examine the given triangular waveform for **symmetry conditions**, and we find that the **odd symmetry** of the waveform is more clearly brought out by **shifting the vertical axis** by $\frac{T_0}{4}$ to the left resulting in periodic triangular waveform $\tilde{f}(t)$ as shown in Figure 2.16 (*b*).

We have
$$\tilde{f}(t) = f\left(t - \frac{T_0}{4}\right)$$

or,
$$f(t) = \tilde{f}\left(t + \frac{T_0}{4}\right)$$

Further,
$$\tilde{f}(t) = -\tilde{f}(-t) \quad ; \quad \text{Odd Symmetry}$$

Therefore, the **d.c.** and **all** the **cosine terms** in the Fourier series **vanish**, *i.e.*, $a_0 = a_n = 0$ and the Fourier series consists of **sine terms only.**

Also,
$$\tilde{f}\left(t \pm \frac{T_0}{2}\right) = -\tilde{f}(t) \quad ; \quad \text{Half-wave (or Mirror or Rotation) Symmetry}$$

Therefore, all the coefficients of **even harmonics** in the Fourier series **vanish**, *i.e.*, the Fourier series contains **only odd harmonics**. We have

$$\tilde{f}(t) = \sum_{n=1}^{\infty} b_n \sin n\omega_0 t \quad ; \quad -\infty < t < +\infty \qquad ...(2.50)$$

The periodic waveform shown in Figure 2.16 (*b*) can be described as

$$\tilde{f}(t) = \begin{cases} \dfrac{4A}{T_0}t & ; \quad 0 \le t \le \dfrac{T_0}{4} \\ -\dfrac{4A}{T_0}t + 2A & ; \quad \dfrac{T_0}{4} \le t \le \dfrac{3T_0}{4} \\ \dfrac{4A}{T_0}t - 4A & ; \quad \dfrac{3T_0}{4} \le t \le T_0 \end{cases}$$

Choosing the interval of integration $[0, T_0]$ and recognizing that $\omega_0 T_0 = 2\pi$, the Fourier coefficient b_n is evaluated as follows

$$b_n = \frac{2}{T_0}\int_0^{T_0} \tilde{f}(t) \sin n\omega_0 t \, dt$$

$$= \frac{4}{T_0}\int_0^{T_0/2} \tilde{f}(t)\sin n\omega_0 t\, dt$$ [Refer Section 2.4]

$$= \frac{16A}{T_0^2}\int_0^{T_0/4} t\,\sin n\omega_0 t\, dt - \frac{16A}{T_0^2}\int_{T_0/4}^{T_0/2} t\,\sin n\omega_0 t\, dt + \frac{8A}{T_0}\int_{T_0/4}^{T_0/2} \sin n\omega_0 t\, dt$$

$$= \frac{16A}{T_0^2}\left[\frac{1}{n^2\omega_0^2}\sin n\omega_0 t - \frac{t}{n\omega_0}\cos n\omega_0 t\right]_0^{T_0/4}$$

$$-\frac{16A}{T_0^2}\left[\frac{1}{n^2\omega_0^2}\sin n\omega_0 t - \frac{t}{n\omega_0}\cos n\omega_0 t\right]_{T_0/4}^{T_0/2}$$

$$+\frac{8A}{T_0}\left[-\frac{\cos n\omega_0 t}{n\omega_0}\right]_{T_0/4}^{T_0/2}$$

$$= \frac{16A}{T_0^2}\left[\frac{1}{n^2\omega_0^2}\sin\left(\frac{n\pi}{2}\right) - \frac{T_0}{4n\omega_0}\cos\left(\frac{n\pi}{2}\right)\right]$$

$$-\frac{16A}{T_0^2}\left[\frac{1}{n^2\omega_0^2}\sin n\pi - \frac{T_0}{2n\omega_0}\cos n\pi\right]$$

$$+\frac{16A}{T_0^2}\left[\frac{1}{n^2\omega_0^2}\sin\left(\frac{n\pi}{2}\right) - \frac{T_0}{4n\omega_0}\cos\left(\frac{n\pi}{2}\right)\right]$$

$$+\frac{8A}{n\omega_0 T_0}\left[-\cos n\pi + \cos\left(\frac{n\pi}{2}\right)\right]$$

$$= \frac{4A}{n^2\pi^2}\sin\left(\frac{n\pi}{2}\right) - \frac{2A}{n\pi}\cos\left(\frac{n\pi}{2}\right) - 0 + \frac{4A}{n\pi}\cos n\pi$$

$$+\frac{4A}{n^2\pi^2}\sin\left(\frac{n\pi}{2}\right) - \frac{2A}{n\pi}\cos\left(\frac{n\pi}{2}\right)$$

$$-\frac{4A}{n\pi}\cos n\pi + \frac{4A}{n\pi}\cos\left(\frac{n\pi}{2}\right)$$

$$= \frac{8A}{n^2\pi^2}\sin\left(\frac{n\pi}{2}\right)$$

Putting above value in equation (2.50), we have

$$\tilde{f}(t) = \sum_{n=1}^{\infty}\left[\frac{8A}{n^2\pi^2}\sin\left(\frac{n\pi}{2}\right)\sin n\omega_0 t\right] \quad ; \quad -\infty < t < +\infty \qquad \text{...(2.51 A)}$$

Equation (2.51 A) gives the **trigonometric Fourier series** representing the periodic triangular waveform $\tilde{f}(t)$. In an expanded form, we have

$$\tilde{f}(t) = \frac{8A}{\pi^2}\sin\omega_0 t - \frac{8A}{9\pi^2}\sin 3\omega_0 t + \frac{8A}{25\pi^2}\sin 5\omega_0 t - \frac{8A}{49\pi^2}\sin 7\omega_0 t + \ldots\ldots \quad ; \quad -\infty < t < +\infty \qquad \ldots(2.51\ B)$$

However,

$$f(t) = \tilde{f}\left(t + \frac{T_0}{4}\right)$$

$$\therefore \quad f(t) = \frac{8A}{\pi^2}\sin\omega_0\left(t + \frac{T_0}{4}\right) - \frac{8A}{9\pi^2}\sin 3\omega_0\left(t + \frac{T_0}{4}\right)$$

$$+ \frac{8A}{25\pi^2}\sin 5\omega_0\left(t + \frac{T_0}{4}\right) - \frac{8A}{49\pi^2}\sin 7\omega_0\left(t + \frac{T_0}{4}\right) + \ldots\ldots$$

or,
$$f(t) = \frac{8A}{\pi^2}\left[\sin\omega_0 t\cos\left(\frac{\omega_0 T_0}{4}\right) + \cos\omega_0 t\sin\left(\frac{\omega_0 T_0}{4}\right)\right]$$

$$- \frac{8A}{9\pi^2}\left[\sin 3\omega_0 t\cos\left(\frac{3\omega_0 T_0}{4}\right) + \cos 3\omega_0 t\sin\left(\frac{3\omega_0 T_0}{4}\right)\right] + \ldots\ldots$$

or,
$$f(t) = \frac{8A}{\pi^2}\left[\sin\omega_0 t\cos\left(\frac{\pi}{2}\right) + \cos\omega_0 t\sin\left(\frac{\pi}{2}\right)\right]$$

$$- \frac{8A}{9\pi^2}\left[\sin 3\omega_0 t\cos\left(\frac{3\pi}{2}\right) + \cos 3\omega_0 t\sin\left(\frac{3\pi}{2}\right)\right] + \ldots\ldots$$

or,
$$f(t) = \frac{8A}{\pi^2}\cos\omega_0 t + \frac{8A}{9\pi^2}\cos 3\omega_0 t + \frac{8A}{25\pi^2}\cos 5\omega_0 t + \frac{8A}{49\pi^2}\cos 7\omega_0 t + \ldots\ldots \quad ; \quad -\infty < t < +\infty \qquad \ldots(2.52)$$

Equation (2.52) gives the **trigonometric Fourier series** representing the periodic triangular waveform $f(t)$.

PROBLEMS

1. Obtain the trigonometric Fourier series for the wave shown in Figure 2.17.

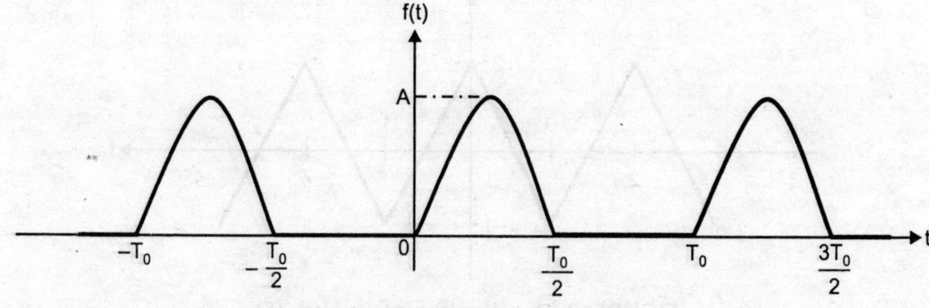

FIGURE 2.17 *Waveform of Problem (1).*

2. Find the Fourier coefficients for the waveform shown in Figure 2.18.

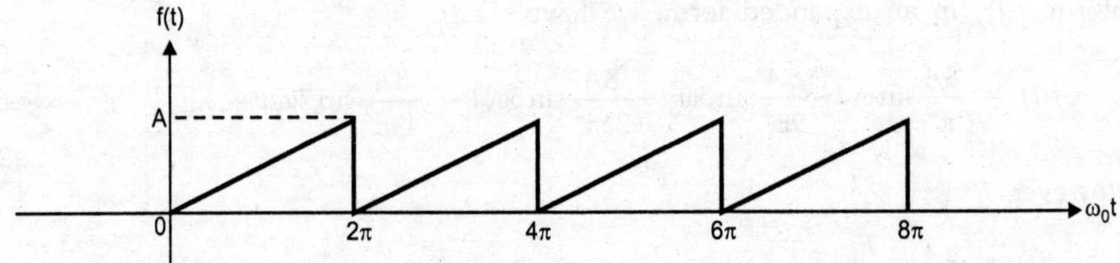

FIGURE 2.18 *Waveform of Problem (2).*

3. Obtain the trigonometric Fourier series for the wave shown in Figure 2.19.

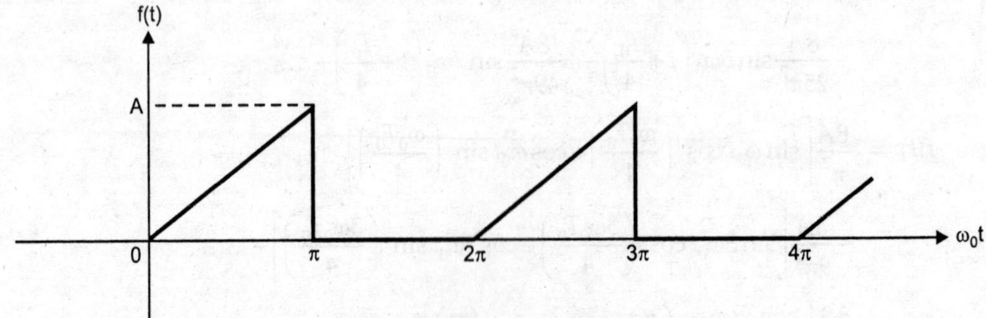

FIGURE 2.19 *Periodic signal waveform of Problem (3).*

4. Find the trigonometric Fourier coefficients for the waveform shown in Figure 2.20.

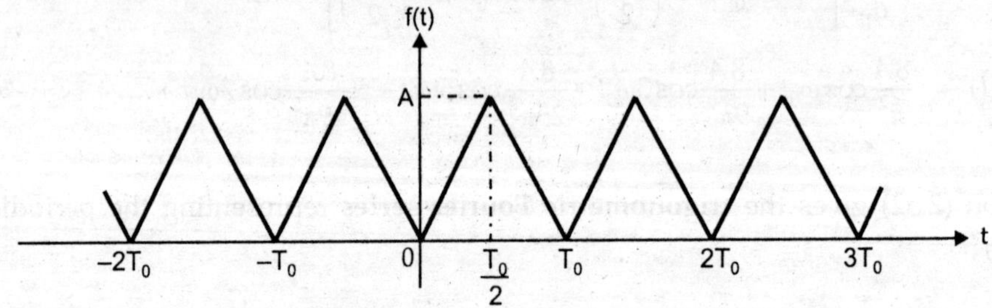

FIGURE 2.20 *Waveform of Problem (4).*

5. Obtain the trigonometric Fourier series for the wave shown in Figure 2.21.

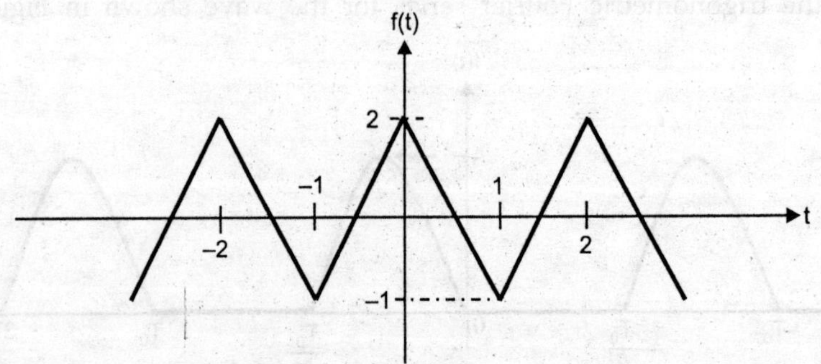

FIGURE 2.21 *Waveform of Problem (5).*

Chapter 3

FOURIER TRANSFORM ANALYSIS

3.1 INTRODUCTION

In previous chapter, we have discussed how a signal $f(t)$ can be expanded into an infinite series of sinusoids of harmonically related frequencies over any interval of duration T_0. We have also discussed that if the signal $f(t)$ happens itself to be **periodic** with period T_0, the Fourier series representing this periodic signal $f(t)$ over a finite interval $[t_1, t_1 + T_0]$ will also represent $f(t)$ for the entire interval $(-\infty, +\infty)$. However, one may pose a logical question that how a **general signal**, whether periodic or not, can be represented in terms of its **frequency components** over the **entire interval** $(-\infty, +\infty)$.

A **non-periodic signal** can be viewed as a **limiting case** of a periodic signal where the period of the signal approaches **infinity**. We, therefore, use this approach to extend the concepts of Fourier series analysis to non-periodic signals using **Fourier transform**. The Fourier transform is the most suitable technique for finding the **response** of a physical system to **non-periodic excitations** in terms of its response to sinusoidal signals and is being widely used in the analysis of communication systems, sampling processes and filters *etc.*

3.2 THE FOURIER TRANSFORM

As already stated, we may derive the **Fourier transform** from the Fourier series by describing a non-periodic signal as the **limiting case** of a periodic signal where **period** T_0 approaches **infinity**. The **exponential Fourier series** representing a periodic signal $f(t)$ over an interval $\left[-\frac{T_0}{2}, \frac{T_0}{2}\right]$ is given by

$$f(t) = \sum_{n=-\infty}^{\infty} F_n e^{jn\omega_0 t}\,;\; -\frac{T_0}{2} < t < \frac{T_0}{2} \qquad \text{...(3.1 A)}$$

where, $$F_n = \frac{1}{T_0}\int_{-T_0/2}^{T_0/2} f(t)e^{-jn\omega_0 t}\,dt \;;\; n = 0, \pm 1, \pm 2, \pm 3, \ldots, \pm\infty \qquad \text{...(3.1 B)}$$

and the **spacing** between two **spectral lines** in the Fourier spectra is given by

$$\Delta\omega = \omega_0 = \frac{2\pi}{T_0}$$

Then $$T_0 = \frac{2\pi}{\Delta\omega}$$

Let us now consider the **limiting case** of a periodic signal where period T_0 approaches **infinity,** *i.e.,* $T_0 \to \infty$, resulting in the case for a **non-periodic signal**. For the case, $\omega_0 \to 0$ and therefore, **spectral lines** in the discrete Fourier spectra **merge** to form the **continuous Fourier spectra.** Also, the **discrete frequency variable** $n\omega_0$ of the discrete Fourier spectra becomes a **continuous frequency variable** ω. We, however, assume that when T_0 becomes **very large,** ω_0 becomes **infinitesimally small** and may be represented by $d\omega$. We may write

$$F_n = \frac{d\omega}{2\pi}\int_{-\infty}^{\infty} f(t)e^{-j\omega t}dt \qquad \text{...(3.2)}$$

Substituting this value of F_n in equation (3.1 A), we have

$$f(t) = \sum_{n=-\infty}^{\infty}\left[\frac{d\omega}{2\pi}\int_{-\infty}^{\infty} f(t)e^{-j\omega t}dt\right]e^{jn\omega_0 t}$$

Under the **limiting process,** with the discrete frequency variable $n\omega_0$ of the discrete Fourier spectra altering to a continuous frequency variable ω, the **summation** must become **integration** so that

$$f(t) = \frac{1}{2\pi}\int_{-\infty}^{\infty}\left[\int_{-\infty}^{\infty} f(t)e^{-j\omega t}dt\right]e^{j\omega t}d\omega$$

or, $$f(t) = \frac{1}{2\pi}\int_{-\infty}^{\infty} F(j\omega)e^{j\omega t}d\omega \;;\; -\infty < t < \infty \qquad \text{...(3.3 A)}$$

where $$F(j\omega) = \int_{-\infty}^{\infty} f(t)e^{-j\omega t}dt \;;\; -\infty < \omega < \infty \qquad \text{...(3.3 B)}$$

Equations (3.3) together form the **Fourier transform pair**. This pair is represented by $f(t) \overset{\mathscr{F}}{\longleftrightarrow} F(j\omega)$, meaning that $F(j\omega)$ is the **Fourier transform** of $f(t)$ and $f(t)$ is the **inverse Fourier transform** of $F(j\omega)$. This relationship may also be written as

$$F(j\omega) = \mathscr{F}[f(t)] \text{ and } f(t) = \mathscr{F}^{-1}[F(j\omega)]$$

3.2.1 Existence of the Fourier Transform: Dirichlet's Conditions

The **sufficient conditions** for the **uniform convergence** of a Fourier transform are called **Dirichlet's conditions.** The Dirichlet conditions are as follows:

Condition (1) : *The signal f(t) must be* ***absolutely integrable****, i.e.,*

$$\int_{-\infty}^{\infty} |f(t)|\,dt < \infty$$

Condition (2) : *The signal f(t) can have only a* ***finite number of maxima and minima*** *in the range* $(-\infty, +\infty)$.

Condition (3) : *The signal f(t) can have only a* ***finite number of discontinuities*** *in the range* $(-\infty, +\infty)$.

Though, almost all physical signals encountered in engineering practice **satisfy** the **second** and **third Dirichlet conditions,** the existence of the Fourier transform is assured if **first Dirichlet condition** is satisfied. Otherwise there is **no guarantee.**

Any periodic signal or signals like step, ramp, *etc.,* **do not satisfy** the first Dirichlet condition and, **strictly speaking,** do not possess the Fourier transform. These signals, however, **do have Fourier transforms in the limit** when we make T_0 **very large but finite.** We can define a **transform pair** that satisfies Fourier transform properties through the use of **impulses** and, therefore, we can still use the Fourier transform as a tool for signal analysis **even though** the Fourier transform **does not converge** for such signals in a strict sense. **Absolute integrability** of $f(t)$ is, thus, a **sufficient but not a necessary condition** for the existence of the Fourier transform of $f(t)$.

Any signal that can be generated in practice satisfies the **Dirichlet conditions** and, therefore, has a Fourier transform. Thus, the **physical existence** of a signal is a **sufficient condition** for the **existence** of its transform.

3.2.2 Fourier Transform in a Real Sense

Let us examine the **connection** between the **line spectrum** of the **Fourier series** and the **continuous spectrum** of the **Fourier transform**. The complex quantity F_n gives the **magnitude** and **phase angle** of the nth **harmonic component** of a periodic signal $f(t)$ and is defined as

$$F_n = \frac{1}{T_0}\int_{-T_0/2}^{T_0/2} f(t)e^{-jn\omega_0 t}\,dt\,;\, n = 0, \pm 1, \pm 2, \pm 3, \ldots\ldots, \pm\infty$$

where ω_0 is the **fundamental frequency** of the periodic signal $f(t)$. While we consider the **limiting case** of a periodic signal where period T_0 approaches **infinity,** *i.e.,* $T_0 \to \infty$, resulting in the case for a **non-periodic signal** ; the discrete variable $n\omega_0$ becomes a **continuous variable** ω. But because of multiplication by $\frac{1}{T_0}$ in the above expression, the magnitude $|F_n| \to 0$. However, the product $F_n \cdot T_0$ **does approach** a finite non-zero value $F(j\omega)$, *i.e.,*

$$\lim_{T_0\to\infty}\left[F_n \cdot T_0\right] = \int_{-\infty}^{\infty} f(t)e^{-j\omega t}\,dt = F(j\omega)$$

where $f(t)$ satisfies Dirichlet's conditions already stated in Section 3.2.1. The **physical significance** to $F(j\omega)$ can easily be assigned by rewriting above equation from equation (3.2) as

$$F(j\omega) = \lim_{T_0\to\infty}\left[F_n \cdot T_0\right] = 2\pi \cdot \frac{F_n}{d\omega}$$

Since this expression involves division by $d\omega$, we may call $F(j\omega)$ as the **frequency density** of the signal $f(t)$. Another explanatory term used for $F(j\omega)$ is **relative frequency distribution.** This term arises because although the **absolute magnitude** of each frequency component is **infinitely small,** their **relative magnitudes** are displayed by a **plot** of $|F(j\omega)|$ *vs.* ω. The magnitude of the components of any frequency ω is proportional to $F(j\omega)$. Therefore, $F(j\omega)$ represents the **frequency spectrum** of $f(t)$ and is called **spectral density function**. Note, however, that the frequency spectrum now is **continuous** and exists at all values of ω. Though $F(j\omega)$ is now called the **continuous spectrum** of $f(t)$, **it in no way implies that the magnitude** $|F(j\omega)|$ **gives the amplitude of the frequency component** ω. The **spectral density function** $F(j\omega)$ can be evaluated from equation (3.3 B).

Example 3.1 *Find the Fourier transforms of the following continuous-time signals*

(a) $f(t) = e^{-at}\ u(t)$

(b) $f(t) = e^{at}\ u(-t)$; *for* $a > 0$

(c) $f(t) = e^{-a|t|}$; *for* $a > 0$

Solution. (a) Since

$$\int_{-\infty}^{\infty} e^{-at}u(t)\,dt = \infty \quad ; \quad \text{for } a \le 0$$

i.e., $f(t)$ is **not absolutely integrable**. Therefore, the Fourier transform **does not converge** for $a \le 0$. However, for $a > 0$, we have

$$F(j\omega) = \int_{-\infty}^{\infty} f(t)e^{-j\omega t}\,dt = \int_{-\infty}^{\infty} e^{-at}\,u(t)e^{-j\omega t}dt$$

$$= \int_{0}^{\infty} e^{-(a+j\omega)t}\,dt = -\left(\frac{1}{a+j\omega}\right)e^{-(a+j\omega)t}\Big|_0^{\infty}$$

$$= \left(\frac{1}{a+j\omega}\right)$$

When writing in **polar form**, the **magnitude** and **phase** of $F(j\omega)$are given by

$$|F(j\omega)| = \frac{1}{\sqrt{(a^2+\omega^2)}}$$

and

$$\arg\,[F(j\omega)] = -\tan^{-1}\left(\frac{\omega}{a}\right)$$

The **magnitude** and **phase spectra** are depicted in Figure 3.1.

(b) For $a > 0$, we have

$$F(j\omega) = \int_{-\infty}^{\infty} f(t)e^{-j\omega t}\,dt = \int_{-\infty}^{0} e^{at}\;u(-t)\,e^{-j\omega t}\,dt$$

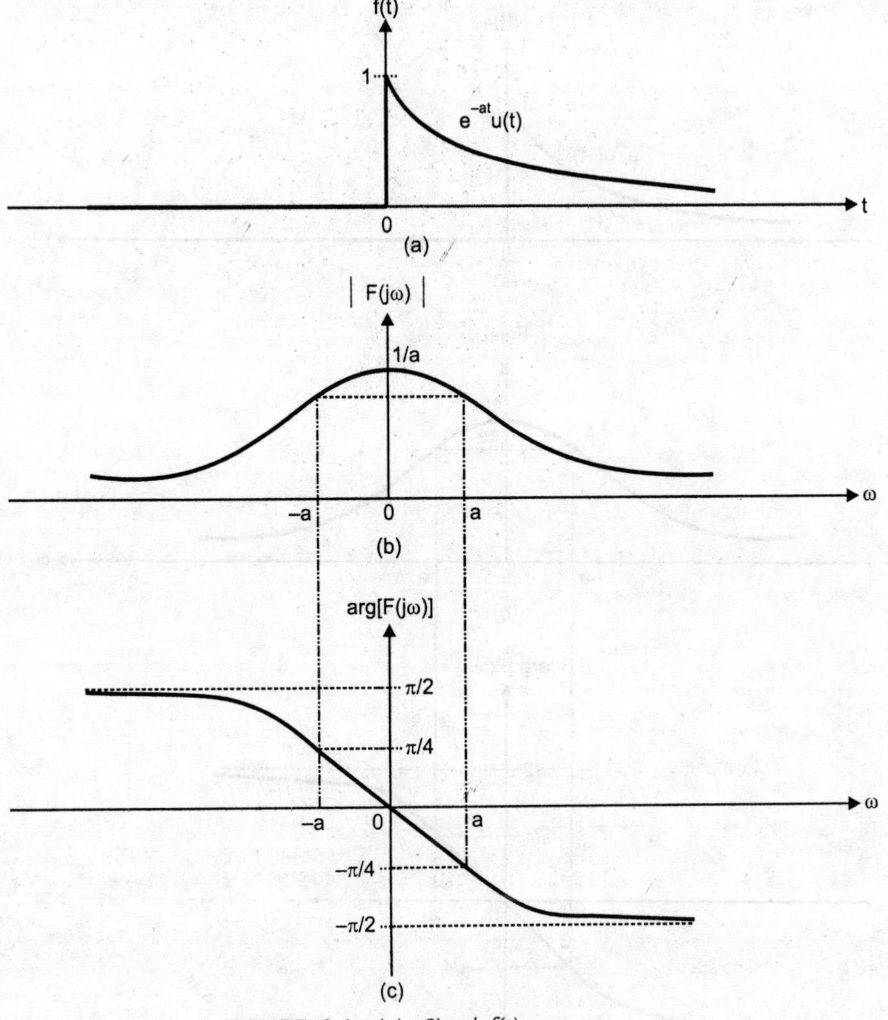

FIGURE 3.1 (*a*) *Signal f(t)*
(*b*) *Magnitude spectrum*
(*c*) *Phase spectrum.*

or,
$$F(j\omega) = \int_{-\infty}^{0} e^{(a-j\omega)t}\,dt = \left(\frac{1}{a-j\omega}\right)e^{(a-j\omega)t}\Big|_{-\infty}^{0}$$

$$= \left(\frac{1}{a-j\omega}\right)$$

When writing in **polar form**, the **magnitude** and **phase** of $F(j\omega)$ are given by

$$|F(j\omega)| = \frac{1}{\sqrt{(a^2+\omega^2)}}$$

and
$$\arg\,[F(j\omega)] = \tan^{-1}\left(\frac{\omega}{a}\right)$$

The **magnitude** and **phase spectra** are depicted in Figure 3.2.

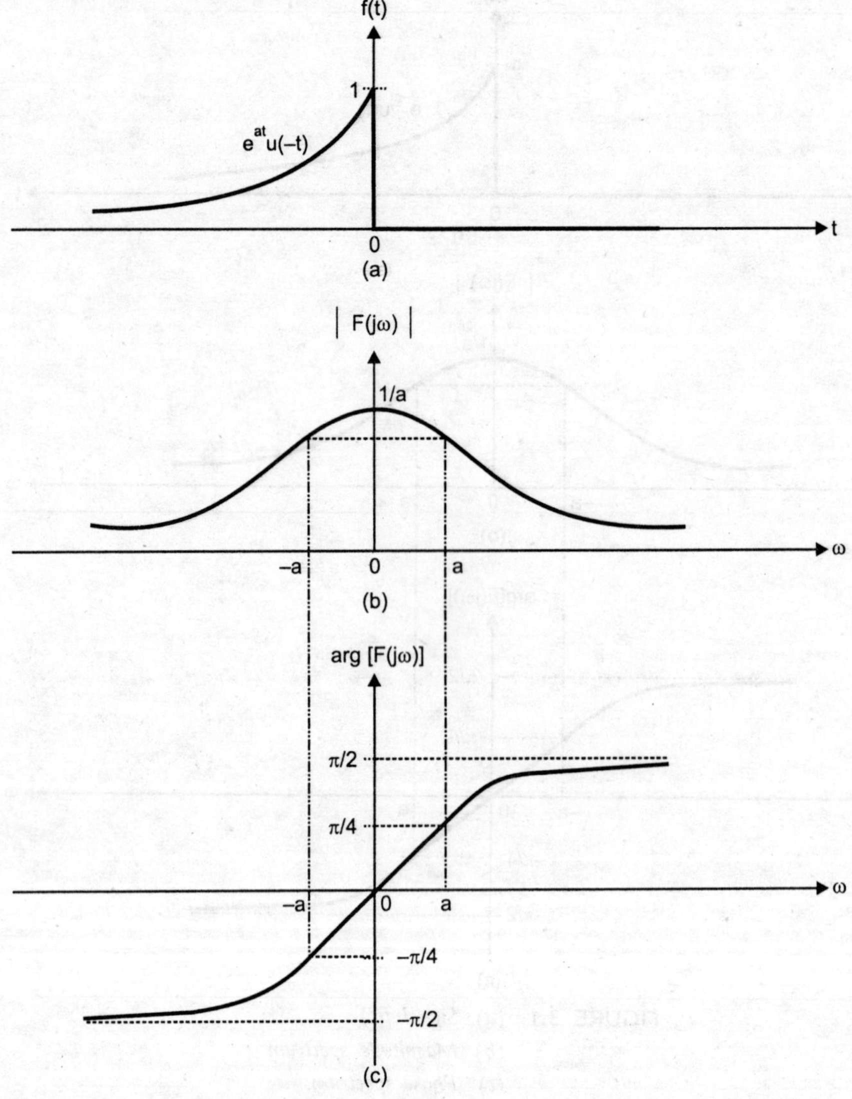

FIGURE 3.2 *(a) Signal f(t)*
(b) Magnitude spectrum
(c) Phase spectrum.

(*c*) For $a > 0$, we have

$$F(j\omega) = \int_{-\infty}^{\infty} f(t)e^{-j\omega t}\,dt = \int_{-\infty}^{\infty} e^{-a|t|}e^{-j\omega t}\,dt$$

$$= \int_{-\infty}^{0} e^{at}e^{-j\omega t}\,dt + \int_{0}^{\infty} e^{-at}e^{-j\omega t}\,dt = \int_{-\infty}^{0} e^{(a-j\omega)t}\,dt + \int_{0}^{\infty} e^{-(a+j\omega)t}\,dt$$

$$= \left[\left(\frac{1}{a-j\omega}\right)e^{(a-j\omega)t}\Big|_{-\infty}^{0}\right] + \left[-\left(\frac{1}{a+j\omega}\right)e^{-(a+j\omega)t}\Big|_{0}^{\infty}\right]$$

$$= \left(\frac{1}{a - j\omega}\right) + \left(\frac{1}{a + j\omega}\right)$$

$$= \frac{2a}{a^2 + \omega^2}$$

In this case $F(j\omega)$ is **real** and the **phase spectrum** is **zero**. We can sketch $F(j\omega)$ directly as shown in Figure 3.3.

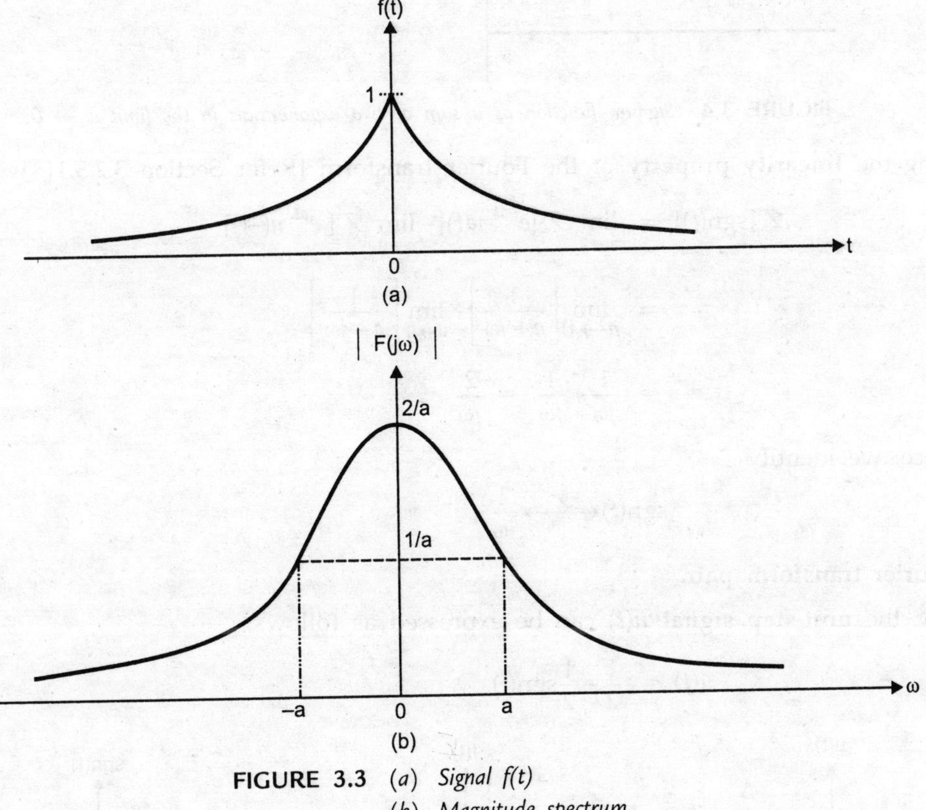

FIGURE 3.3 *(a) Signal f(t)*
(b) Magnitude spectrum.

Example 3.2. *Find the Fourier transform of the signal $f(t)$ = sgn(t) and hence obtain the Fourier transform of $u(t)$.*

Solution. The **signum function** is defined as

$$\text{sgn}(t) = \begin{cases} 1 \;;\; t > 0 \\ 0 \;;\; t = 0 \\ -1 \;;\; t < 0 \end{cases}$$

We can represent $f(t) = \text{sgn}(t)$ as a sum of two **exponentials,** as illustrated in Figure 3.4, in the limit $a \to 0$ as follows

$$\text{sgn}(t) = \lim_{a \to 0} [e^{-at}\, u(t) - e^{at}\, u(-t)]$$

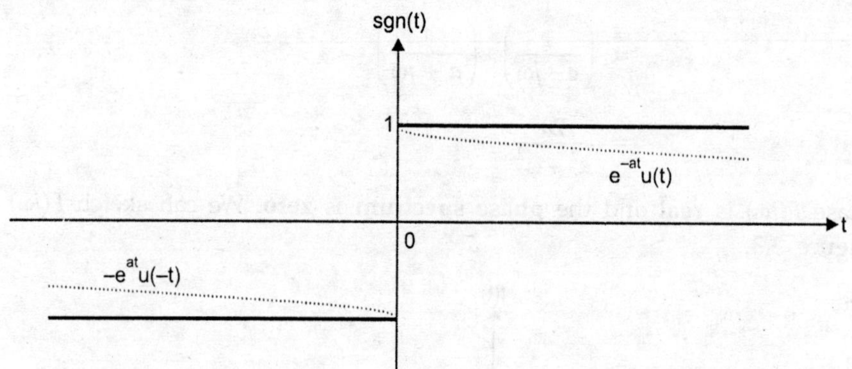

FIGURE 3.4 *Signum function as a sum of two exponentials in the limit $a \to 0$.*

Using the **linearity property** of the Fourier transform [Refer Section 3.2.3.1], we have

$$\mathcal{F}[\text{sgn}(t)] = \lim_{a\to 0} \mathcal{F}[e^{-at}\, u(t)] - \lim_{a\to 0} \mathcal{F}[\, e^{at}\, u(-t)]$$

$$= \lim_{a\to 0}\left[\frac{1}{a+j\omega}\right] - \lim_{a\to 0}\left[\frac{1}{a-j\omega}\right]$$

$$= \frac{1}{j\omega} + \frac{1}{j\omega} = \frac{2}{j\omega}$$

Hence, we identify

$$\text{sgn}(t) \xleftrightarrow{\mathcal{F}} \frac{2}{j\omega} \qquad \text{...(3.4)}$$

as a **Fourier transform pair.**

Now, the **unit-step signal** $u(t)$ can be expressed as follows

$$u(t) = \frac{1}{2} + \frac{1}{2}\text{sgn}(t)$$

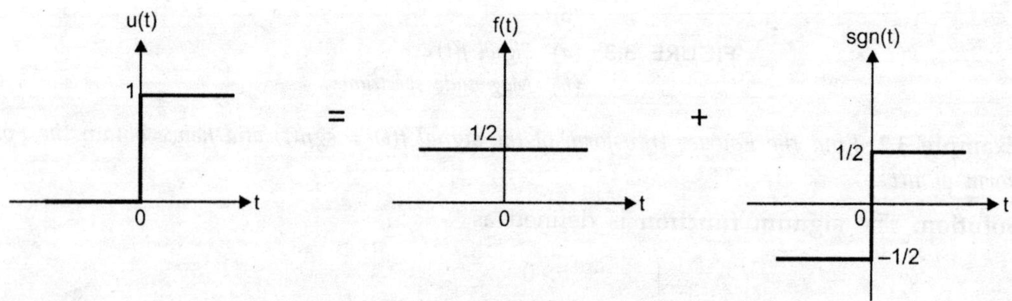

FIGURE 3.5 *Representation of $u(t) = \frac{1}{2} + \frac{1}{2}\,sgn(t)$.*

Thus, $$\mathcal{F}[u(t)] = \mathcal{F}\left[\frac{1}{2} + \frac{1}{2}\text{sgn}(t)\right]$$

Using the result of Example 3.8 and relation (3.4) so as to obtain

$$\mathcal{F}[u(t)] = \pi\, \delta(\omega) + \frac{1}{j\omega}$$

Hence, we identify

$$u(t) \stackrel{\mathcal{F}}{\longleftrightarrow} \pi\,\delta(\omega) + \frac{1}{j\omega}$$

as a **Fourier transform pair.**

Example 3.3. *Find the Fourier transform of the signal $f(t) = \delta(t)$.*

Solution. The signal $f(t) = \delta(t)$ **does not satisfy** the **Dirichlet's conditions,** since the impulse $\delta(t)$ is only defined within an integral. In spite of this potential problem, we attempt to go on with equation (3.3 B) so as to write

$$F(j\omega) = \int_{-\infty}^{\infty} f(t)e^{-j\omega t}\,dt = \int_{-\infty}^{\infty} \delta(t)e^{-j\omega t}\,dt = 1$$

Hence, we identify

$$\delta(t) \stackrel{\mathcal{F}}{\longleftrightarrow} 1$$

as a **Fourier transform pair** and the impulse contains **unity contributions** from **complex sinusoids** of all frequencies, from $\omega = -\infty$ to $\omega = \infty$.

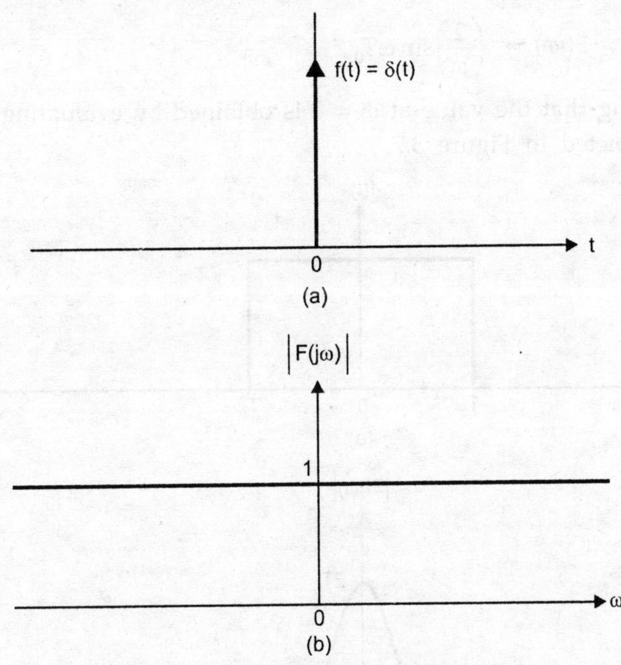

FIGURE 3.6 (*a*) *Signal $f(t) = \delta(t)$*
(*b*) *Magnitude spectrum.*

Example 3.4. *Find the Fourier transform of the rectangular pulse defined as*

$$f(t) = \begin{cases} 1 \ ; & -T_0 \le t \le T_0 \\ 0 \ ; & \textit{Otherwise} \end{cases}$$

Solution. The **rectangular pulse** $f(t)$ is **absolutely integrable provided** $T_0 < \infty$. For $\omega \neq 0$, we have

$$\begin{aligned} F(j\omega) &= \int_{-\infty}^{\infty} f(t)e^{-j\omega t}\,dt \\ &= \int_{-T_0}^{T_0} e^{-j\omega t}\,dt \\ &= -\left(\frac{1}{j\omega}\right)e^{-j\omega t}\Big|_{-T_0}^{T_0} \\ &= \left(\frac{2}{\omega}\right)\sin\omega T_0 \quad ; \quad \omega \neq 0 \end{aligned}$$

For $\omega = 0$, using **L' Hopital's Rule,** we have

$$\lim_{\omega \to 0}\left(\frac{2}{\omega}\right)\sin\omega T_0 = 2T_0$$

Thus, we write for all ω

$$F(j\omega) = \left(\frac{2}{\omega}\right)\sin\omega T_0$$

with the understanding that the value at $\omega = 0$ is obtained by evaluating a limit. In this case $F(j\omega)$ is **real**. It is depicted in Figure 3.7.

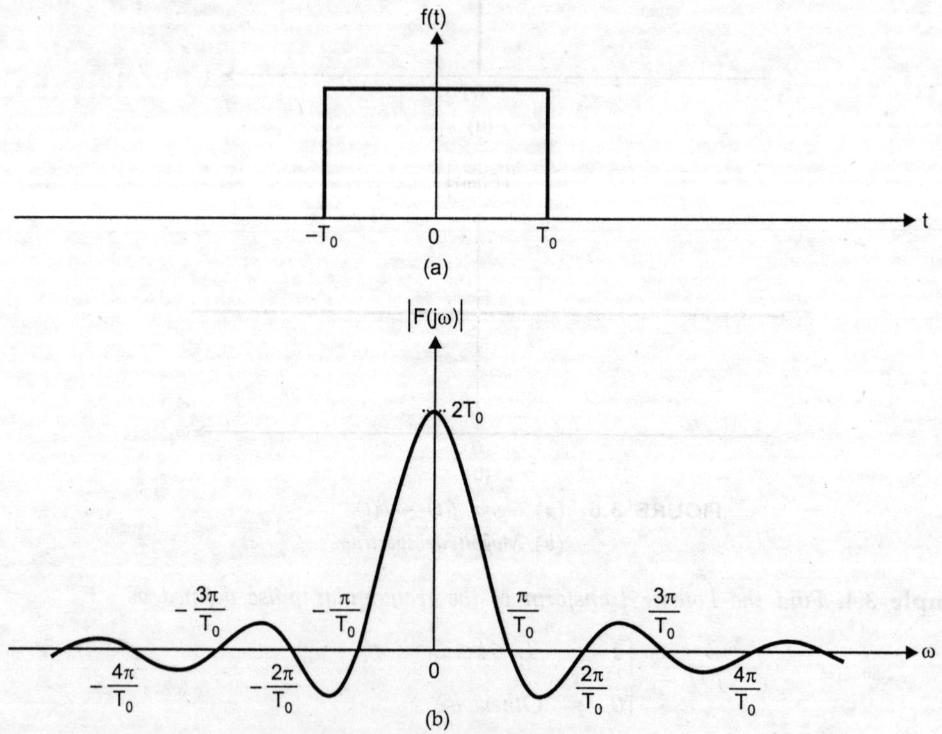

FIGURE 3.7 *(a) Rectangular pulse*
(b) Magnitude spectrum.

The **magnitude** and **phase spectra** are given by

$$|F(j\omega)| = 2\left|\frac{\sin\omega T_0}{\omega}\right|$$

and
$$\arg\,[F(j\omega)] = \begin{cases} 0 \;\; ; & \dfrac{\sin\omega T_0}{\omega} > 0 \\ \pi \;\; ; & \text{Otherwise} \end{cases}$$

We may also write $F(j\omega)$ using the **sinc function notation** as

$$F(j\omega) = 2T_0 \text{ sinc}\left(\frac{\omega T_0}{\pi}\right)$$

3.2.3 Properties of the Fourier Transform

Let $\mathscr{F}[f(t)] = F(j\omega)$

and $\mathscr{F}[g(t)] = G(j\omega)$

Now, the properties of the Fourier transform are summarized in following sections one by one :

3.2.3.1 Linearity

$$\mathscr{F}\left[af(t) \pm bg(t)\right] = aF(j\omega) \pm bG(j\omega) \qquad \text{...(3.5)}$$

Proof.
$$\mathscr{F}\left[af(t) \pm bg(t)\right] = \int_{-\infty}^{\infty}\left[af(t) \pm bg(t)\right] e^{-j\omega t}\, dt$$

$$= a\int_{-\infty}^{\infty} f(t)\, e^{-j\omega t}\, dt \pm b\int_{-\infty}^{\infty} g(t)\, e^{-j\omega t}\, dt$$

$$= aF(j\omega) \pm bG(j\omega)$$

3.2.3.2 Scaling

$$\mathscr{F}[f(at)] = \frac{1}{|a|} F\left(\frac{j\omega}{a}\right) \qquad \text{...(3.6)}$$

where, a is a **non-zero real number.**

Proof.
$$\mathscr{F}[f(at)] = \int_{-\infty}^{\infty} f(at)\, e^{-j\omega t}\, dt$$

Let $\lambda = at$, we have

$$\mathscr{F}[f(at)] = \begin{cases} \dfrac{1}{a}\displaystyle\int_{-\infty}^{\infty} f(\lambda)\, e^{-\left(\frac{j\omega}{a}\right)\lambda}\, d\lambda \;\; ; & a > 0 \\ -\dfrac{1}{a}\displaystyle\int_{-\infty}^{\infty} f(\lambda)\, e^{-\left(\frac{j\omega}{a}\right)\lambda}\, d\lambda \;\; ; & a < 0 \end{cases}$$

In combined form, we can write

$$\mathscr{F}[f(at)] = \frac{1}{|a|}\int_{-\infty}^{\infty} f(\lambda)\, e^{-\left(\frac{j\omega}{a}\right)\lambda}\, d\lambda$$

$$= \frac{1}{|a|} F\left(\frac{j\omega}{a}\right)$$

3.2.3.3 *Time Shifting*

$$\mathscr{F}[f(t-\tau)] = e^{-j\omega\tau} F(j\omega) \qquad \text{...(3.7)}$$

Proof. $$\mathscr{F}[f(t-\tau)] = \int_{-\infty}^{\infty} f(t-\tau)\, e^{-j\omega t}\, dt$$

Let $\lambda = t - \tau$, we have

$$\mathscr{F}[f(t-\tau)] = \int_{-\infty}^{\infty} f(\lambda)\, e^{-j\omega(\tau+\lambda)}\, d\lambda$$

$$= \int_{-\infty}^{\infty} f(\lambda)\, e^{-j\omega\tau}\, e^{-j\omega\lambda}\, d\lambda$$

$$= e^{-j\omega\tau}\int_{-\infty}^{\infty} f(\lambda)\, e^{-j\omega\lambda}\, d\lambda$$

$$= e^{-j\omega\tau} F(j\omega)$$

This result shows that delaying a signal by τ **does not change its amplitude spectrum.** The **phase spectrum, however, is changed by** $-\omega\tau$.

3.2.3.4 *Frequency Shifting*

$$\mathscr{F}[e^{j\alpha t} f(t)] = F(j(\omega-\alpha)) \qquad \text{...(3.8)}$$

Proof. To establish this property, consider the expression

$$g(t) = \frac{1}{2\pi}\int_{-\infty}^{\infty} G(j\omega)\, e^{j\omega t}\, d\omega$$

Assuming $G(j\omega) = F(j(\omega - \alpha))$, we may write

$$g(t) = \frac{1}{2\pi}\int_{-\infty}^{\infty} F(j(\omega-\alpha))\, e^{j\omega t}\, d\omega$$

Let $\lambda = \omega - \alpha$, we have

$$g(t) = \frac{1}{2\pi}\int_{-\infty}^{\infty} F(j\lambda)\, e^{j(\alpha+\lambda)t}\, d\lambda$$

$$= \frac{1}{2\pi}\int_{-\infty}^{\infty} F(j\lambda)\, e^{j\alpha t}\, e^{j\lambda t}\, d\lambda$$

$$= e^{j\alpha t}\left[\frac{1}{2\pi}\int_{-\infty}^{\infty} F(j\lambda)\, e^{j\lambda t}\, d\lambda\right]$$

$$= e^{j\alpha t} f(t)$$

Thus, $\mathscr{F}[e^{j\alpha t} f(t)] = F(j(\omega - \alpha))$

3.2.3.5 Convolution

$$\mathscr{F}[f(t) * g(t)] = F(j\omega)\, G(j\omega) \qquad ...(3.9)$$

Proof. $\mathscr{F}[f(t) * g(t)] = \int_{-\infty}^{\infty} [f(t) * g(t)]\, e^{-j\omega t}\, dt$

$$= \int_{-\infty}^{\infty}\left[\int_{-\infty}^{\infty} f(\tau) g(t-\tau)\, d\tau\right] e^{-j\omega t}\, dt$$

Interchanging the order of integration and noting that $f(\tau)$ does not depend on t, we may write

$$\mathscr{F}[f(t) * g(t)] = \int_{-\infty}^{\infty} f(\tau)\left[\int_{-\infty}^{\infty} g(t-\tau)\, e^{-j\omega t}\, dt\right] d\tau$$

Let $\lambda = t - \tau$, we have

$$\mathscr{F}[f(t) * g(t)] = \int_{-\infty}^{\infty} f(\tau)\left[\int_{-\infty}^{\infty} g(\lambda)\, e^{-j\omega(\tau+\lambda)}\, d\lambda\right] d\tau$$

$$= \int_{-\infty}^{\infty} f(\tau)\, e^{-j\omega\tau}\left[\int_{-\infty}^{\infty} g(\lambda)\, e^{-j\omega\lambda}\, d\lambda\right] d\tau$$

$$= \left[\int_{-\infty}^{\infty} f(\tau)\, e^{-j\omega\tau}\, d\tau\right] G(j\omega)$$

$$= F(j\omega)\, G(j\omega)$$

3.2.3.6 Modulation

$$\mathscr{F}[f(t)\, g(t)] = \frac{1}{2\pi}[F(j\omega) * G(j\omega)] \qquad ...(3.10)$$

Proof

$$\mathscr{F}^{-1}\left[\frac{1}{2\pi}[F(j\omega) * G(j\omega)]\right] = \frac{1}{2\pi}\int_{-\infty}^{\infty}\left[\frac{1}{2\pi}[F(j\omega) * G(j\omega)]\right] e^{j\omega t}\, d\omega$$

$$= \frac{1}{2\pi}\int_{-\infty}^{\infty}\left[\frac{1}{2\pi}\int_{-\infty}^{\infty}F(j\lambda)\,G(j(\omega-\lambda))\,d\lambda\right]e^{j\omega t}\,d\omega$$

$$= \frac{1}{(2\pi)^2}\int_{-\infty}^{\infty}\int_{-\infty}^{\infty}F(j\lambda)\,G(j(\omega-\lambda))\,e^{j\omega t}\,d\lambda\,d\omega$$

Let $\eta = \omega - \lambda$, we have

$$\mathcal{F}^{-1}\left[\frac{1}{2\pi}[F(j\omega)*G(j\omega)]\right] = \frac{1}{(2\pi)^2}\int_{-\infty}^{\infty}\int_{-\infty}^{\infty}F(j\lambda)\,G(j\eta)\,e^{j(\lambda+\eta)t}d\lambda\,d\eta$$

$$= \frac{1}{(2\pi)^2}\int_{-\infty}^{\infty}\int_{-\infty}^{\infty}F(j\lambda)\,G(j\eta)\,e^{j\lambda t}\,e^{j\eta t}\,d\lambda\,d\eta$$

$$= \left[\frac{1}{2\pi}\int_{-\infty}^{\infty}F(j\lambda)\,e^{j\lambda t}\,d\lambda\right]\left[\frac{1}{2\pi}\int_{-\infty}^{\infty}G(j\eta)\,e^{\eta t}\,d\eta\right]$$

$$= f(t)\,g(t)$$

which implies that

$$\mathcal{F}[f(t)\,g(t)] = \frac{1}{2\pi}[F(j\omega)*G(j\omega)]$$

3.2.3.7 *Differentiation in Frequency-Domain*

$$\mathcal{F}[-jt\,f(t)] = \frac{d}{d\omega}F(j\omega) \qquad \text{...(3.11)}$$

Proof.
$$\mathcal{F}[-jt\,f(t)] = \int_{-\infty}^{\infty}[-jt\,f(t)]\,e^{-j\omega t}\,dt = \int_{-\infty}^{\infty}f(t)[-jt\,e^{-j\omega t}]\,dt$$

$$= \int_{-\infty}^{\infty}f(t)\left[\frac{d}{d\omega}e^{-j\omega t}\right]dt = \frac{d}{d\omega}\left[\int_{-\infty}^{\infty}f(t)\,e^{-j\omega t}\,dt\right]$$

$$= \frac{d}{d\omega}F(j\omega)$$

3.2.3.8 *Differentiation in Time-Domain*

$$\mathcal{F}\left[\frac{d}{dt}f(t)\right] = j\omega\,F(j\omega) \qquad \text{...(3.12)}$$

Proof. To establish this property, consider the expression

$$f(t) = \frac{1}{2\pi}\int_{-\infty}^{\infty}F(j\omega)\,e^{j\omega t}\,d\omega$$

Differentiating both sides of this equation with respect to t yields

$$\frac{d}{dt}f(t) = \frac{1}{2\pi}\int_{-\infty}^{\infty} F(j\omega)\left[\frac{d}{dt}e^{j\omega t}\right]d\omega = \frac{1}{2\pi}\int_{-\infty}^{\infty} F(j\omega)\,[j\omega\, e^{j\omega t}]\,d\omega$$

$$= \frac{1}{2\pi}\int_{-\infty}^{\infty} [j\omega\, F(j\omega)]\, e^{j\omega t}\, d\omega = \frac{j\omega}{2\pi}\int_{-\infty}^{\infty} F(j\omega)\, e^{j\omega t}\, d\omega$$

which implies that

$$\mathscr{F}\left[\frac{d}{dt}f(t)\right] = j\omega F(j\omega)$$

3.2.3.9 *Integration in Time-Domain*

$$\mathscr{F}\left[\int_{-\infty}^{t} f(\tau)d\tau\right] = \frac{F(j\omega)}{j\omega} + \pi\, F(j0)\,\delta(\omega) \qquad \text{...(3.13))}$$

Proof. The **convolution sum** between a signal $f(t)$ and a unit-step signal $u(t)$ is given by

$$f(t) * u(t) = \int_{-\infty}^{\infty} f(\tau)\, u(t-\tau)\, d\tau$$

Again, a **shifted unit-step signal** $u(t - \tau)$ is defined as

$$u(t-\tau) = \begin{cases} 1 \;;\; t \geq \tau \\ 0 \;;\; t < \tau \end{cases}$$

It follows that

$$f(t) * u(t) = \int_{-\infty}^{t} f(\tau)\, d\tau \qquad \text{...(3.14)}$$

Again, from the **convolution property** of the Fourier transform, we have

$$\mathscr{F}[f(t) * u(t)] = F(j\omega)\, U(j\omega)$$

Using result of Example 3.2, we obtain

$$\mathscr{F}[f(t) * u(t)] = F(j\omega)\left[\frac{1}{j\omega} + \pi\,\delta(\omega)\right] = \frac{F(j\omega)}{j\omega} + \pi\, F(j\omega)\,\delta(\omega)$$

Since the impulse exists only at $\omega = 0$ and the value of the Fourier transform $F(j\omega)$ at $\omega = 0$ is $F(j0)$, we may write

$$F(j\omega)\,\delta(\omega) = F(j0)\,\delta(\omega)$$

Therefore,

$$\mathscr{F}[f(t) * u(t)] = \frac{F(j\omega)}{j\omega} + \pi\, F(j0)\,\delta(\omega) \qquad \text{...(3.15)}$$

Substituting the value of $f(t)*u(t)$ from equation (3.14) into equation (3.15), we get

$$\mathcal{F}\left[\int_{-\infty}^{t} f(\tau)d\tau\right] = \frac{F(j\omega)}{j\omega} + \pi F(j0)\,\delta(\omega)$$

3.2.4 Fourier Transform for Periodic Signals

Strictly speaking, the Fourier transform of a periodic signal **does not exist**, since it **fails** to satisfy the condition of **absolute integrability**. We, however, may express a periodic signal by its **Fourier series**. The **Fourier transform** of the **periodic signal** is then the **sum** of Fourier transforms of its individual components. We can express a periodic signal $f(t)$ with period T_0 as

$$f(t) = \sum_{n=-\infty}^{\infty} F_n\, e^{jn\omega_0 t} \quad ; \quad \omega_0 = \frac{2\pi}{T_0}$$

Taking the Fourier transform of both sides, we have

$$\mathcal{F}[f(t)] = \mathcal{F}\left[\sum_{n=-\infty}^{\infty} F_n\, e^{jn\omega_0 t}\right]$$

or,

$$\mathcal{F}[f(t)] = \sum_{n=-\infty}^{\infty} F_n\, [\mathcal{F}(e^{jn\omega_0 t})]$$

Now, consider the expression

$$\mathcal{F}^{-1}[2\pi\,\delta(\omega-\omega_0)] = \frac{1}{2\pi}\int_{-\infty}^{\infty} [2\pi\,\delta(\omega-\omega_0)]\, e^{j\omega t}\, d\omega$$

or,

$$\mathcal{F}^{-1}[2\pi\,\delta(\omega-\omega_0)] = \int_{-\infty}^{\infty} e^{j\omega t}\,\delta(\omega-\omega_0)\, d\omega$$

Since the impulse $\delta(\omega - \omega_0)$ exist only at $\omega = \omega_0$, and the value of $e^{j\omega t}$ at $\omega = \omega_0$ is $e^{j\omega_0 t}$, we may write

$$e^{j\omega t}\,\delta(\omega-\omega_0) = e^{j\omega_0 t}\,\delta(\omega-\omega_0)$$

Thus, we have

$$\mathcal{F}^{-1}[2\pi\,\delta(\omega-\omega_0)] = e^{j\omega_0 t}\left[\int_{-\infty}^{\infty}\delta(\omega-\omega_0)\, d\omega\right]$$

or,

$$\mathcal{F}^{-1}[2\pi\,\delta(\omega-\omega_0)] = e^{j\omega_0 t}$$

Hence we identify

$$e^{j\omega_0 t} \stackrel{\mathcal{F}}{\longleftrightarrow} 2\pi\,\delta(\omega-\omega_0)$$

as a **Fourier transform pair with the understanding that** the spectrum of the **everlasting exponential** $e^{j\omega_0 t}$ is a **single impulse** of strength 2π at $\omega = \omega_0$.

Using this relation, we have

$$\mathscr{F}[f(t)] = \sum_{n=-\infty}^{\infty} F_n [\mathscr{F}(e^{jn\omega_0 t})]$$

or,

$$\mathscr{F}[f(t)] = 2\pi \sum_{n=-\infty}^{\infty} F_n \, \delta(\omega - n\omega_0) \qquad ...(3.16)$$

Thus, the resulting Fourier transform consists of impulses located at the **harmonic frequencies** $n\omega_0$ of the signal $f(t)$, with the strength of each impulse equal to 2π times the value of corresponding Fourier series coefficient F_n. This will turn out to be a very useful result. The result should, of course, be no surprise, since we know that a periodic signal contains components only of **discrete harmonic frequencies.**

Example 3.5 *A rectangular pulse f(t) is defined as*

$$f(t) = \begin{cases} 1 & ; \quad -1 \le t \le 1 \\ 0 & ; \quad \text{Otherwise} \end{cases}$$

Using the scaling property and the Fourier transform of f(t), determine the Fourier transform of the scaled rectangular pulse defined as

$$g(t) = \begin{cases} 1 & ; \quad -2 \le t \le 2 \\ 0 & ; \quad \text{Otherwise} \end{cases}$$

Solution. Substituting $T_0 = 1$ into the result obtained in Example 3.4, we have

$$F(j\omega) = \left(\frac{2}{\omega}\right) \sin \omega$$

The magnitude spectrum is given by

$$|F(j\omega)| = 2\left|\frac{\sin \omega}{\omega}\right|$$

The signal $f(t)$ and its magnitude spectrum are shown in Figure 3.8.

Note that $g(t) = f\left(\frac{t}{2}\right)$. Using **scaling property** with $a = \frac{1}{2}$, we obtain

$$G(j\omega) = 2F(j2\omega)$$

or,

$$G(j\omega) = \left(\frac{2}{\omega}\right) \sin 2\omega$$

The magnitude spectrum is given by

$$|G(j\omega)| = 2\left|\frac{\sin 2\omega}{\omega}\right|$$

The signal $g(t)$ and its magnitude spectrum are shown in Figure 3.9.

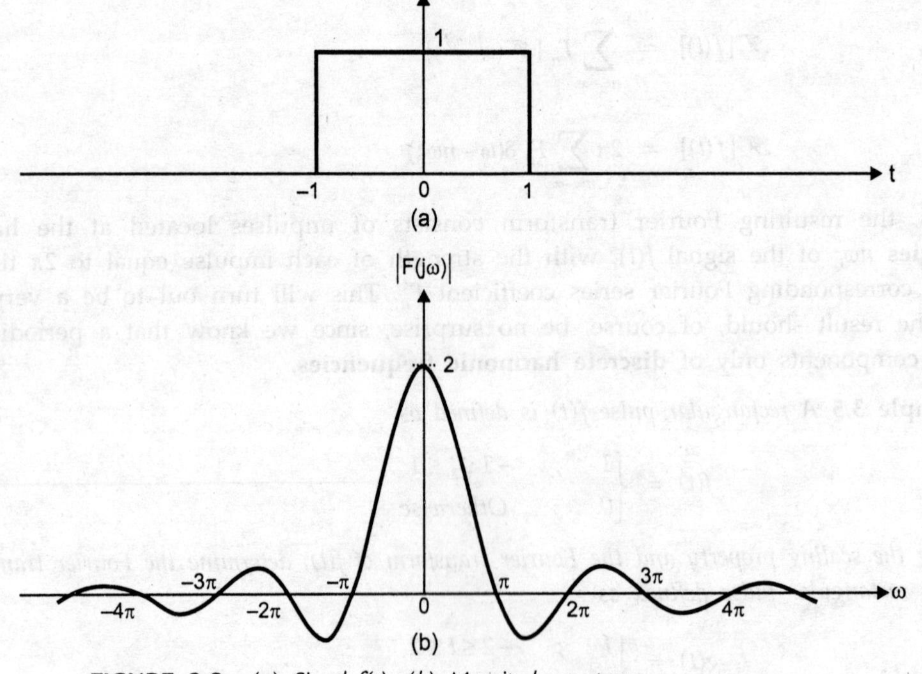

FIGURE 3.8 *(a) Signal f(t), (b) Magnitude spectrum.*

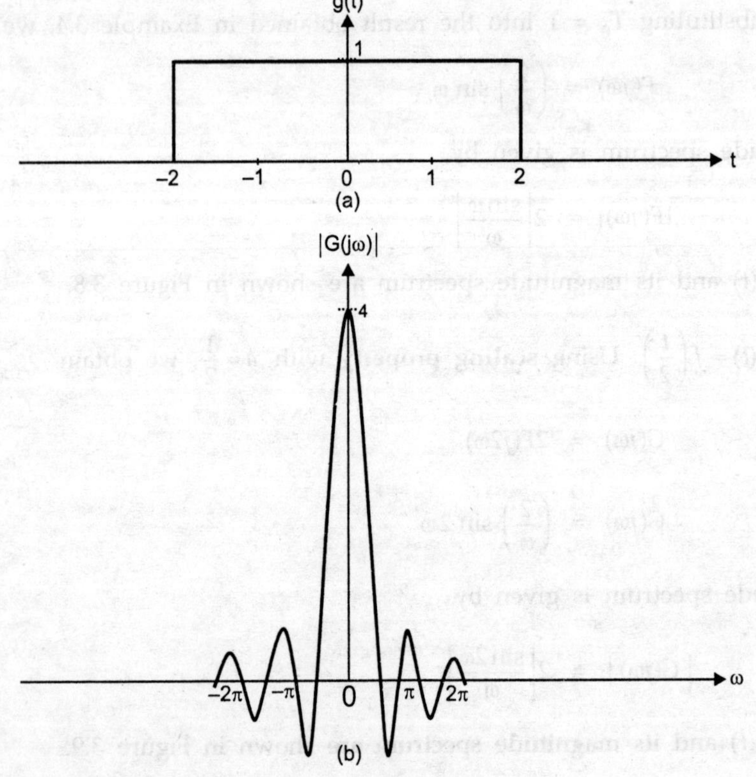

FIGURE 3.9 *(a) Signal g(t)*
(b) Magnitude spectrum.

Example 3.6 *Find the Fourier transform for the periodic impulse train* $f(t) = \delta_{T_0}(t)$ *as shown in Figure 3.10 (a). Also sketch the corresponding spectrum.*

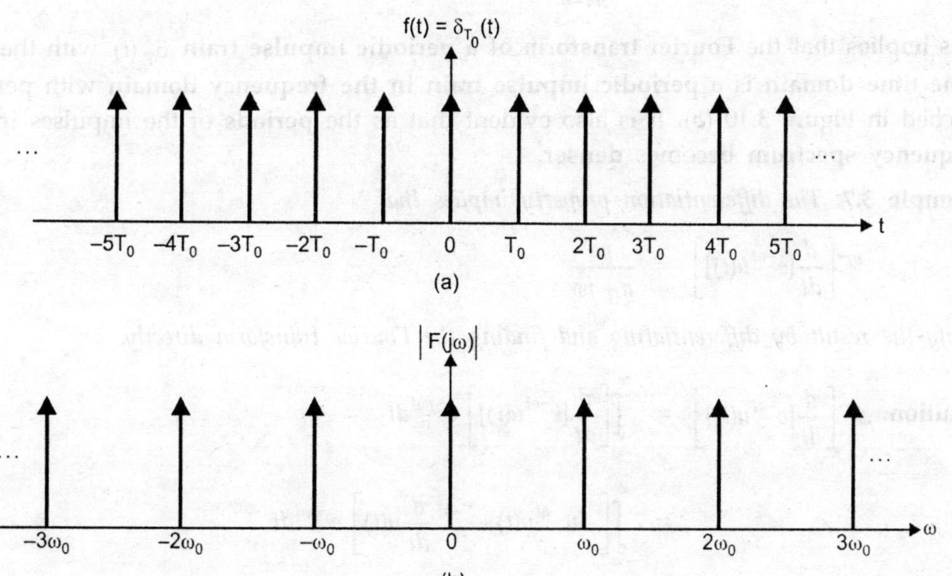

FIGURE 3.10 *(a) Periodic impulse-train*
(b) Magnitude spectrum.

Solution. The **exponential Fourier series** representing the periodic signal $\delta_{T_0}(t)$ is given by

$$\delta_{T_0}(t) = \sum_{n=-\infty}^{\infty} F_n\, e^{jn\omega_0 t} \quad ; \quad -\infty < t < +\infty \qquad \text{...(3.17)}$$

so that

$$\mathscr{F}[\delta_{T_0}(t)] = 2\pi \sum_{n=-\infty}^{\infty} F_n\, \delta(\omega - n\omega_0) \qquad \text{...(3.18)}$$

where,

$$F_n = \frac{1}{T_0}\int_{T_0} \delta_{T_0}(t) e^{-jn\omega_0 t} dt$$

Choosing the interval of integration $\left[-\frac{T_0}{2}, \frac{T_0}{2}\right]$ and recognizing that over this interval $\delta_{T_0}(t) = \delta(t)$, we have

$$F_n = \frac{1}{T_0}\int_{-T_0/2}^{T_0/2} \delta(t) e^{-jn\omega_0 t} dt = \frac{1}{T_0}$$

Putting F_n in equation (3.18), we get the Fourier transform for the **periodic impulse train** $\delta_{T_0}(t)$ as

$$\mathscr{F}[\delta_{T_0}(t)] = \frac{2\pi}{T_0}\sum_{n=-\infty}^{\infty} \delta(\omega - n\omega_0)$$

or
$$\mathcal{F}[\delta_{T_0}(t)] = \omega_0 \sum_{n=-\infty}^{\infty} \delta(\omega - n\omega_0) \qquad ...(3.19)$$

This implies that the **Fourier transform of a periodic impulse train** $\delta_{T_0}(t)$ with the period T_0 **in the time domain is a periodic impulse train in the frequency domain** with period ω_0, as sketched in Figure 3.10 (*b*). It is also evident that as the **periods** of the impulses **increase,** the **frequency spectrum** becomes **denser.**

Example 3.7. *The differentiation property implies that*

$$\mathcal{F}\left[\frac{d}{dt}[e^{-at}u(t)]\right] = \frac{j\omega}{a+j\omega}$$

Verify the result by differentiating and finding the Fourier transform directly.

Solution.
$$\mathcal{F}\left[\frac{d}{dt}[e^{-at}u(t)]\right] = \int_{-\infty}^{\infty}\left[\frac{d}{dt}[e^{-at}u(t)]\right] e^{-j\omega t}\,dt$$

$$= \int_{-\infty}^{\infty}\left[-ae^{-at}u(t) + e^{-at}\frac{d}{dt}u(t)\right] e^{-j\omega t}\,dt$$

$$= \int_{-\infty}^{\infty}[-ae^{-at}u(t) + e^{-at}\delta(t)]\, e^{-j\omega t}\,dt$$

$$= \int_{-\infty}^{\infty}[-ae^{-at}u(t) + \delta(t)]\, e^{-j\omega t}\,dt$$

$$= \int_{-\infty}^{\infty}[-ae^{-at}u(t)]\, e^{-j\omega t}\,dt + \int_{-\infty}^{\infty}\delta(t)\, e^{-j\omega t}\,dt$$

$$= -\frac{a}{a+j\omega} + 1$$

$$= \frac{j\omega}{a+j\omega}$$

Which is the same as mentioned in the question.

3.3 INVERSE FOURIER TRANSFORM

The process for transforming $F(j\omega)$ back to a continuous-time signal $f(t)$ is called the **inverse Fourier transformation.** The inverse Fourier transformation can be accomplished by two methods. These are :

1. Direct Integration Method
2. Partial-Fraction Expansion Method

Now, we shall discuss these methods one by one :

3.3.1 Direct Integration Method

If $F(j\omega)$ is the Fourier transform of $f(t)$, we have the relationship

$$f(t) = \frac{1}{2\pi}\int_{-\infty}^{\infty} F(j\omega)\, e^{j\omega t}\, d\omega$$

This relationship directly gives the inverse Fourier transform of $F(j\omega)$. To illustrate how a direct integration method is employed to obtain the inverse Fourier transform, let us consider the following examples:

Example 3.8. *Find the inverse Fourier transform of the spectrum* $F(j\omega) = 2\pi\ \delta(\omega)$.

Solution. Here again we may expect **convergence irregularities** since $F(j\omega)$ has an **infinite discontinuity at the origin**. Using equation (3.3 B), we have

$$\begin{aligned} f(t) &= \int_{-\infty}^{\infty} F(j\omega)\, e^{j\omega t}\, d\omega \\ &= \int_{-\infty}^{\infty} 2\pi\ \delta(\omega) e^{j\omega t}\, d\omega \\ &= 1 \end{aligned}$$

Hence we identify

$$1 \stackrel{\mathcal{F}}{\longleftrightarrow} 2\pi\ \delta(\omega)$$

as a **Fourier transform** pair. This implies that the frequency content of a **d.c. signal** is **concentrated entirely at** $\omega = 0$.

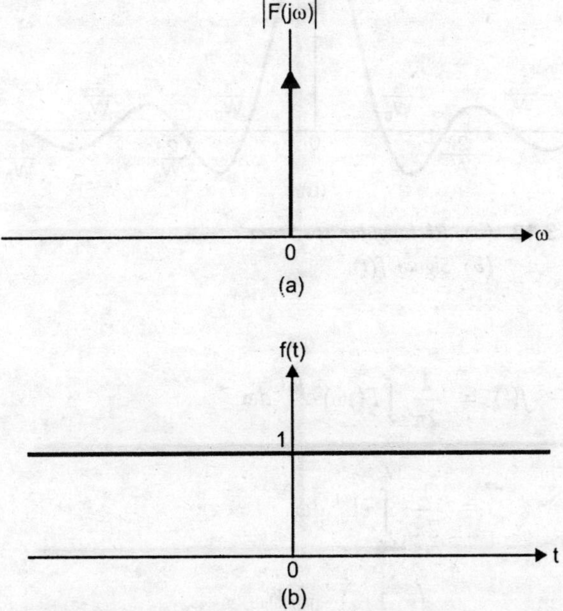

FIGURE 3.11 *(a) Magnitude spectrum*
(b) Signal f(t).

Comment : While convergence of the Fourier transform **cannot be guaranteed** in examples (3.3) and (3.8), the transform pairs **do satisfy** the properties of a Fourier transform pair and are, therefore, useful for analysis. In both cases the transform pairs are the **consequences** of the properties of the **impulse function**. By permitting the use of impulse function we greatly expand the **class of signals** that are represented by the Fourier transform and, thus, **enhance** the power of the Fourier transform as a **problem solving tool**.

Example 3.9. *Find the inverse Fourier transform of the rectangular spectrum as depicted in Figure 3.12 (a).*

Solution. The rectangular spectrum $F(j\omega)$ is defined as

$$F(j\omega) = \begin{cases} 1 & ; \quad -W_0 \le \omega \le W_0 \\ 0 & ; \quad \text{Otherwise} \end{cases}$$

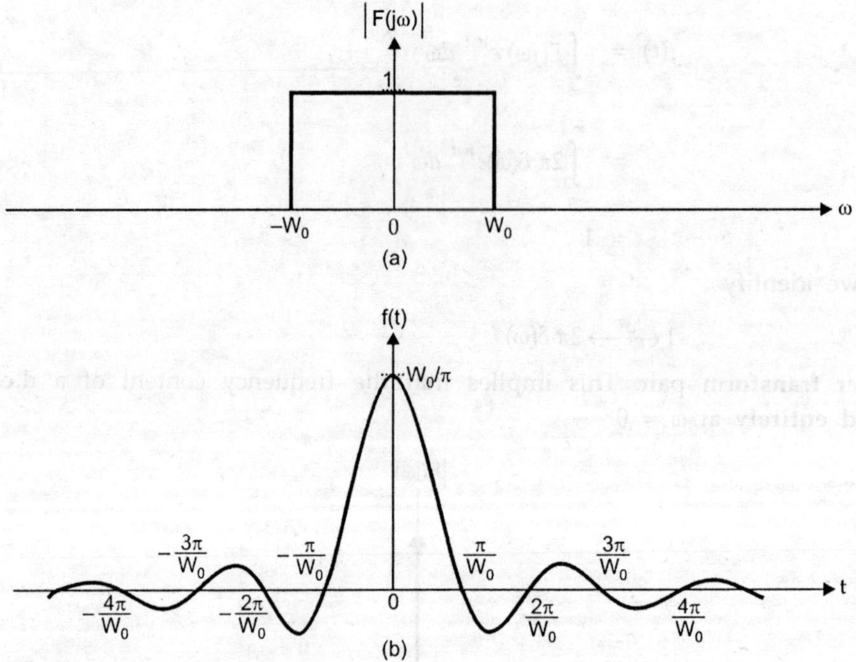

FIGURE 3.12 *(a) Rectangular spectrum*
(b) Signal f(t).

For $t \neq 0$, we have

$$f(t) = \frac{1}{2\pi}\int_{-\infty}^{\infty} F(j\omega)\, e^{j\omega t}\, d\omega$$

$$= \frac{1}{2\pi}\int_{-W_0}^{W_0} e^{j\omega t}\, d\omega$$

$$= -\left(\frac{1}{2j\pi t}\right) e^{j\omega t}\Big|_{-W_0}^{W_0}$$

$$= \left(\frac{1}{\pi t}\right) \sin W_0 t \quad ; \quad t \neq 0$$

For $t = 0$, using **L' Hopital's Rule**, we have

$$\lim_{t \to 0} \left(\frac{1}{\pi t}\right) \sin W_0 t = \frac{W_0}{\pi}$$

Thus, we write for all t

$$f(t) = \left(\frac{1}{\pi t}\right) \sin W_0 t$$

with the understanding that the value at $t = 0$ is obtained by evaluating a limit. The signal $f(t)$ is depicted in Figure 3.12 (*b*). We may also write $f(t)$ using the **sinc function notation** as

$$f(t) = \frac{W_0}{\pi} \text{ sinc}\left(\frac{W_0 t}{\pi}\right).$$

3.3.2 Partial-Fraction Expansion Method

This method is particularly suited to **rational Fourier transforms**. In this method, we determine inverse Fourier transforms by inspection. Given knowledge of several basic transform pairs and the Fourier transform properties; we are able to invert a very large class of rational Fourier transforms. Basically, the procedure consists of expanding the rational Fourier transform expression into a **linear combination of simpler terms** using **partial-fraction expansion method.** The inverse Fourier transform of each term can then be obtained by means of a list of **Fourier transform pairs** (given in **Appendix-B**) in conjunction with the **properties of the Fourier transform** (already discussed in **Section 3.2.3**). To illustrate how a **partial-fraction expansion** can be used to obtain the inverse Fourier transform, let us consider the following example:

Example 3.10. *Find the inverse Fourier transform of*

$$F(j\omega) = \frac{j\omega + 1}{(j\omega)^2 + 5j\omega + 6}$$

Solution.

$$F(j\omega) = \frac{j\omega + 1}{(j\omega)^2 + 5j\omega + 6}$$

or,

$$F(j\omega) = \frac{j\omega + 1}{(j\omega + 2)(j\omega + 3)}$$

or,

$$F(j\omega) = \left.\frac{j\omega + 1}{(j\omega + 3)}\right|_{j\omega = -2} \frac{1}{(j\omega + 2)} + \left.\frac{j\omega + 1}{(j\omega + 2)}\right|_{j\omega = -3} \frac{1}{(j\omega + 3)}$$

or,

$$F(j\omega) = \frac{-1}{(j\omega + 2)} + \frac{2}{(j\omega + 3)} \quad ...(3.20)$$

Thus,

$$f(t) = f_1(t) + f_2(t)$$

where, $$f_1(t) \overset{\mathcal{F}}{\longleftrightarrow} \frac{-1}{j\omega+2}$$

and $$f_1(t) \overset{\mathcal{F}}{\longleftrightarrow} \frac{2}{j\omega+3}$$

By inspection, we can identify

$$f_1(t) = -e^{-2t}\, u(t)$$

and $$f_2(t) = 2e^{-3t}\, u(t)$$

so that $$f(t) = [2e^{-3t} - e^{-2t}]\, u(t).$$

3.4 DUALITY

While observing the examples that have been dealt in this chapter, we find a **consistent symmetry between the time and frequency domain representations of signals.** We can also observe the **symmetries in the properties of the Fourier transform** as well. For example, convolution in one domain corresponds to modulation in the other domain and so on. These symmetries exist because of the **symmetry in the definitions of time and frequency domain representations of signals.** If we are careful, we may **interchange** time and frequency variables. This interchangeability property is termed as **duality.**

Example 3.11. *Use duality to evaluate the Fourier transform of*

$$f(t) = \frac{1}{1+jt}$$

Solution. First recognize that

$$e^{-t}\, u(t) = \frac{1}{2\pi}\int_{-\infty}^{\infty}\left[\frac{1}{1+j\omega}\right] e^{j\omega t}\, d\omega$$

Multiplying this equation by 2π and replacing t by $-t$, we get

$$2\pi\, e^{t}\, u(-t) = \int_{-\infty}^{\infty}\left[\frac{1}{1+j\omega}\right] e^{-j\omega t}\, d\omega$$

Using **duality property,** we have

$$2\pi\, e^{\omega}\, u(-\omega) = \int_{-\infty}^{\infty}\left[\frac{1}{1+jt}\right] e^{-j\omega t}\, dt$$

which implies that

$$\frac{1}{1+jt} \overset{\mathcal{F}}{\longleftrightarrow} 2\pi\, e^{\omega}\, u(-\omega)$$

We conclude that

$$F(j\omega) = \mathcal{F}\left[\frac{1}{1+jt}\right]$$

or, $$F(j\omega) = 2\pi\, e^{\omega}\, u(-\omega)$$

Example 3.12. *Use duality to evaluate the inverse Fourier transform of the step function in frequency,* $F(j\omega) = u(\omega)$.

Solution. First recognize that

$$\frac{1}{j\omega} + \pi\,\delta(\omega) = \int_{-\infty}^{\infty} u(t)\, e^{-j\omega t}\, dt$$

Dividing this equation by 2π and replacing ω by $-\omega$, we get

$$\frac{1}{2\pi}\left[-\frac{1}{j\omega} + \pi\,\delta(-\omega)\right] = \frac{1}{2\pi}\int_{-\infty}^{\infty} u(t)\, e^{j\omega t}\, dt$$

or,
$$\frac{1}{2}\left[-\frac{1}{j\pi\omega} + \delta(-\omega)\right] = \frac{1}{2\pi}\int_{-\infty}^{\infty} u(t)\, e^{j\omega t}\, dt$$

Using **duality property,** we have

$$\frac{1}{2}\left[-\frac{1}{j\pi t} + \delta(-t)\right] = \frac{1}{2\pi}\int_{-\infty}^{\infty} u(\omega)\, e^{j\omega t}\, d\omega$$

which implies that

$$\frac{1}{2}\left[-\frac{1}{j\pi t} + \delta(-t)\right] \xleftrightarrow{\mathscr{F}} u(\omega)$$

We conclude that

$$f(t) = \mathscr{F}^{-1}[u(\omega)]$$

or,
$$f(t) = \frac{1}{2}\left[\delta(-t) - \frac{1}{j\pi t}\right]$$

3.5 PARSEVAL'S RELATION

The **Parseval's relationships** state that the **energy or power** in the **time-domain representation** of a signal is **equal** to the **energy or power** in the **frequency-domain representation.** Hence, energy or power is **conserved** in the **Fourier representation.**

The energy of a **continuous-time non-periodic signal** is

$$E_{\infty} = \int_{-\infty}^{\infty} |f(t)|^2\, dt$$

where it is assumed that $f(t)$ may be **complex-valued** in general. Note that

$$|f(t)|^2 = f(t)\, f^*(t)$$

and that $f^*(t)$ is expressed in terms of its Fourier transform $F^*(j\omega)$ as

$$f^*(t) = \frac{1}{2\pi}\int_{-\infty}^{\infty} F^*(j\omega)e^{-j\omega t}d\omega$$

Substitute this in the expression for E_∞ so as to obtain

$$E_\infty = \int_{-\infty}^{\infty} f(t)\frac{1}{2\pi}\int_{-\infty}^{\infty} F^*(j\omega)e^{-j\omega t}\, d\omega\, dt$$

Interchanging the order of integration

$$E_\infty = \frac{1}{2\pi}\int_{-\infty}^{\infty} F^*(j\omega)\left[\int_{-\infty}^{\infty} f(t)\, e^{-j\omega t}\, dt\right] d\omega$$

or,

$$E_\infty = \frac{1}{2\pi}\int_{-\infty}^{\infty} F^*(j\omega)F(j\omega)\, d\omega$$

and so conclude

$$\int_{-\infty}^{\infty} |f(t)|^2\, dt = \frac{1}{2\pi}\int_{-\infty}^{\infty} |F(j\omega)|^2\, d\omega \qquad ...(3.21)$$

Hence the energy in the time-domain representation of the signal is equal to the energy in the frequency-domain representation **normalized by** 2π. The quantity $|F(j\omega)|^2$ is termed as the **energy spectrum** of the signal. Equation (3.21) is also referred to as **Rayleigh's energy theorem.**

3.6 FREQUENCY RESPONSE

Figure 3.13 shows a linear **continuous-time system** wherein the output signal $y(t)$ of the system is given by the **convolution sum** between the **input signal** $x(t)$ and the **impulse response function** $g(t)$. We may write

$$y(t) = x(t) * g(t)$$

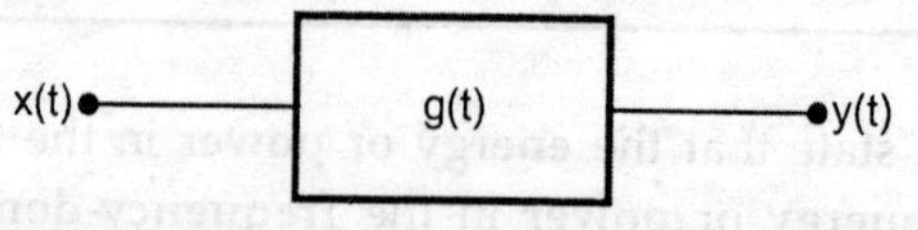

x(t) g(t) y(t)

FIGURE 3.13 *Linear continuous-time system.*

Using the **convolution property** of the Fourier transform, we have

$$Y(j\omega) = X(j\omega)\, G(j\omega) \qquad ...(3.22)$$

This is an important result wherein $G(j\omega)$ is interpreted as the **frequency response** of the linear continuous-time system as shown in Figure 3.14.

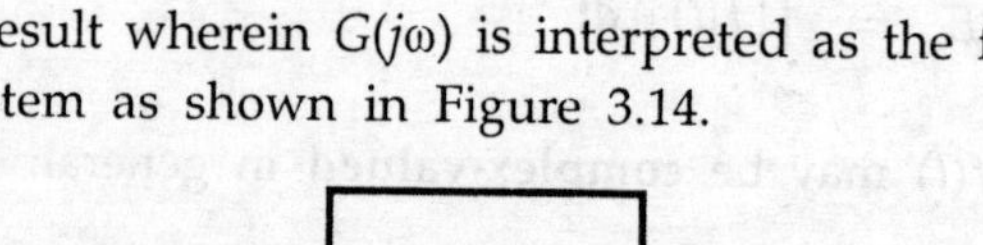

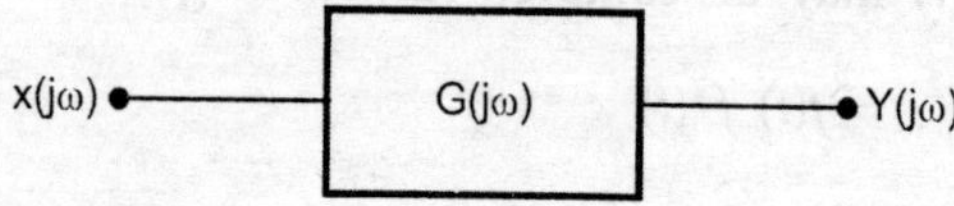

FIGURE 3.14 *Block-diagram for a frequency response system.*

Thus, from equation (3.22), the **frequency response** is given by

$$G(j\omega) = \frac{Y(j\omega)}{X(j\omega)} \qquad ...(3.23)$$

It is to be noted that $G(j\omega)$ is the Fourier transform of the **impulse response function** $g(t)$ and **captures the change in complex amplitude** of the Fourier transform of the input at each frequency ω.

Example 3.13. *Find out the impulse response function g(t) of an LTI system described by the following differential equation.*

$$\frac{d^2}{dt^2}y(t) + 4\frac{d}{dt}y(t) + 3y(t) = \frac{d}{dt}x(t) + 2x(t)$$

Solution. Taking the Fourier transforms of both sides, we have

$$(j\omega)^2 Y(j\omega) + 4j\omega Y(j\omega) + 3Y(j\omega) = j\omega X(j\omega) + 2X(j\omega)$$

or, $$[(j\omega)^2 + 4j\omega + 3]\, Y(j\omega) = [j\omega + 2]\, X(j\omega)$$

or, $$\frac{Y(j\omega)}{X(j\omega)} = \frac{j\omega + 2}{(j\omega)^2 + 4j\omega + 3}$$

or, $$G(j\omega) = \frac{j\omega + 2}{(j\omega + 1)(j\omega + 3)}$$

or, $$G(j\omega) = \left.\frac{j\omega + 2}{(j\omega + 3)}\right|_{j\omega=-1} \frac{1}{(j\omega + 1)} + \left.\frac{j\omega + 2}{(j\omega + 1)}\right|_{j\omega=-3} \frac{1}{(j\omega + 3)}$$

ro, $$G(j\omega) = \frac{0.5}{(j\omega + 1)} + \frac{0.5}{(j\omega + 3)}$$

Thus, $$g(t) = g_1(t) + g_2(t)$$

where, $$g_1(t) \xleftrightarrow{\mathcal{F}} \frac{0.5}{j\omega + 1}$$

and $$g_2(t) \xleftrightarrow{\mathcal{F}} \frac{0.5}{j\omega + 3}$$

By inspection, we can identify

$$g_1(t) = \frac{1}{2}\, e^{-t}\, u(t)$$

and $$g_2(t) = \frac{1}{2}\, e^{-3t}\, u(t)$$

so that $$g(t) = \frac{1}{2}\, [e^{-t} + e^{-3t}]\, u(t)$$

PROBLEMS

1. Determine the Fourier transform of

$$f(t) = \begin{cases} t & ; \ |t| \le 1 \\ 0 & ; \ |t| > 1 \end{cases}$$

2. Obtain the Fourier transform of

$$f(t) = te^{-at}\, u(t)$$

[Hint: Use the frequency differentiation property]

3. Determine the Fourier transform of the complex sinusoidal pulse

$$f(t) = \begin{cases} e^{10jt} & ; \ |t| \le \pi \\ 0 & ; \ \text{otherwise} \end{cases}$$

[Hint: Use the frequency shift property]

4. The output of a system in response to an input $f(t) = e^{-2t}\, u(t)$ is given by

$$y(t) = e^{-t}\, u(t)$$

Determine the frequency response and the impulse response of this system.

5. The impulse response of a system is given by

$$g(t) = e^{-at}\, u(t) \ ; \ a > 0$$

Use Fourier transform method so as to determine the response of this system to the input signal

$$x(t) = e^{-bt}\, u(t) \ ; \ b > 0$$

Assume that :

(*a*) $a \ne b$

(*b*) $a = b$

6. Find the inverse Fourier transform of

(*a*) $$F(j\omega) = \begin{cases} 2\cos\omega & ; \ |\omega| \le \pi \\ 0 & ; \ |\omega| > \pi \end{cases}$$

(*b*) $$F(j\omega) = \frac{-j\omega}{(j\omega)^2 + 3j\omega + 2}$$

7. An LTI system is described by the following differential equation.

$$\frac{d^2}{dt^2}y(t) + 4\frac{d}{dt}y(t) + 3y(t) = \frac{d}{dt}x(t) + 2x(t)$$

Use Fourier transform method so as to determine the response of this system to the input signal

$$x(t) = e^{-t}\, u(t)$$

Chapter 4

LAPLACE TRANSFORM ANALYSIS

4.1 INTRODUCTION

In previous chapter, we have discussed that a signal $f(t)$ can be represented in the **frequency-domain** by its **Fourier transform** $F(j\omega)$ provided it is **absolutely integrable** in the sense

$$\int_{-\infty}^{\infty} |f(t)| \, dt < \infty.$$

However, one may encounter with a large class of problems involving signals that are **not absolutely integrable**. The Fourier transform does not exist for signals that are not absolutely integrable and, therefore, Fourier transform based methods cannot be used in this class of problems. This difficulty is overcome by **generalizing** the representation of a signal offered by the Fourier transform to a representation in terms of **complex exponential signals** which is termed as the **Laplace transform**, named after **Pierre Simon de Laplace** (1749–1827).

Use of the Laplace transform for **system analysis** is advantageous in that it provides a broader **characterization** of systems and their **interaction** with signals than that is possible with the Fourier methods. The existence of the Laplace transform for signals that have no Fourier transform is a **significant advantage** of using the complex exponential representation. Furthermore, the Laplace transform converts the governing differential equation for physical systems to an **algebraic equation** and **initial conditions** are included **automatically** as a part of the solution.

4.2 THE LAPLACE TRANSFORM

As already stated, the Fourier transform does not exist for signals that are not absolutely integrable. In order to analyze such signals, we **generalize** the Fourier transform representation of a signal $f(t)$ by multiplying it with a **real exponential convergence factor** $e^{-\sigma t}$ and then **Fourier transforming** the product. The value of the **real constant** σ is such that

$$\int_{-\infty}^{\infty} \left| f(t)\, e^{-\sigma t} \right| dt < \infty.$$

We, therefore, may write the Fourier transform of $f(t)e^{-\sigma t}$ as

$$\begin{aligned} F(\sigma + j\omega) &= \mathscr{F}[f(t)\, e^{-\sigma t}] \\ &= \int_{-\infty}^{\infty} [f(t)\, e^{-\sigma t}]\, e^{-j\omega t}\, dt \\ &= \int_{-\infty}^{\infty} f(t)\, e^{-(\sigma+j\omega)t}\, dt \end{aligned}$$

With $s = \sigma + j\omega$, we have

$$F(s) = \int_{-\infty}^{\infty} f(t)\, e^{-st}\, dt \qquad \text{...(4.1)}$$

Conversely, we can invert the relationship using the inverse Fourier transform as given in equation (3.3 A) so as to get

$$\begin{aligned} f(t)\, e^{-\sigma t} &= \mathscr{F}^{-1}[F(\sigma + j\omega)] \\ &= \frac{1}{2\pi} \int_{-\infty}^{\infty} F(\sigma + j\omega)\, e^{j\omega t}\, d\omega \end{aligned}$$

Multiplying both sides by $e^{\sigma t}$, we get

$$f(t) = \frac{1}{2\pi} \int_{-\infty}^{\infty} F(\sigma + j\omega)\, e^{(\sigma+j\omega)t}\, d\omega \qquad \text{...(4.2)}$$

Changing the variable of integration from ω to $s = \sigma + j\omega$ so as to obtain

$$d\omega = \left(\frac{1}{j}\right) ds$$

Recognizing the new limits of integration from $s = \sigma - j\infty$ to $s = \sigma + j\infty$, equation (4.2) can be rewritten as

$$f(t) = \frac{1}{2\pi j} \int_{\sigma-j\infty}^{\sigma+j\infty} F(s)\, e^{st}\, ds \qquad \text{...(4.3)}$$

Equations (4.1) and (4.3) together form the **Laplace transform pair**. This pair is represented by $f(t) \xleftrightarrow{\mathscr{L}} F(s)$, meaning that $F(s)$ is the **Laplace transform** of $f(t)$, and $f(t)$ is the **inverse Laplace transform** of $F(s)$. This relationship may also be written as

$$F(s) = \mathscr{L}[f(t)] \text{ and } f(t) = \mathscr{L}^{-1}[F(s)]$$

We recognize that the **Laplace transform** of $f(t)$ can be interpreted as the **Fourier transform** of $f(t)$ **after multiplication** by a **real exponential convergence factor** $e^{-\sigma t}$. Therefore, **absolute**

integrability of $f(t)e^{-\sigma t}$ becomes a **necessary condition** for convergence of the Laplace transform. It should not be believed that one can always find some σ which ensure the absolute integrability of $f(t)e^{-\sigma t}$. An example is $f(t) = e^{e^t}$; $t > 0$, which **grows faster than** $e^{-\sigma t}$ **decays.** Therefore, this signal is **not Laplace transformable**. However, in linear systems analysis, we seldom encounter signals which are not Laplace transformable.

For **causal signals,** we have

$$f(t) = 0; \; t < 0$$

Therefore, the Laplace transform $F(s)$ of continuous-time signal $f(t)$, given by equation (4.1) reduces to

$$F(s) = \int_{0^-}^{\infty} f(t)\, e^{-st}\, dt \qquad \text{...(4.4)}$$

The Laplace transform $F(s)$, defined by equation (4.1) is called **bilateral (or two-sided) Laplace transform** and the same defined by equation (4.4) is called **unilateral (or one-sided) Laplace transform.**

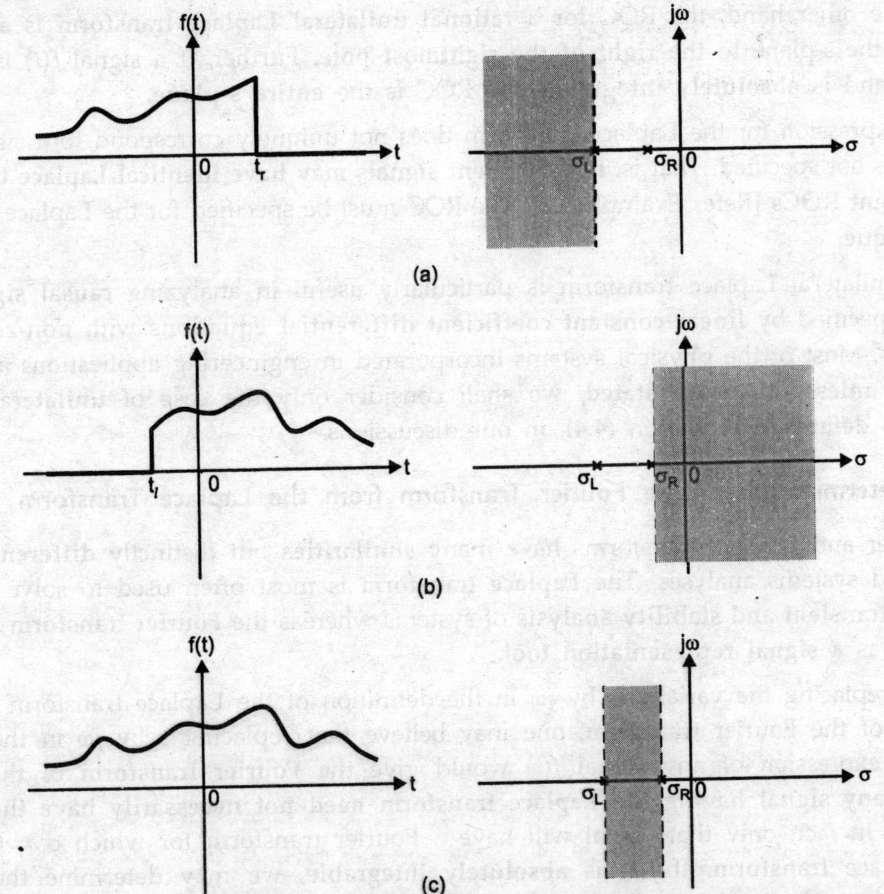

FIGURE 4.1 *(a) A left-sided signal has ROC to the left of the leftmost pole in the s-plane*
(b) A right-sided signal has ROC to the right of the rightmost pole in the s-plane
(c) The ROC for a two-sided signal is a strip of finite width in the s-plane.

The **difference** between bilateral Laplace transform and unilateral Laplace transform is that the lower limit of integration is $-\infty$ for bilateral Laplace transform and 0^- for unilateral Laplace transform. The lower limit of integration, *i.e.*, 0^- implies that we also take account of any impulse **concentrated** at $t = 0$. It should be noted that two signals that **differ** for $t < 0$, but are **identical** for $t \geq 0$, will have **different bilateral Laplace transforms but identical unilateral transforms**. Similarly, any signal that is **identically zero** for $t < 0$ has **identical bilateral and unilateral Laplace transforms.**

The range of values of s, for which the Laplace integral **converges**, is called the **region of convergence (ROC)** of the Laplace transform. Define a **left-sided signal** as one for which $f(t) = 0$ for $t > t_r$; a **right-sided signal** as one for which; $f(t) = 0$ for $t < t_l$; and a **two-sided signal** as one that is infinite in extent in both directions. Then, the ROC for a rational bilateral Laplace transform of a **right-sided signal** is the region in the s-plane to the **right of the rightmost pole** whereas the ROC for a rational bilateral Laplace transform of a **left-sided signal** is the region in the s-plane to the **left of the leftmost pole.** The ROC for a rational bilateral Laplace transform of a **two-sided signal** is a **single strip in the s-plane** as illustrated in Figure 4.1.

On the other hand, the ROC for a **rational unilateral Laplace transform** is **always** the region in the s-plane to the **right of the rightmost pole.** Further, if a signal $f(t)$ is of **finite duration** and is **absolutely integrable**, the ROC is the **entire s-plane.**

The expression for the Laplace transform does not uniquely correspond to a signal $f(t)$ if the ROC is not specified. That is, **two different signals may have identical Laplace transforms but different ROCs** [Refer Example 4.2]. The ROC must be specified for the Laplace transform to be **unique.**

The unilateral Laplace transform is particularly useful in analyzing **causal signals and systems** specified by **linear constant coefficient differential equations** with **non-zero initial conditions.** Most of the physical systems incorporated in engineering applications are **causal,** therefore, **unless otherwise stated,** we shall consider only the case of **unilateral Laplace transform,** defined by equation (4.4), in our discussions.

4.2.1 Determination of the Fourier Transform from the Laplace Transform

The Fourier and Laplace transforms have many **similarities** but distinctly **different roles** in signals and systems analyses. The **Laplace transform** is most often used to solve problems involving **transient and stability analysis** of systems whereas the **Fourier transform** is usually employed as a **signal representation tool.**

Since replacing the variable s by $j\omega$ in the definition of the Laplace transform gives the definition of the Fourier transform, one may believe that replacing s by $j\omega$ in the Laplace transform expression of any signal $f(t)$ would give the **Fourier transform** of that signal. However, **any signal having the Laplace transform need not necessarily have the Fourier transform.** In fact, only that signal will have a Fourier transform for which $\sigma = 0$ gives a valid **Laplace transform.** If $f(t)$ is **absolutely integrable,** we may determine the Fourier transform from the Laplace transform by setting $\sigma = 0$ as follows:

$$F(j\omega) = F(s)|_{\sigma = 0}$$

4.2.2 Properties of the Laplace Transform

Let $\mathscr{L}[f(t)] = F(s)$

and $\mathscr{L}[g(t)] = G(s)$

Now, the properties of the Laplace transform are **summarized** in following sections one by one:

4.2.2.1 Linearity

$$\mathscr{L}[af(t) \pm bg(t)] = aF(s) \pm bG(s) \qquad ...(4.5)$$

Proof.
$$\mathscr{L}[af(t) \pm bg(t)] = \int_{0^-}^{\infty}[af(t) \pm bg(t)]\, e^{-st}\, dt$$

$$= a\int_{0^-}^{\infty} f(t)\, e^{-st}\, dt \pm b\int_{0^-}^{\infty} g(t)\, e^{-st}\, dt$$

$$= aF(s) \pm bG(s)$$

4.2.2.2 Scaling

$$\mathscr{L}[f(at)] = \frac{1}{|a|}F\left(\frac{s}{a}\right) \qquad ...(4.6)$$

Proof.
$$\mathscr{L}[f(at)] = \int_{0^-}^{\infty} f(at)\, e^{-st}\, dt$$

Let $\lambda = at$, we have

$$\mathscr{L}[f(at)] = \begin{cases} \dfrac{1}{a}\displaystyle\int_{0^-}^{\infty} f(\lambda)\, e^{-\left(\frac{s}{a}\right)\lambda}\, d\lambda & ; \; a > 0 \\ -\dfrac{1}{a}\displaystyle\int_{0^-}^{\infty} f(\lambda)\, e^{-\left(\frac{s}{a}\right)\lambda}\, d\lambda & ; \; a < 0 \end{cases}$$

In combined form, we can write

$$\mathscr{L}[f(at)] = \frac{1}{|a|}\int_{0^-}^{\infty} f(\lambda)\, e^{-\left(\frac{s}{a}\right)\lambda}\, d\lambda$$

$$= \frac{1}{|a|}F\left(\frac{s}{a}\right)$$

4.2.2.3 Time Shifting

$$\mathscr{L}[f(t-\tau)] = e^{-s\tau} F(s) \text{ for all } \tau > 0 \text{ such that } f(t-\tau)\, u(t) = f(t-\tau)\, u(t-\tau) \qquad (4.7)$$

Proof. $\mathscr{L}\,[f(t-\tau)] = \int_{0^-}^{\infty} f(t-\tau)\, e^{-st}\, dt$

Let $\lambda = t - \tau$, we have

$$\mathscr{L}\,[f(t-\tau)] = \int_{0^-}^{\infty} f(\lambda)\, e^{-s(\tau+\lambda)}\, d\lambda$$

$$= \int_{0^-}^{\infty} f(\lambda)\, e^{-s\tau}\, e^{-s\lambda}\, d\lambda$$

$$= e^{-s\tau} \int_{0^-}^{\infty} f(\lambda)\, e^{-s\lambda}\, d\lambda$$

$$= e^{-s\tau} F(s)$$

Here, the **restriction** on the shift arises because the unilateral Laplace transform is defined solely in terms of the **positive-time portions** of the signal $f(t)$.

4.2.2.4 Shifting in s-Domain

$$\mathscr{L}\,[e^{\alpha t} f(t)] = F(s-\alpha) \qquad \text{...(4.8)}$$

Proof. To establish this property, consider the expression

$$g(t) = \frac{1}{2\pi j} \int_{\sigma-j\infty}^{\sigma+j\infty} G(s)\, e^{st}\, ds$$

Assuming $G(s) = F(s-\alpha)$, we may write

$$g(t) = \frac{1}{2\pi j} \int_{\sigma-j\infty}^{\sigma+j\infty} F(s-\alpha)\, e^{st}\, ds$$

Let $\lambda = s - \alpha$, we have

$$g(t) = \frac{1}{2\pi j} \int_{\sigma-j\infty}^{\sigma+j\infty} F(\lambda)\, e^{(\alpha+\lambda)t}\, d\lambda$$

$$= \frac{1}{2\pi j} \int_{\sigma-j\infty}^{\sigma+j\infty} F(\lambda)\, e^{\alpha t}\, e^{\lambda t}\, d\lambda$$

$$= e^{\alpha t} \left[\frac{1}{2\pi j} \int_{\sigma-j\infty}^{\sigma+j\infty} F(\lambda)\, e^{\lambda t}\, d\lambda \right]$$

$$= e^{\alpha t} f(t)$$

Thus, $\mathscr{L}[e^{\alpha t} f(t)] = F(s-\alpha)$

4.2.2.5 Convolution

$$\mathscr{L}[f(t)^* g(t)] = F(s)G(s) \quad ...(4.9)$$

Proof. $$\mathscr{L}[f(t)^* g(t)] = \int_{0^-}^{\infty}[f(t) * g(t)]\, e^{-st}\, dt$$

$$= \int_{t=0^-}^{\infty}\left[\int_{\tau=0^-}^{t} f(\tau)g(t-\tau)\, d\tau\right] e^{-st}\, dt$$

The upper limit of the integral inside the bracket can be made ∞ without altering its value, because $g(t) = 0$; $t < 0$ or equivalently $g(t - \tau) = 0$; $t < \tau$.

Thus, $$\mathscr{L}[f(t)^* g(t)] = \int_{t=0^-}^{\infty}\left[\int_{\tau=0^-}^{\infty} f(\tau)g(t-\tau)\, d\tau\right] e^{-st}\, dt$$

Interchanging the order of integration and noting that $f(\tau)$ **does not depend** on t, we may write

$$\mathscr{L}[f(t)* g(t)] = \int_{\tau=0^-}^{\infty} f(\tau)\left[\int_{t=0^-}^{\infty} g(t-\tau)\, e^{-st}\, dt\right] d\tau$$

Let $\lambda = t - \tau$, we have

$$\mathscr{L}[f(t)* g(t)] = \int_{\tau=0^-}^{\infty} f(\tau)\left[\int_{t=0^-}^{\infty} g(\lambda)\, e^{-s(\tau+\lambda)}\, d\lambda\right] d\tau$$

$$= \int_{\tau=0^-}^{\infty} f(\tau)\, e^{-s\tau}\left[\int_{t=0^-}^{\infty} g(\lambda)\, e^{-s\lambda}\, d\lambda\right] d\tau$$

$$= \left[\int_{\tau=0^-}^{\infty} f(\tau)\, e^{-s\tau}\, d\tau\right] G(s)$$

$$= F(s)G(s)$$

4.2.2.6 Differentiation in s-Domain (Multiplication by t)

$$\mathscr{L}[t\ f(t)] = -\frac{d}{ds}F(s) \quad ...(4.10)$$

Proof. $$\mathscr{L}[t\ f(t)] = \int_{0^-}^{\infty}[t\, f(t)]\, e^{-st}\, dt$$

$$= -\int_{0^-}^{\infty} f(t)[-t\, e^{-st}]\, dt$$

$$= -\int_{0^-}^{\infty} f(t)\left[\frac{d}{ds}\, e^{-st}\right] dt$$

$$= -\frac{d}{ds}\left[\int_{0^-}^{\infty} f(t)\, e^{-st}\, dt\right]$$

$$= -\frac{d}{ds} F(s)$$

4.2.2.7 *Differentiation in Time-Domain*

$$\mathcal{L}\left[\frac{d}{dt} f(t)\right] = sF(s) - f(0^-) \qquad \text{...(4.11)}$$

Proof. $$\mathcal{L}\left[\frac{d}{dt} f(t)\right] = \int_{0^-}^{\infty} \left[\frac{d}{dt} f(t)\right] e^{-st}\, dt$$

Integrating by parts, we obtain

$$\mathcal{L}\left[\frac{d}{dt} f(t)\right] = [e^{-st} f(t)]\Big|_{0^-}^{\infty} + s\int_{0^-}^{\infty} f(t)\, e^{-st}\, dt = s\int_{0^-}^{\infty} f(t)\, e^{-st}\, dt + [e^{-st} f(t)]\Big|_{0^-}^{\infty}$$

$$= sF(s) - f(0^-)$$

Provided that

$$\lim_{t\to\infty} [e^{-st} f(t)] = 0$$

4.2.2.8 *Integration in s-Domain (Division by t)*

$$\mathcal{L}\left[\frac{f(t)}{t}\right] = \int_{s}^{\infty} F(s) ds$$

Proof. We have

$$F(s) = \mathcal{L}[f(t)]$$

$$= \int_{0^-}^{\infty} f(t)\, e^{-st}\, dt$$

Integrating both sides by taking limits from s to ∞, we have

$$\int_{s}^{\infty} F(s)\, ds = \int_{s}^{\infty} \left[\int_{0^-}^{\infty} f(t)\, e^{-st}\, dt\right] ds$$

Interchanging the order of integration, we may write

$$\int_{s}^{\infty} F(s)\, ds = \int_{0^-}^{\infty} f(t) \left[\int_{s}^{\infty} e^{-st}\, ds\right] dt$$

$$= \int_{0^-}^{\infty} f(t)\left[\frac{e^{-st}}{t}\right] dt$$

$$= \int_{0^-}^{\infty} \left[\frac{f(t)}{t}\right] e^{-st}\, dt$$

Thus, $$\mathscr{L}\left[\frac{f(t)}{t}\right] = \int_{s}^{\infty} F(s)ds$$

4.2.2.9 Integration in Time-Domain

$$\mathscr{L}\left[\int_{-\infty}^{t} f(\tau)d\tau\right] = \frac{F(s)}{s} + \frac{1}{s}\left[\int_{-\infty}^{0^-} f(\tau)\, d\tau\right] \qquad ...(4.12)$$

Proof. $$\mathscr{L}\left[\int_{-\infty}^{t} f(\tau)d\tau\right] = \int_{0^-}^{\infty}\left[\int_{-\infty}^{t} f(\tau)d\tau\right] e^{-st}\, dt$$

Integrating by parts, we obtain

$$\mathscr{L}\left[\int_{-\infty}^{t} f(\tau)d\tau\right] = \left[\left(\int_{-\infty}^{t} f(\tau)d\tau\right)\left(-\frac{e^{-st}}{s}\right)\right]_{0^-}^{\infty} + \frac{1}{s}\left[\int_{0^-}^{\infty} f(t)\, e^{-st} dt\right]$$

$$= \frac{1}{s}\left[\int_{0^-}^{\infty} f(t)\, e^{-st} dt\right] + \left[\left(\int_{-\infty}^{t} f(\tau)d\tau\right)\left(-\frac{e^{-st}}{s}\right)\right]_{0^-}^{\infty}$$

$$= \frac{F(s)}{s} + \frac{1}{s}\left[\int_{-\infty}^{0^-} f(\tau)\, d\tau\right]$$

Provided that

$$\lim_{t\to\infty}\left[e^{-st}\int_{-\infty}^{t} f(\tau)d\tau\right] = 0$$

4.2.2.10 Initial Value Theorem

$$f(0^+) = \lim_{s\to\infty}\,[sF(s)] \qquad ...(4.13)$$

Proof. $$\mathscr{L}\left[\frac{d}{dt} f(t)\right] = sF(s) - f(0^-)$$

Taking the limit $s \to \infty$ on both sides, we obtain

$$\lim_{s\to\infty}\left\{\mathscr{L}\left[\frac{d}{dt} f(t)\right]\right\} = \lim_{s\to\infty}\,[sF(s) - f(0^-)]$$

or,
$$\lim_{s\to\infty}\left\{\int_{0^-}^{\infty}\left[\frac{d}{dt}f(t)\right]e^{-st}\,dt\right\} = \lim_{s\to\infty}[sF(s)] - f(0^-) \qquad \text{...(4.14)}$$

Case 1: If $f(t)$ is **continuous** at $t = 0$ then $\frac{d}{dt}f(t)$ **does not contain any impulse** at $t = 0$. It, therefore, follows that

$$f(0^-) = f(0) = f(0^+)$$

and
$$\lim_{s\to\infty}\left\{\int_{0^-}^{\infty}\left[\frac{d}{dt}f(t)\right]e^{-st}\,dt\right\} = 0$$

Thus, from equation (4.14), we have

$$\lim_{s\to\infty}[sF(s)] = f(0^-) = f(0^+) \qquad \text{...(4.15 A)}$$

provided that $\lim_{s\to\infty}[sF(s)]$ exists.

Case 2: If $f(t)$ is **discontinuous** at $t = 0$ then $\frac{d}{dt}f(t)$ **contains an impulse** $[f(0^+) - f(0^-)]\delta(t)$ at $t = 0$, and

$$\lim_{s\to\infty}\left\{\int_{0^-}^{\infty}\left[\frac{d}{dt}f(t)\right]e^{-st}\,dt\right\} = f(0^+) - f(0^-)$$

Thus, from equation (4.14), we have

$$f(0^+) = \lim_{s\to\infty}[sF(s)] \qquad \text{...(4.15 B)}$$

provided that $\lim_{s\to\infty}[sF(s)]$ exists.

It must be noted that $\lim_{s\to\infty}[sF(s)]$, if exists, **always gives the initial value** of $f(t)$ as $t \to 0^+$ **regardless of the lower limit** used on the Laplace integral.

4.2.2.11 Final Value Theorem

$$f(\infty) = \lim_{t\to\infty} f(t) = \lim_{s\to 0}[sF(s)] \qquad \text{...(4.16)}$$

Proof.
$$\mathcal{L}\left[\frac{d}{dt}f(t)\right] = sF(s) - f(0^-)$$

Taking the limit $s \to 0$ on both sides, we obtain

$$\lim_{s\to 0}\left\{\mathcal{L}\left[\frac{d}{dt}f(t)\right]\right\} = \lim_{s\to 0}[sF(s) - f(0^-)]$$

or,
$$\lim_{s\to 0}\left\{\int_{0^-}^{\infty}\left[\frac{d}{dt}f(t)\right]e^{-st}\,dt\right\} = \lim_{s\to 0}[sF(s)] - f(0^-)$$

or, $$\int_{0^-}^{\infty}\left[\frac{d}{dt}f(t)\right]dt = \lim_{s\to 0}[sF(s)] - f(0^-)$$

or, $$f(t)\Big|_{0^-}^{\infty} = \lim_{s\to 0}[sF(s)] - f(0^-)$$

or, $$\lim_{t\to\infty}[f(t)] - \lim_{t\to 0^-}[f(t)] = \lim_{s\to 0}[sF(s)] - f(0^-)$$

or, $$\lim_{t\to\infty}[f(t)] - f(0^-) = \lim_{s\to 0}[sF(s)] - f(0^-)$$

or, $$\lim_{t\to\infty}[f(t)] = \lim_{s\to 0}[sF(s)]$$

or, $$f(\infty) = \lim_{t\to\infty} f(t) = \lim_{s\to 0}[sF(s)]$$

provided that these limits exist.

4.2.3 Laplace Transforms of Some Common Continuous-Time Signals

4.2.3.1 Unit-Impulse Signal

It is defined as

$$\delta(t) = 0 \;\; ; \;\; t \neq 0 \qquad \text{...(4.17 A)}$$

and $$\int_{-\infty}^{\infty}\delta(t)\,dt = 1 \qquad \text{...(4.17 B)}$$

Using definition (4.1), we have

$$\mathscr{L}[\delta(t)] = \int_{-\infty}^{\infty}\delta(t)\,e^{-st}\,dt$$

or, $$\mathscr{L}[\delta(t)] = 1 \qquad \text{ROC: Entire } s\text{-plane.} \qquad \text{...(4.18)}$$

4.2.3.2 Unit-Step Signal

It is defined as

$$u(t) = \begin{cases} 1 & ; \; t \geq 0 \\ 0 & ; \; t < 0 \end{cases}$$

Using definition (4.4), we have

$$\mathscr{L}[u(t)] = \int_{0^-}^{\infty} u(t)\,e^{-st}\,dt$$

or, $$\mathscr{L}[u(t)] = \left[-\frac{e^{-st}}{s}\right]_{0^-}^{\infty}$$

or, $$\mathscr{L}[u(t)] = \frac{1}{s} \qquad \text{ROC: } \mathscr{Re}(s) > 0 \Rightarrow \sigma > 0 \qquad \text{...(4.19)}$$

4.2.3.3 Unit-Ramp Signal

A unit-ramp is obtained by multiplying a unit-step signal by t,

i.e.,
$$r(t) = t\ u(t)$$

$$\mathscr{L}[r(t)] = \mathscr{L}[t\ u(t)]$$

$$\therefore \quad = -\frac{d}{ds}U(s) \qquad \text{[Property 4.2.2.6]}$$

Since
$$U(s) = \frac{1}{s}$$

Therefore,
$$\mathscr{L}[r(t)] = -\frac{d}{ds}\left(\frac{1}{s}\right)$$

or,
$$\mathscr{L}[r(t)] = \frac{1}{s^2} \qquad \text{ROC: } \mathscr{R}e\,(s) > 0 \Rightarrow \sigma > 0 \qquad ...(4.20)$$

4.2.3.4 Exponential Signal

It is defined as

$$f(t) = \begin{cases} e^{-at} & ; \quad t \geq 0 \\ 0 & ; \quad t < 0 \end{cases}$$

Using definition (4.4), we have

$$\mathscr{L}[f(t)] = \int_{0^-}^{\infty} f(t)\, e^{-st}\, dt$$

or,
$$\mathscr{L}[f(t)] = \int_{0^-}^{\infty} e^{-at}\, e^{-st}\, dt$$

or,
$$\mathscr{L}[f(t)] = \int_{0^-}^{\infty} e^{-(s+a)\,t}\, dt$$

or,
$$\mathscr{L}[f(t)] = \left[-\frac{e^{-(s+a)\,t}}{s+a}\right]_{0^-}^{\infty}$$

or,
$$\mathscr{L}[f(t)] = \frac{1}{s+a} \qquad \text{ROC: } \mathscr{R}e\,(s) > -a \Rightarrow \sigma > -a \qquad ...(4.21)$$

4.2.3.5 Sinusoidal Signals

$$\mathscr{L}[\sin \omega t] = \mathscr{L}\left[\frac{e^{j\omega t} - e^{-j\omega t}}{2j}\right]$$

$$= \frac{1}{2j}\mathscr{L}[e^{j\omega t}] - \frac{1}{2j}\mathscr{L}[e^{-j\omega t}]$$

$$= \frac{1}{2j}\left[\frac{1}{s-j\omega}\right] - \frac{1}{2j}\left[\frac{1}{s+j\omega}\right]$$

or, $$\mathscr{L}[\sin \omega t] = \frac{\omega}{s^2+\omega^2} \qquad \text{ROC: } \mathscr{Re}(s) > 0 \Rightarrow \sigma > 0 \qquad ...(4.22)$$

Similarly, $$\mathscr{L}[\cos \omega t] = \mathscr{L}\left[\frac{e^{j\omega t}+e^{-j\omega t}}{2}\right]$$

$$= \frac{1}{2}\mathscr{L}[e^{j\omega t}] + \frac{1}{2}\mathscr{L}[e^{-j\omega t}]$$

$$= \frac{1}{2}\left[\frac{1}{s-j\omega}\right] + \frac{1}{2}\left[\frac{1}{s+j\omega}\right]$$

or, $$\mathscr{L}[\cos \omega t] = \frac{s}{s^2+\omega^2} \qquad \text{ROC: } \mathscr{Re}(s) > 0 \Rightarrow \sigma > 0 \qquad ...(4.23)$$

4.2.4 Laplace Transform of Periodic Signals

Let $f(t)$ be a **periodic signal** with period T_0.

Then, $$f(t) = f(t + T_0) \qquad \text{for all } t$$

Now, $$\mathscr{L}[f(t)] = \int_{0^-}^{\infty} f(t)\, e^{-st}\, dt$$

$$= \int_{0^-}^{T_0} f(t)\, e^{-st}\, dt + \int_{T_0}^{2T_0} f(t)\, e^{-st}\, dt + \ldots\ldots + \int_{nT_0}^{(n+1)T_0} f(t)\, e^{-st}\, dt + \ldots\ldots$$

$$= \int_{0^-}^{T_0} f(t)\, e^{-st}\, dt + e^{-sT_0}\int_{0^-}^{T_0} f(t)\, e^{-st}\, dt + \ldots\ldots + e^{-nsT_0}\int_{0^-}^{T_0} f(t)\, e^{-st}\, dt + \ldots\ldots$$

[Refer Section 4.2.2.3]

$$= [1+e^{-sT_0}+e^{-sT_0}\ldots\ldots+e^{-sT_0}]\int_{0^-}^{T_0} f(t)\, e^{-st}\, dt$$

$$= \frac{1}{1-e^{-sT_0}}\left[\int_{0^-}^{T_0} f(t)\, e^{-st}\, dt\right]$$

This implies that the **Laplace transform** of a **periodic signal** $f(t)$ with time period T_0 is equal to the **product** of $\frac{1}{1-e^{-sT_0}}$ and the **Laplace transform** of the **first cycle** of the **periodic waveform**.

Example 4.1. *Determine the Laplace transforms of the following continuous-time signals.*

(a) t^n	(b) $t^n e^{at}$	(c) sinh ωt
(d) cosh ωt	(e) e^{-at} sin ωt	(f) e^{-at} cos ωt

Solution. (*a*) Using definition (4.4), we have

$$\mathscr{L}[t^n] = \int_{0^-}^{\infty} t^n e^{-st}\, dt$$

Integrating by parts, we have

$$\mathscr{L}[t^n] = \left[t^n\left(-\frac{e^{-st}}{s}\right)\right]_{0^-}^{\infty} - \int_{0^-}^{\infty}\left[nt^{n-1}\left(-\frac{e^{-st}}{s}\right)\right] dt$$

$$= 0+\left(\frac{n}{s}\right)\int_{0^-}^{\infty} t^{n-1}\, e^{-st}\, dt$$

$$= \left(\frac{n}{s}\right)\mathscr{L}[t^{n-1}]$$

$$= \left(\frac{n}{s}\right)\cdot\left(\frac{n-1}{s}\right)\cdot\left(\frac{n-2}{s}\right)\cdots\cdots\left(\frac{2}{s}\right)\cdot\left(\frac{1}{s}\right)\mathscr{L}[t^0]$$

$$= \left(\frac{n}{s}\right)\cdot\left(\frac{n-1}{s}\right)\cdot\left(\frac{n-2}{s}\right)\cdots\cdots\left(\frac{2}{s}\right)\cdot\left(\frac{1}{s}\right)\mathscr{L}[1]$$

$$= \left(\frac{n}{s}\right)\cdot\left(\frac{n-1}{s}\right)\cdot\left(\frac{n-2}{s}\right)\cdots\cdots\left(\frac{2}{s}\right)\cdot\left(\frac{1}{s}\right)\cdot\left(\frac{1}{s}\right)$$

or, $$\mathscr{L}[t^n] = \frac{n!}{s^{n+1}} \qquad \text{ROC: } \mathscr{R}(s) > 0 \Rightarrow \sigma > 0 \qquad ...(4.24)$$

(*b*) We have the result

$$\mathscr{L}[t^n] = \frac{n!}{s^{n+1}}$$

Using **shifting property**, we obtain

$$\mathscr{L}[t^n t^{at}] = \frac{n!}{(s-a)^{n+1}} \qquad \text{ROC: } \mathscr{R}(s) > a \Rightarrow \sigma > a \qquad ..(4.25)$$

(*c*) $$\mathscr{L}[\sinh \omega t] = \mathscr{L}\left[\frac{e^{\omega t}-e^{-\omega t}}{2}\right]$$

$$= \frac{1}{2}\mathscr{L}[e^{\omega t}]-\frac{1}{2}\mathscr{L}[e^{-\omega t}]$$

$$= \frac{1}{2}\left[\frac{1}{s-\omega}\right]-\frac{1}{2}\left[\frac{1}{s+\omega}\right]$$

or, $$\mathscr{L}[\sinh \omega t] = \frac{\omega}{s^2-\omega^2} \qquad \text{ROC: } \mathscr{R}(s) > 0 \Rightarrow \sigma > 0 \qquad ...(4.26)$$

(*d*) $$\mathscr{L}[\cosh \omega t] = \mathscr{L}\left[\frac{e^{\omega t}+e^{-\omega t}}{2}\right]$$

$$= \frac{1}{2}\mathcal{L}[e^{\omega t}] + \frac{1}{2}\mathcal{L}[e^{-\omega t}]$$

$$= \frac{1}{2}\left[\frac{1}{s-\omega}\right] + \frac{1}{2}\left[\frac{1}{s+\omega}\right]$$

or, $$\mathcal{L}[\cosh \omega t] = \frac{s}{s^2 - \omega^2} \qquad \text{ROC: } \mathcal{R}e\,(s) > 0 \Rightarrow \sigma > 0 \qquad ..(4.27)$$

(*e*) We have the result

$$\mathcal{L}[\sin \omega t] = \frac{s}{s^2 - \omega^2}$$

Using **shifting property**, we obtain

$$\mathcal{L}[e^{-at} \sin \omega t] = \frac{\omega}{(s+a)^2 + \omega^2} \qquad \text{ROC: } \mathcal{R}e\,(s) > -a \Rightarrow \sigma > -a \qquad ...(4.28)$$

(*f*) We have the result

$$\mathcal{L}[\cos \omega t] = \frac{s}{s^2 + \omega^2}$$

Using **shifting property**, we obtain

$$\mathcal{L}[e^{-at} \cos \omega t] = \frac{s}{(s+a)^2 + \omega^2} \qquad \text{ROC: } \mathcal{R}e\,(s) > -a \Rightarrow \sigma > -a \qquad ...(4.29)$$

Example 4.2. *Determine the Laplace transforms of the following continuous-time signals.*

(a) $e^{-at}\, u(t)$

(b) $-e^{-at}\, u(-t)$

Compare and comment on the results obtained.

Solution. (*a*) Unit-step signal $u(t)$ is defined as

$$u(t) = \begin{cases} 1 & ; \quad t \geq 0 \\ 0 & ; \quad t < 0 \end{cases}$$

Since $e^{-at}\, u(t)$ is a **causal signal**, therefore, using definition (4.4), we have

$$\mathcal{L}[e^{-at}\, u(t)] = \int_{0^-}^{\infty} e^{-at}\, u(t)\, e^{-st}\, dt$$

or, $$\mathcal{L}[e^{-at}\, u(t)] = \int_{0^-}^{\infty} e^{-at}\, e^{-st}\, dt$$

or, $$\mathcal{L}[e^{-at}\, u(t)] = \int_{0^-}^{\infty} e^{-(s+a)\,t}\, dt$$

or, $$\mathcal{L}[e^{-at}\, u(t)] = \left[-\frac{e^{-(s+a)t}}{s+a}\right]_{0^-}^{\infty}$$

or, $$\mathcal{L}[e^{-at}\ u(t)] = \frac{1}{s+a} \qquad \text{ROC: } \mathcal{R}(s) > -a \Rightarrow \sigma > -a \qquad ...(4.30)$$

(*b*) $u(-t)$ is defined as

$$u(-t) = \begin{cases} 1 & ; \ t \le 0 \\ 0 & ; \ t > 0 \end{cases}$$

Since $-e^{-at}\ u(-t)$ is a **non-causal signal**, therefore, using definition (4.1), we have

$$\mathcal{L}[-e^{-at}\ u(-t)] = -\int_{-\infty}^{\infty} e^{-at}\ u(-t)\, e^{-st}\, dt$$

or, $$\mathcal{L}[-e^{-at}\ u(-t)] = -\int_{-\infty}^{0^-} e^{-at}\ e^{-st}\, dt$$

or, $$\mathcal{L}[-e^{-at}\ u(-t)] = -\int_{-\infty}^{0^-} e^{-(s+a)\,t}\, dt$$

or, $$\mathcal{L}[-e^{-at}\ u(-t)] = \left[\frac{e^{-(s+a)t}}{s+a}\right]_{-\infty}^{0^-}$$

or, $$\mathcal{L}[-e^{-at}\ u(-t)] = \frac{1}{s+a} \qquad \text{ROC: } \mathcal{R}(s) < -a \Rightarrow \sigma < -a \qquad ...(4.31)$$

From equations (4.30) and (4.31), it is clear that two signals may have **identical Laplace transforms but different ROCs**. The expression for the Laplace transform does not uniquely correspond to a signal if the ROC is not specified. Thus, the ROC must be specified for the Laplace transform to be **unique**.

Example 4.3. *Find the Laplace transform of the signal* $e^{-a|t|}$.

Solution. Since $e^{-a|t|}$ is a **two-sided signal**, we can represent it as the **sum of a right-sided and a left-sided signal** as follows

$$e^{-a|t|} = e^{-at}\ u(t) + e^{at}\ u(-t) \qquad ...(4.32)$$

Using **linearity property**, we may write

$$\mathcal{L}[e^{-a|t|}] = \mathcal{L}[e^{-at}\ u(t)] + \mathcal{L}[e^{at}\ u(-t)]$$

Now, $$e^{-at}\ u(t) \xleftrightarrow{\mathcal{L}} \frac{1}{s+a} \quad ; \quad \text{ROC: } \mathcal{R}(s) > -a \Rightarrow \sigma > -a \qquad ...(4.33\text{ A})$$

and $$e^{at}\ u(-t) \xleftrightarrow{\mathcal{L}} \frac{-1}{s-a} \quad ; \quad \text{ROC: } \mathcal{R}(s) < a \Rightarrow \sigma < a \qquad ...(4.33\text{ B})$$

Although the Laplace transforms of each individual terms in equation (4.33) have a region of convergence, there is **no common region of convergence** when $a \le 0$, and therefore, for $a \le 0$ the Laplace transform of $e^{-a|t|}$ **does not exist**. However, when $a > 0$, the Laplace transform of $e^{-a|t|}$ is given by

$$\mathcal{L}[e^{-a|t|}] = \frac{1}{s+a} - \frac{1}{s-a}$$

or, $$\mathscr{L}[e^{-a|t|}] = \frac{-2a}{s^2 - a^2} \quad ; \quad -a < \mathscr{R}(s) < a \Rightarrow -a < \sigma < a \qquad ...(4.34)$$

Example 4.4. *Find out the initial and final values of the continuous-time signal f(t), whose Laplace transform is given by*

$$F(s) = \frac{5s+3}{s(s+1)}$$

Solution. $$f(0^+) = \lim_{s\to\infty} [sF(s)] = \lim_{s\to\infty} \left[\frac{5s+3}{s+1}\right]$$

$$= \lim_{s\to\infty} \left[\frac{5+\frac{3}{s}}{1+\frac{1}{s}}\right] = 5$$

and, $$f(\infty) = \lim_{s\to 0} [sF(s)] = \lim_{s\to 0} \left[\frac{5s+3}{s+1}\right] = 3$$

Example 4.5. *Obtain the Laplace transform of a gate signal.*

Solution. A **gate signal** is defined as

$$G_{T_1,T_2}(t) = u(t-T_1) - u(t-T_2) \qquad ...(4.35)$$

and is illustrated in Figure 4.2. This signal represents a **rectangular pulse of unit height** starting at $t = T_1$ and ends at $t = T_2$.

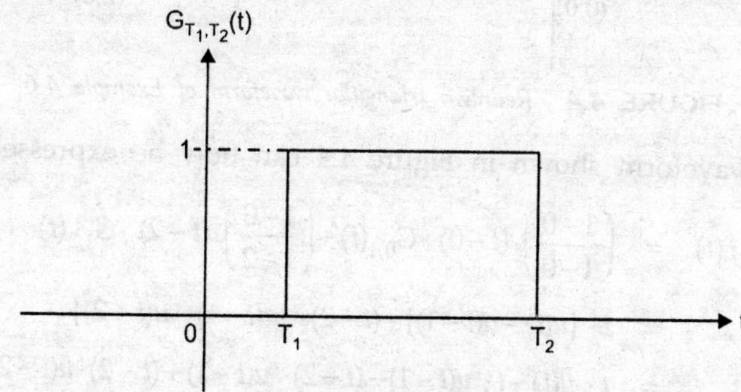

FIGURE 4.2 *Gate signal.*

$$\therefore \quad \mathscr{L}[G_{T_1,T_2}(t)] = \mathscr{L}[u(t-T_1) - u(t-T_2)]$$

Using **linearity property**, we have

$$\mathscr{L}[G_{T_1,T_2}(t)] = \mathscr{L}[u(t-T_1)] - \mathscr{L}[u(t-T_2)]$$

$$= e^{-sT_1}\mathscr{L}[u(t)] - e^{-sT_2}\mathscr{L}[u(t)]$$

$$= \frac{1}{s}[e^{-sT_1} - e^{-sT_2}]$$

Example 4.6. *Determine the Laplace transform of a triangular waveform as shown in Figure 4.3.*

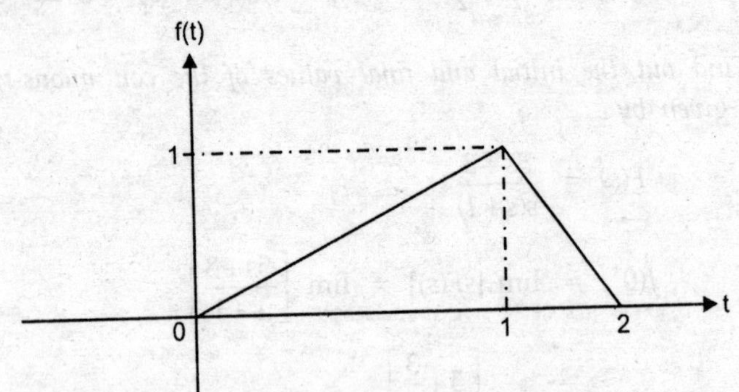

FIGURE 4.3 *Triangular waveform of Example 4.6.*

Solution. Before writing the synthesis equation, we first redraw the triangular waveform indicating the **coordinates** of every point as shown in Figure 4.4.

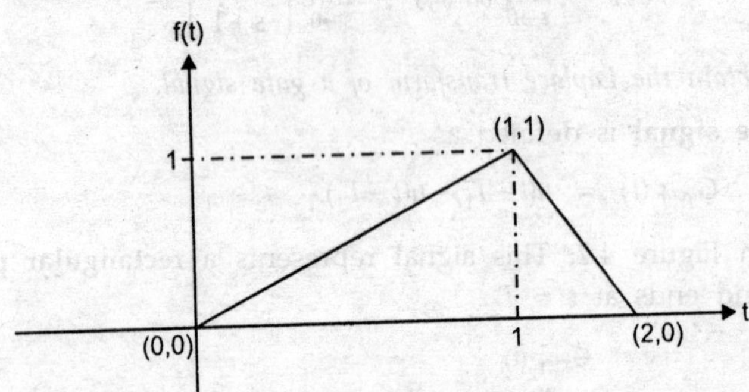

FIGURE 4.4 *Redrawn triangular waveform of Example 4.6.*

The triangular waveform shown in Figure 4.4 can now be expressed as follows:

$$\begin{aligned}
f(t) &= \left(\frac{1-0}{1-0}\right)\cdot(t-0)\cdot G_{0,1}(t)+\left(\frac{1-0}{1-2}\right)\cdot(t-2)\cdot G_{1,2}(t) \\
&= t\cdot[u(t)-u(t-1)]-(t-2)\cdot[u(t-1)-u(t-2)] \\
&= t\cdot u(t)-t\cdot u(t-1)-(t-2)\cdot u(t-1)+(t-2)\cdot u(t-2) \\
&= t\cdot u(t)-(t-1)\cdot u(t-1)-u(t-1)-(t-1-1)\cdot u(t-1)+(t-2)\cdot u(t-2) \\
&= t\cdot u(t)-(t-1)\cdot u(t-1)-u(t-1)-(t-1)\cdot u(t-1)+u(t-1)+(t-2)\cdot u(t-2) \\
&= t\cdot u(t)-(t-1)\cdot u(t-1)-(t-1)\cdot u(t-1)+(t-2)\cdot u(t-2) \\
&= t\cdot u(t)-2(t-1)\cdot u(t-1)+(t-2)\cdot u(t-2)
\end{aligned}$$

Thus,
$$\begin{aligned}
\mathscr{L}[f(t)] &= \frac{1}{s^2}-\frac{2e^{-s}}{s^2}+\frac{2e^{-2s}}{s^2} \\
&= \frac{1}{s^2}[1-2e^{-s}+2e^{-2s}]
\end{aligned}$$

Example 4.7. *Find out the Laplace transform of a single half-wave rectified sinusoidal signal as shown in Figure 4.5.*

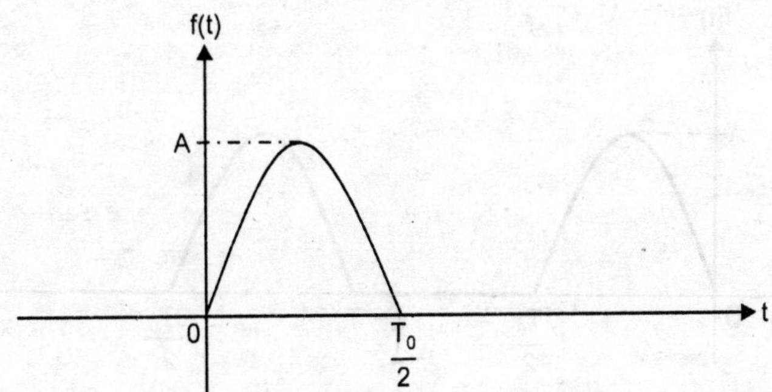

FIGURE 4.5 *Single half-wave rectified sinusoidal waveform of Example 4.7.*

Solution. The single half-wave rectified sinusoidal waveform shown in Figure 4.5 can be expressed as follows:

$$
\begin{aligned}
f(t) &= A\sin\left(\frac{2\pi}{T_0}\cdot t\right)\cdot G_{0,T_0/2}(t) \\
&= A\sin\left(\frac{2\pi}{T_0}\cdot t\right)\cdot\left[u(t)-u\left(t-\frac{T_0}{2}\right)\right] \\
&= A\sin\left(\frac{2\pi}{T_0}\cdot t\right)\cdot u(t)-A\sin\left(\frac{2\pi}{T_0}\cdot t\right)\cdot u\left(t-\frac{T_0}{2}\right) \\
&= A\sin\left(\frac{2\pi}{T_0}\cdot t\right)\cdot u(t)-A\sin\left[\frac{2\pi}{T_0}\left(t-\frac{T_0}{2}+\frac{T_0}{2}\right)\right]\cdot u\left(t-\frac{T_0}{2}\right) \\
&= A\sin\left(\frac{2\pi}{T_0}\cdot t\right)\cdot u(t)-A\sin\left[\frac{2\pi}{T_0}\left(t-\frac{T_0}{2}\right)+\pi\right]\cdot u\left(t-\frac{T_0}{2}\right) \\
&= A\sin\left(\frac{2\pi}{T_0}\cdot t\right)\cdot u(t)+A\sin\left[\frac{2\pi}{T_0}\left(t-\frac{T_0}{2}\right)\right]\cdot u\left(t-\frac{T_0}{2}\right)
\end{aligned}
$$

Thus,

$$
\begin{aligned}
\mathscr{L}[f(t)] &= A\left[\frac{\left(\frac{2\pi}{T_0}\right)}{s^2+\left(\frac{2\pi}{T_0}\right)^2}\right]+Ae^{-\frac{sT_0}{2}}\left[\frac{\left(\frac{2\pi}{T_0}\right)}{s^2+\left(\frac{2\pi}{T_0}\right)^2}\right] \\
&= A\left(1+e^{-\frac{sT_0}{2}}\right)\left[\frac{\left(\frac{2\pi}{T_0}\right)}{s^2+\left(\frac{2\pi}{T_0}\right)^2}\right]
\end{aligned}
$$

Example 4.8. *Determine the Laplace transform of a periodic half-wave rectified sinusoidal signal as shown in Figure 4.6.*

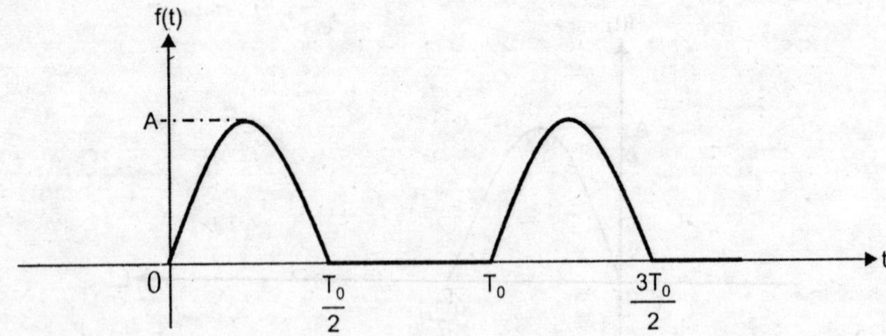

FIGURE 4.6 *Periodic half-wave rectified sinusoidal waveform of Example 4.8.*

Solution. The Laplace transform of a periodic signal $f(t)$ with time period T_0 is equal to the product of $\dfrac{1}{1-e^{-sT_0}}$ and the Laplace transform of the first cycle of the periodic waveform.

From Example 4.7, we have the **Laplace transform** of the **first cycle of the periodic waveform** given by

$$F_1(s) = A\left(1+e^{-\frac{sT_0}{2}}\right)\left[\frac{\left(\frac{2\pi}{T_0}\right)}{s^2+\left(\frac{2\pi}{T_0}\right)^2}\right]$$

so that

$$\mathscr{L}[f(t)] = \left(\frac{1}{1-e^{-sT_0}}\right)\cdot F_1(s)$$

$$= A\frac{\left(1+e^{-\frac{sT_0}{2}}\right)}{\left(\frac{1}{1-e^{-sT_0}}\right)}\cdot\left[\frac{\left(\frac{2\pi}{T_0}\right)}{s^2+\left(\frac{2\pi}{T_0}\right)^2}\right]$$

$$= A\frac{\left(1+e^{-\frac{sT_0}{2}}\right)}{\left(\frac{1}{1-e^{-sT_0}}\right)}\cdot\frac{\left(1-e^{-\frac{sT_0}{2}}\right)}{\left(1-e^{-\frac{sT_0}{2}}\right)}\cdot\left[\frac{\left(\frac{2\pi}{T_0}\right)}{s^2+\left(\frac{2\pi}{T_0}\right)^2}\right]$$

$$= \frac{A}{\left(1-e^{-\frac{sT_0}{2}}\right)}\cdot\left[\frac{\left(\frac{2\pi}{T_0}\right)}{s^2+\left(\frac{2\pi}{T_0}\right)^2}\right]$$

4.3 THE INVERSE LAPLACE TRANSFORM

If $F(s)$ is the Laplace transform of $f(t)$ then

$$f(t) = \frac{1}{2\pi j} \int_{\sigma - j\infty}^{\sigma + j\infty} F(s)\, e^{st}\, ds$$

But, we usually do not evaluate this integral directly since this integral is usually carried out as a **contour integral** in the complex plane by using **Cauchy's integral theorem**. Instead, we determine inverse Laplace transforms by inspection utilizing the **one-to-one relationship** between a signal $f(t)$ and its Laplace transform $F(s)$.

Given knowledge of several **basic transform pairs** and the Laplace transform properties; we are able to invert a very large class of rational Laplace transforms in this manner. Basically, the procedure consists of expanding the rational Laplace transform expression into a linear combination of simpler terms using **partial-fraction expansion method**. The inverse Laplace transform of each term can then be obtained by inspection. In particular, suppose that the partial-fraction expansion of $F(s)$ is of the form

$$F(s) = \sum_{i=1}^{n} \frac{A_i}{s + a_i} \qquad \text{...(4.36)}$$

so that the inverse Laplace transform of $F(s)$ equals the **sum** of the inverse Laplace transforms of the individual terms in the equation (4.36).

Case 1: If ROC: $\mathcal{R}e(s) > -a_i \Rightarrow \sigma > -a_i$, then the inverse Laplace transform of the corresponding term in equation (4.36) is $A_i\, e^{-a_i t}\, u(t)$, which is a **right-sided signal**.

Case 2: If ROC: $\mathcal{R}e(s) < -a_i \Rightarrow \sigma < -a_i$, then the inverse Laplace transform of the corresponding term in equation (4.36) is $-A_i\, e^{-a_i t}\, u(-t)$, which is a **left-sided signal**.

In general, the partial-fraction expansion of a rational Laplace transform may include **higher-order terms in addition to the first-order terms** in equation. The inverse Laplace transform of each term can then be obtained by means of a list of **Laplace transform pairs** (given in **Appendix-C**) in conjunction with the **properties of the Laplace transform** (already discussed in **Section 4.2.2**). To illustrate how a partial-fraction expansion can be used to obtain the inverse Laplace transform, let us consider the following examples.

Example 4.9 *Determine the inverse Laplace transform of*

$$F(s) = \frac{10}{s^2 + 10s + 16}$$

ROC: *(a)* $\mathcal{R}e(s) > -2$

(b) $\mathcal{R}e(s) < -8$

(c) $-8 < \mathcal{R}e(s) < -2$

Solution. Given,

$$F(s) = \frac{10}{s^2 + 10s + 16}$$

or, $$F(s) = \frac{10}{(s+2)(s+8)}$$

Here, $F(s)$ has two poles; one at $s = -2$ and other at $s = -8$.

Now, $$F(s) = \left.\frac{10}{s+8}\right|_{s=-2} \frac{1}{(s+2)} + \left.\frac{10}{s+2}\right|_{s=-8} \frac{1}{(s+8)}$$

or, $$F(s) = \frac{5}{3} \cdot \frac{1}{(s+2)} - \frac{5}{3} \cdot \frac{1}{(s+8)} \qquad \text{...(4.37)}$$

In order to determine the inverse Laplace transform of each of these individual terms, we must specify the ROC associated with each.

(*a*) ROC: $\mathcal{R}_e(s) > -2$ (Given)

Here, the ROC for $F(s)$ is the region in the s-plane to the **right of the rightmost pole** at $s = -2$. Consequently, the ROC for each individual term in equation (4.37) must also be the region in the s-plane to the **right of the pole** associated with that term.

Thus, $$f(t) = f_1(t) - f_2(t)$$

where, $$f_1(t) \overset{\mathcal{L}}{\longleftrightarrow} \frac{5}{3} \cdot \frac{1}{(s+2)} \quad ; \quad \mathcal{R}_e(s) > -2$$

and $$f_2(t) \overset{\mathcal{L}}{\longleftrightarrow} \frac{5}{3} \cdot \frac{1}{(s+8)} \quad ; \quad \mathcal{R}_e(s) > -8$$

By inspection, we can identify

$$f_1(t) = \frac{5}{3} e^{-2t} u(t)$$

and $$f_2(t) = \frac{5}{3} e^{-8t} u(t)$$

so that $$f(t) = \frac{5}{3} e^{-2t} u(t) - \frac{5}{3} e^{-8t} u(t)$$

(*b*) ROC: $\mathcal{R}_e(s) < -8$ (Given)

Here, the ROC for $F(s)$ is the region in the s-plane to the **left of the leftmost pole** at $s = -8$. Consequently, the ROC for each individual term in equation (4.37) must also be the region in the s-plane to the **left of the pole** associated with that term.

Thus, $$f(t) = f_1(t) - f_2(t)$$

where, $$f_1(t) \overset{\mathcal{L}}{\longleftrightarrow} \frac{5}{3} \cdot \frac{1}{(s+2)} \quad ; \quad \mathcal{R}_e(s) < -2$$

and $$f_2(t) \overset{\mathcal{L}}{\longleftrightarrow} \frac{5}{3} \cdot \frac{1}{(s+8)} \quad ; \quad \mathcal{R}_e(s) < -8$$

By inspection, we can identify

$$f_1(t) = -\frac{5}{3} e^{-2t} u(-t)$$

and $$f_2(t) = -\frac{5}{3}e^{-8t}\,u(-t)$$

so that $$f(t) = -\frac{5}{3}e^{-2t}\,u(-t) + \frac{5}{3}e^{-8t}\,u(-t)$$

(c) ROC: $-8 < \mathcal{R}e(s) < -2$ (Given)

Here, the ROC for $F(s)$ is the region in the s-plane to the **right of the leftmost pole** at $s = -8$ and to the **left of the rightmost pole** at $s = -2$. Consequently, the ROC corresponding to first term in equation (4.37) must also be the region in the s-plane to the **left of the pole** at $s = -2$ and the ROC corresponding to second term in equation (4.37) must also be the region in the s-plane to the **right of the pole** at $s = -8$.

Thus, $$f(t) = f_1(t) - f_2(t)$$

where, $$f_1(t) \xleftrightarrow{\mathcal{L}} \frac{5}{3}\cdot\frac{1}{(s+2)} \quad ; \quad \mathcal{R}e(s) < -2$$

and $$f_2(t) \xleftrightarrow{\mathcal{L}} \frac{5}{3}\cdot\frac{1}{(s+8)} \quad ; \quad \mathcal{R}e(s) > -8$$

By inspection, we can identify

$$f_1(t) = -\frac{5}{3}e^{-2t}\,u(-t)$$

and $$f_2(t) = \frac{5}{3}e^{-8t}\,u(t)$$

so that $$f(t) = -\frac{5}{3}e^{-2t}\,u(-t) - \frac{5}{3}e^{-8t}\,u(-t)$$

Example 4.10. *Obtain the inverse Laplace transform of*

$$F(s) = \frac{2s^2+5s+5}{(s+1)^2(s+2)}$$

Solution. Given,

$$F(s) = \frac{2s^2+5s+5}{(s+1)^2(s+2)}$$

Here, $F(s)$ has three poles; two at $s = -1$ and one at $s = -2$.

Now, $$F(s) = \left.\frac{2s^2+5s+5}{(s+2)}\right|_{s=-1}\frac{1}{(s+1)^2} + \frac{d}{ds}\left[\frac{2s^2+5s+5}{(s+2)}\right]_{s=-1}\frac{1}{(s+1)} + \left.\frac{2s^2+5s+5}{s+1}\right|_{s=-2}\frac{1}{(s+2)}$$

or, $$F(s) = \frac{2}{(s+1)^2} - \frac{1}{(s+1)} + \frac{3}{(s+2)} \qquad ...(4.38)$$

In order to determine the inverse Laplace transform of each of these individual terms, we must specify the ROC associated with each.

Thus, $$f(t) = f_1(t) + f_2(t) + f_3(t)$$

where, $$f_1(t) \xleftrightarrow{\mathcal{L}} \frac{2}{(s+1)^2} \quad ; \quad \mathcal{R}e(s) > -1$$

$$f_2(t) \xleftrightarrow{\mathscr{L}} \frac{1}{(s+1)} \quad ; \quad \mathscr{R}e(s) > -1$$

and

$$f_3(t) \xleftrightarrow{\mathscr{L}} \frac{3}{(s+2)} \quad ; \quad \mathscr{R}e(s) > -2$$

By inspection, we can identify

$$f_1(t) = 2t\, e^{-t}\, u(t)$$

$$f_2(t) = e^{-t}\, u(t)$$

and

$$f_3(t) = 3e^{-2t}\, u(t)$$

so that

$$f(t) = 2t\, e^{-t}\, u(t) - e^{-t}\, u(t) + 3e^{-2t}\, u(t)$$

4.4 SOLUTION OF DIFFERENTIAL EQUATIONS USING THE LAPLACE TRANSFORM

The operation of continuous-time systems are described by a set of **differential equations.** The analysis and design of continuous-time systems may effectively be carried out in terms of differential equations. Essentially by using the Laplace transform method, the solution of differential equations progresses systematically and involves only **algebraic manipulations.** In order to solve the differential equations, we shall utilize the property of **differentiation in time-domain** of the Laplace transform, written as

$$\mathscr{L}\left[\frac{d^n}{dt^n} f(t)\right] = s^n F(s) - s^{n-1} f(0^-) - s^{n-2}\frac{d}{dt} f(0^-) - s^{n-3}\frac{d^2}{dt^2} f(0^-) - \;\ldots\ldots\; - \frac{d^{n-1}}{dt^{n-1}} f(0^-)$$

It is to be noted that when any differential equation is transformed into an algebraic equation in *s* by the Laplace transform method, the **initial data are automatically included** in the algebraic representation.

Example 4.11. *Find out the response of the system described by the differential equation*

$$\frac{d^2}{dt^2} f(t) + 3\frac{d}{dt} f(t) + 2y(t) = 0$$

Given that

$$f(0^-) = 2 \text{ and } \frac{d}{dt} f(0^-) = -3$$

Solution. Taking the Laplace transforms of both sides of the above differential equation, we get

$$s^2 F(s) - s f(0^-) - \frac{d}{dt} f(0^-) + 3sF(s) - 3f(0^-) + 2F(s) = 0$$

or,

$$(s^2 + 3s + 2)F(s) = s f(0^-) + \frac{d}{dt} f(0^-) + 3f(0^-)$$

or,

$$(s^2 + 3s + 2)F(s) = 2s + 3$$

or,

$$F(s) = \frac{2s+3}{s^2+3s+2}$$

or,

$$F(s) = \frac{2s+3}{(s+1)(s+2)}$$

Here, $F(s)$ has two poles; one at $s = -1$ and other at $s = -2$.

Now, $$F(s) = \left.\frac{2s+3}{(s+2)}\right|_{s=-1}\frac{1}{(s+1)} + \left.\frac{2s+3}{(s+1)}\right|_{s=-2}\frac{1}{(s+2)}$$

or, $$F(s) = \frac{1}{(s+1)} + \frac{1}{(s+2)} \qquad \text{...(4.39)}$$

In order to determine the inverse Laplace transform of each of these individual terms, we must specify the ROC associated with each.

Thus, $$f(t) = f_1(t) + f_2(t)$$

where, $$f_1(t) \xleftrightarrow{\mathscr{L}} \frac{1}{(s+1)} \quad ; \quad \mathscr{R}_e(s) > -1$$

and $$f_2(t) \xleftrightarrow{\mathscr{L}} \frac{1}{(s+2)} \quad ; \quad \mathscr{R}_e(s) > -2$$

By inspection, we can identify

$$f_1(t) = e^{-t}\, u(t)$$

and $$f_2(t) = e^{-2t}\, u(t)$$

so that $$f(t) = [e^{-t} + e^{-2t}]\, u(t)$$

4.5 THE TRANSFER FUNCTION

Figure 4.7 shows a **linear continuous-time system** wherein the output signal $y(t)$ of the system is given by the **convolution sum** between the **input signal** $x(t)$ and the **impulse response function** $g(t)$. We may write

$$y(t) = x(t) * g(t)$$

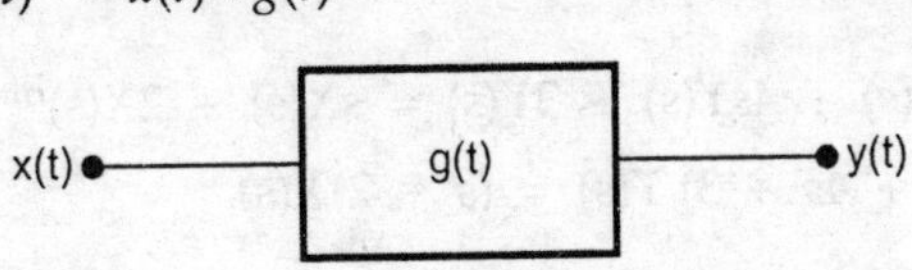

x(t) g(t) y(t)

FIGURE 4.7 *Linear continuous-time system.*

Using **convolution property** of the Laplace transform, we have

$$Y(s) = X(s)G(s) \qquad \text{...(4.40)}$$

This is an important result wherein $G(s)$ is interpreted as the **transfer function** of the linear continuous-time system as shown in Figure 4.8.

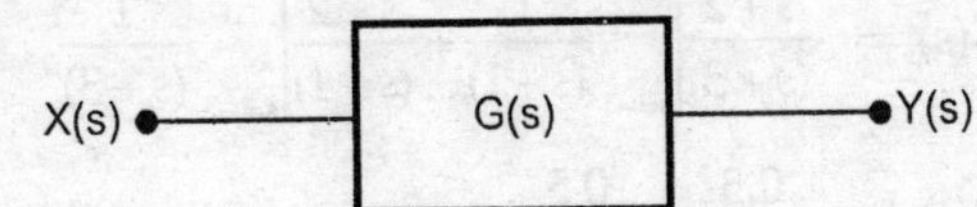

FIGURE 4.8 *Block-diagram for a transfer function system.*

Thus, from equation (4.40), the **transfer function** is expressed as

$$G(s) = \frac{Y(s)}{X(s)} \qquad \text{...(4.41)}$$

It is to be noted that $G(s)$ is the Laplace transform of the **impulse response function** $g(t)$ of an **initially relaxed system, *i.e.*, a system with zero initial conditions.**

As seen from equation (4.41), we can define the transfer function as the ratio of the Laplace transform of the output to the Laplace transform of the input, keeping all initial condition zero,

i.e.,
$$G(s) = \left.\frac{\text{Laplace transform of output}}{\text{Laplace transform of input}}\right|_{\text{Initial conditions}=0}$$

or,
$$G(s) = \left.\frac{Y(s)}{X(s)}\right|_{\text{Initial conditions}=0} \qquad \text{...(4.42)}$$

Hence, the transfer function is a **mathematical model of an initially relaxed system.** From the knowledge of this model the output of the system to any input can be obtained by using the relation (4.42) and then taking the inverse Laplace transform to obtain $y(t)$.

The ROC associated with the transfer function $G(s)$ for a **causal system** is the region in the *s*-plane to the **right of the rightmost pole.** However, the **converse may or may not be true.** Moreover, the ROC associated with the transfer function $G(s)$ can also be related to the **stability** of a system. A **causal system** with **rational transfer function** $G(s)$ is said to be **stable if and only if** all the poles of $G(s)$ lie in the **left-half of the *s*-plane,** *i.e.*, all of the poles have **negative real parts.**

Example 4.12. *Find out the impulse response function g(t) of an LTI system described by the following differential equation.*

$$\frac{d^2}{dt^2}y(t) + 4\frac{d}{dt}y(t) + 3y(t) = \frac{d}{dt}x(t) + 2x(t)$$

Solution. Taking the Laplace transforms of both sides assuming all initial conditions to be zero, we have

$$s^2Y(s) + 4sY(s) + 3Y(s) = sX(s) + 2X(s)$$

or,
$$(s^2 + 4s + 3)\,Y(s) = (s + 2)X(s)$$

or,
$$\frac{Y(s)}{X(s)} = \frac{s+2}{s^2+4s+3}$$

or,
$$G(s) = \frac{s+2}{(s+1)(s+3)}$$

Here, $G(s)$ has two poles; one at $s = -1$ and other at $s = -3$.

Now,
$$G(s) = \left.\frac{s+2}{(s+3)}\right|_{s=-1}\frac{1}{(s+1)} + \left.\frac{s+2}{(s+1)}\right|_{s=-3}\frac{1}{(s+3)}$$

or,
$$G(s) = \frac{0.5}{(s+1)} + \frac{0.5}{(s+3)} \qquad \text{...(4.43)}$$

In order to determine the inverse Laplace transform of each of these individual terms, we must specify the ROC associated with each.

Thus,
$$g(t) = g_1(t) + g_2(t)$$

where, $g_1(t) \overset{\mathscr{L}}{\longleftrightarrow} \dfrac{0.5}{(s+1)}$; $\mathscr{R}e(s) > -1$

and $g_2(t) \overset{\mathscr{L}}{\longleftrightarrow} \dfrac{0.5}{(s+3)}$; $\mathscr{R}e(s) > -3$

By inspection, we can identify

$$g_1(t) = \frac{1}{2} e^{-t} u(t)$$

and $$g_2(t) = \frac{1}{2} e^{-3t} u(t)$$

so that $$g(t) = \frac{1}{2} [e^{-t} + e^{-3t}] u(t)$$

Example 4.13. *Determine the stability of the systems with impulse response functions given as*

(a) $g(t) = e^{-t} u(t)$

(b) $g(t) = e^{t} u(t)$

Solution. (*a*) This system is **causal** and its transfer function $G(s)$ is given by

$$G(s) = \mathscr{L}[e^{-t} u(t)]$$

or, $$G(s) = \int_{0^-}^{\infty} e^{-t} u(t) e^{-st} dt$$

or, $$G(s) = \int_{0^-}^{\infty} e^{-t} e^{-st} dt$$

or, $$G(s) = \int_{0^-}^{\infty} e^{-(s+1)t} dt$$

or, $$G(s) = \left[-\frac{e^{-(s+1)t}}{s+1} \right]_{0^-}^{\infty}$$

or, $$G(s) = \frac{1}{s+1} \qquad \text{ROC: } \mathscr{R}e(s) > -1 \Rightarrow \sigma > -1 \qquad ...(4.44)$$

We find that the transfer function $G(s)$ has only one pole at $s = -1$, which is in the **left-half of the** ***s*****-plane**. Thus, the system under consideration is **stable**.

(*b*) This system is also causal and its transfer function $G(s)$ is given by

$$G(s) = \mathscr{L}[e^{t} u(t)]$$

or, $$G(s) = \int_{0^-}^{\infty} e^{t} u(t) e^{-st} dt$$

or, $$G(s) = \int_{0^-}^{\infty} e^{t} e^{-st} dt$$

or, $$G(s) = \int_{0^-}^{\infty} e^{-(s-1)t}\, dt$$

or, $$G(s) = \left[-\frac{e^{-(s-1)t}}{s-1} \right]_{0^-}^{\infty}$$

or, $$G(s) = \frac{1}{s-1} \qquad \text{ROC: } \mathcal{Re}\,(s) > 1 \Rightarrow \sigma > 1 \qquad ...(4.45)$$

We find that the transfer function $G(s)$ has only one pole at $s = 1$, which is in the **right-half of the s-plane**. Thus, the system under consideration is **unstable**.

4.6 CIRCUIT ANALYSIS USING LAPLACE TRANSFORM

An **electrical system** is described by the **integro-differential equations**. It is very difficult to solve such equations in **time-domain**. The Laplace transform analysis of such electrical networks converts the integro-differential equations into **simple algebraic equations**. Then by taking the inverse Laplace transform, time domain function can be obtained.

Example 4.14. *Consider the circuit shown in Figure 4.9, where the switch S is switched on at t = 0. Obtain the expression for the current. Also find the current through the capacitor at t = 0⁺. Assume the capacitor to be discharged initially.*

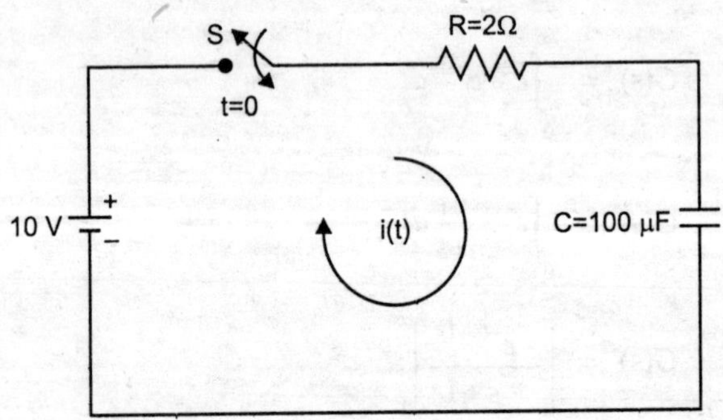

FIGURE 4.9 *Circuit of Example 4.14.*

Solution. The current $i(t)$ after **switching** is governed by the differential equation

$$Ri(t) + \frac{1}{C}\int_0^t i(\tau)\, d\tau = 10$$

Taking the Laplace transform of both sides, we have

$$RI(s) + \frac{1}{C}\left(\frac{I(s)}{s} + \frac{q(0^-)}{s} \right) = \frac{10}{s}$$

or, $$2I(s) + \frac{1}{100 \times 10^{-6}}\left(\frac{I(s)}{s} + \frac{q(0^-)}{s} \right) = \frac{10}{s}$$

As the capacitor is **discharged initially**, we have

$$q(0^-) = 0$$

Thus,
$$2I(s)+\frac{1}{100\times10^{-6}}\frac{I(s)}{s} = \frac{10}{s}$$

or,
$$I(s) = \frac{10}{\left(2s+\frac{1}{100\times10^{-6}}\right)}$$

or,
$$I(s) = \frac{5}{\left(s+\frac{1}{200\times10^{-6}}\right)}$$

Taking the inverse Laplace transform, we obtain

$$i(t) = 5e^{-t/(200\times10^{-6})}\,u(t)$$

Thus, the current through the capacitor at t = 0^+ is given by

$$i(0^+) = \lim_{s\to\infty}[sI(s)]$$

$$= \lim_{s\to\infty}\left[\frac{5s}{\left(s+\frac{1}{200\times10^{-6}}\right)}\right]$$

$$= \lim_{s\to\infty}\left[\frac{5}{\left(1+\frac{1}{200\,s\times10^{-6}}\right)}\right]$$

$$= 5A$$

Example 4.15. *Consider the circuit shown in Figure 4.10, where the switch S is switched from position (1) to position (2) at t = 0. Find the current through the inductor as a function of time.*

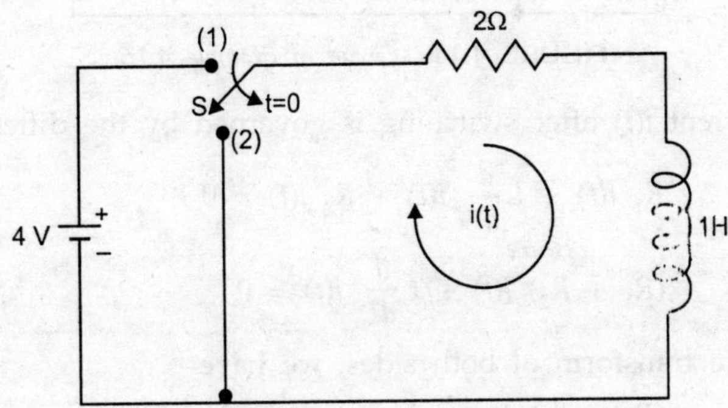

FIGURE 4.10 *Circuit of Example 4.15.*

Solution. The inductor current is governed by the differential equation

$$2i(t) + \frac{d}{dt}i(t) = \begin{cases}4 & ; \quad t<0\\ 0 & ; \quad t\geq 0\end{cases}$$

Taking the Laplace transform of both sides, we have

$$2I(s) + sI(s) - i(0^-) = 0$$

Assuming that the circuit was in **steady-state** for $t < 0$, we observe from the circuit diagram that

$$i(0^-) = \frac{4V}{2\Omega} = 2A \qquad \text{...(4.46 A)}$$

Thus,
$$2I(s) + sI(s) - 2 = 0$$

or,
$$(s + 2)I(s) - 2 = 0$$

or,
$$I(s) = \frac{2}{s+2}$$

Taking the inverse Laplace transform, we obtain

$$i(t) = 2e^{-2t}\, u(t) \qquad \text{...(4.46 B)}$$

While observing equations (4.46), we have the current through the inductor as

$$i(t) = \begin{cases} 2e^{-2t} & ; \quad t \geq 0 \\ 2 & ; \quad t < 0 \end{cases}$$

Example 4.16. *Consider the circuit shown in Figure 4.11, which is initially under steady-state condition. The switch S is switched from position (1) to position (2) at t = 0. Find the current after switching.*

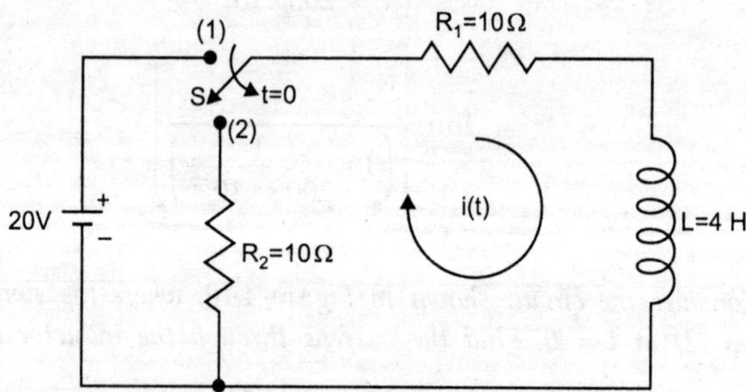

FIGURE 4.11 *Circuit of Example 4.16.*

Solution. The current $i(t)$ after switching is governed by the differential equation

$$R_1\, i(t) + L\frac{d}{dt}\, i(t) + R_2\, i(t) = 0$$

or,
$$(R_1 + R_2)\, i(t) + L\frac{d}{dt}\, i(t) = 0$$

Taking the Laplace transform of both sides, we have

$$(R_1 + R_2)\, I(s) + L[sI(s) - i(0^-)] = 0$$

or,
$$20I(s) + 4[sI(s) - i(0^-)] = 0$$

As the circuit is **initially under steady-state,** we observe from the circuit diagram that

$$i(0^-) = \frac{20\,V}{10\,\Omega} = 2A \qquad \text{...(4.47 A)}$$

Thus, $$20I(s) + 4sI(s) - 8 = 0$$

or, $$4(s + 5)I(s) - 8 = 0$$

or, $$I(s) = \frac{2}{s+5}$$

Taking the inverse Laplace transform, we obtain

$$i(t) = 2e^{-5t}\ u(t) \qquad \text{...(4.47 B)}$$

Thus, the current after switching is given by

$$i(t) = 2e^{-5t}\ \text{A}$$

Example 4.17. *Consider the circuit shown in Figure 4.12, where the switch S is switched from position (1) to position (2) at t = 0. Obtain the expression for the current i(t).*

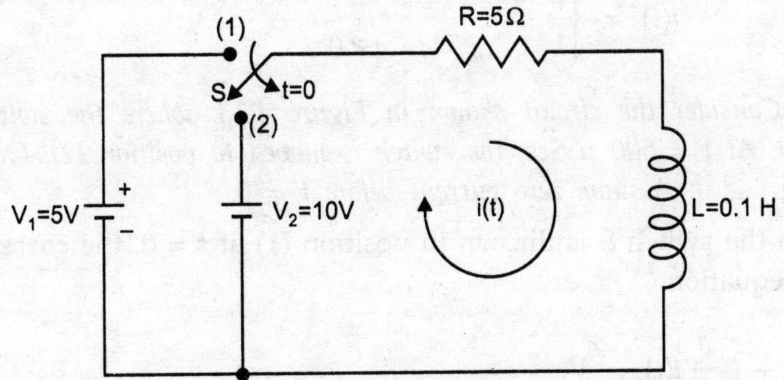

FIGURE 4.12 *Circuit of Example 4.17.*

Solution. The current $i(t)$ is governed by the differential equation

$$Ri(t) + L\frac{d}{dt}i(t) = \begin{cases} V_1 & ; \quad t < 0 \\ V_2 & ; \quad t \geq 0 \end{cases}$$

Taking the Laplace transform of both sides, we have

$$RI(s) + L[sI(s) - i(0^-)] = \frac{V_2}{s}$$

or, $$5I(s) + 0.1[sI(s) - i(0^-)] = \frac{10}{s}$$

Assuming that the circuit was in **steady-state** for $t < 0$, we observe from the circuit diagram that

$$i(0^-) = \frac{5V}{5\Omega} = 1A \qquad \text{...(4.48 A)}$$

Thus, $$5I(s) + 0.1[sI(s) - 1] = \frac{10}{s}$$

or, $$(0.1s + 5)I(s) - 0.1 = \frac{10}{s}$$

or, $$I(s) = \frac{0.1\,s+10}{s\,(0.1\,s+5)}$$

or, $$I(s) = \frac{s+100}{s\,(s+50)}$$

or, $$I(s) = \left.\frac{s+100}{(s+50)}\right|_{s=0}\frac{1}{s}+\left.\frac{s+100}{s}\right|_{s=-50}\frac{1}{(s+50)}$$

or, $$I(s) = \frac{2}{s}-\frac{1}{(s+50)}$$

Taking the inverse Laplace transform, we obtain

$$i(t) = [2 - e^{-50t}]\,u(t) \qquad \text{...(4.48 B)}$$

While observing equations (4.48), we have the expression for the current $i(t)$ as

$$i(t) = \begin{cases} 2-e^{-50t} & ; \quad t \geq 0 \\ 1 & ; \quad t < 0 \end{cases}$$

Example 4.18. *Consider the circuit shown in Figure 4.13, where the switch S is thrown in position (1) at t = 0. At t = 500 μ Sec, the switch is moved to position (2). Obtain the expression for the current i(t) ; t ≥ 0. Assume zero current before t = 0.*

Solution. When the switch S is thrown in position (1) at t = 0, the current $i(t)$ is governed by the differential equation

$$Ri(t) + L\frac{d}{dt}i(t) = V_1$$

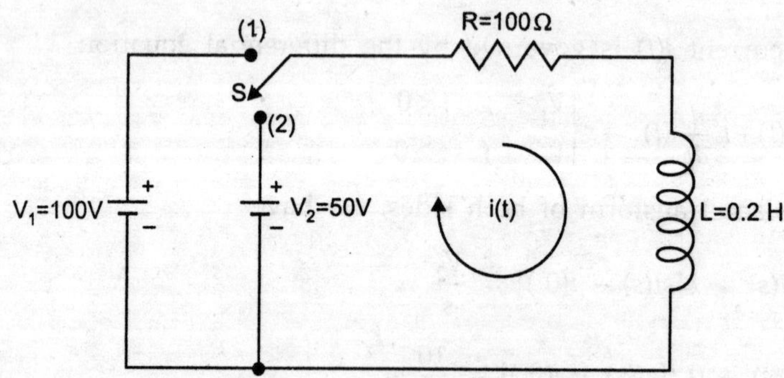

FIGURE 4.13 *Circuit of Example 4.18.*

Taking the Laplace transform of both sides, we have

$$RI(s) + L[sI(s) - i(0^-)] = \frac{V_1}{s}$$

or, $$100\,I(s) + 0.2[sI(s) - i(0^-)] = \frac{100}{s}$$

Assuming zero current before t = 0, we obtain

$$100\,I(s) + 0.2sI(s) = \frac{100}{s}$$

or, $$(0.2s + 100)\, I(s) = \frac{100}{s}$$

or, $$I(s) = \frac{100}{s\,(0.2\,s+100)}$$

or, $$I(s) = \frac{500}{s\,(s+500)}$$

or, $$I(s) = \left.\frac{500}{(s+500)}\right|_{s=0} \frac{1}{s} + \left.\frac{500}{s}\right|_{s=-500} \frac{1}{(s+500)}$$

or, $$I(s) = \frac{1}{s} - \frac{1}{(s+500)}$$

Taking the inverse Laplace transform, we obtain

$$i(t) = [1 - e^{-500t}]\, u(t) \qquad \text{...(4.49 A)}$$

Therefore, the current in the circuit at the **instant just before** the switch S is moved to position (2) is given by

$$i(500^-) = 1 - e^{-500 \times (500 \times 10^{-6})} = 0.22\ A$$

Now, when the switch S is moved to position (2) at $t = 500\ \mu\ Sec$, the current $i(t)$ is governed by the differential equation

$$Ri(t) + L\frac{d}{dt}i(t) = V_2$$

Taking the Laplace transform of both sides, we have

$$RI(s) + L\,[sI(s) - i(500^-)] = \frac{V_2}{s}$$

or, $$100\, I(s) + 0.2[sI(s) - i(500^-)] = \frac{50}{s}$$

or, $$100\, I(s) + 0.2[sI(s) - 0.22)] = \frac{50}{s}$$

or, $$(0.2s + 100)\, I(s) - 0.044 = \frac{50}{s}$$

or, $$I(s) = \frac{0.044\,s+50}{s\,(0.2\,s+100)}$$

or, $$I(s) = \frac{0.22\,s+250}{s\,(s+500)}$$

or, $$I(s) = \left.\frac{0.22\,s+250}{(s+500)}\right|_{s=0} \frac{1}{s} + \left.\frac{0.22\,s+250}{s}\right|_{s=-500} \frac{1}{(s+500)}$$

or, $$I(s) = \frac{0.5}{s} - \frac{0.28}{(s+500)}$$

Taking the inverse Laplace transform, we obtain

$$i(t) = [0.5 - 0.28\, e^{-500t}]\, u(t) \qquad \text{...(4.49 B)}$$

While observing equations (4.49), we have the expression for the current $i(t)$; $t \geq 0$ as

$$i(t) = \begin{cases} 1 - e^{-500t} & ; \quad 0 \leq t < 500\ \mu\ Sec \\ 0.5 - 0.28\, e^{-500t} & ; \quad t > 500\ \mu\ Sec \end{cases}$$

Example 4.19. *Consider the circuit shown in Figure 4.14, where the switch S is thrown from position (1) to position (2) at t = 0. Just before the switch is thrown, the initial conditions are*

$$i_L(0^-) = 2A \text{ and } v_C(0^-) = 2V$$

Find the current after switching. Assume R = 3Ω, L = 1H, C = 0.5F, and V_1 = 5V.

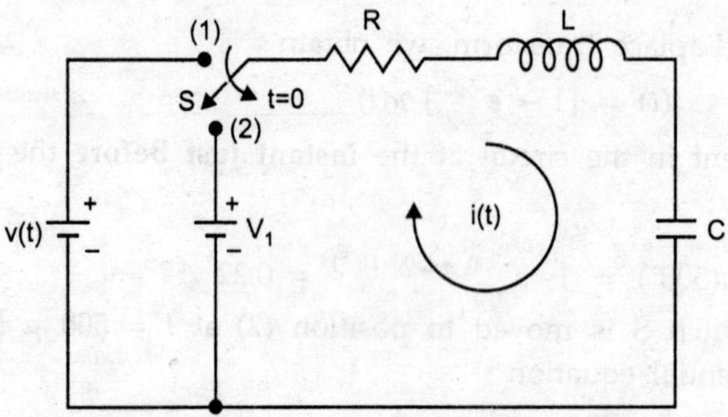

FIGURE 4.14 *Circuit for Example 4.19.*

Solution. The current *i(t)* after switching is governed by the differential equation

$$Ri(t) + L\frac{d}{dt}i(t) + \frac{1}{C}\int_0^t i(\tau)\, d\tau = V_1$$

Taking the Laplace transform of both sides, we have

$$RI(s) + L[sI(s) - i(0^-)] + \frac{1}{C}\left(\frac{I(s)}{s} + \frac{q(0^-)}{s}\right) = \frac{V_1}{s}$$

or, $$RI(s) + L[sI(s) - i_L(0^-)] + \frac{1}{C}\left(\frac{I(s)}{s} + \frac{C\, v_C(0^-)}{s}\right) = \frac{V_1}{s}$$

or, $$3I(s) + [sI(s) - 2] + \frac{1}{0.5}\left(\frac{I(s)}{s} + \frac{0.5 \times 2}{s}\right) = \frac{5}{s}$$

or, $$I(s) = \frac{\left(2 + \frac{3}{s}\right)}{\left(3 + s + \frac{2}{s}\right)}$$

or, $$I(s) = \frac{2s + 3}{s^2 + 3s + 2}$$

or, $$I(s) = \frac{2s + 3}{(s+1)(s+2)}$$

or, $$I(s) = \left.\frac{2s+3}{(s+2)}\right|_{s=-1} \frac{1}{(s+1)} + \left.\frac{2s+3}{(s+1)}\right|_{s=-2} \frac{1}{(s+2)}$$

or, $$I(s) = \frac{1}{(s+1)} + \frac{1}{(s+2)}$$

Taking the inverse Laplace transform, we obtain

$$i(t) = [e^{-t} - e^{-2t}]\ u(t) \quad \text{...(4.50)}$$

Thus, the current after switching is given by

$$i(t) = [e^{-t} - e^{-2t}]\ A$$

PROBLEMS

1. Determine the Laplace transform and ROC of
$$f(t) = u(t - 5)$$
2. Determine the Laplace transform of
   (a) $f(t) = [-e^{3t}\ u(t)]*[t\ u(t)]$
   (b) $f(t) = e^{-t}\ (t - 2)\ u(t - 2)$
3. Find the initial and final values of $f(t)$ when
$$F(s) = -e^{-5s}\left[-\frac{2}{s(s+2)}\right]$$
4. Find out the Laplace transform of saw-tooth waveform as shown in Figure 4.15.

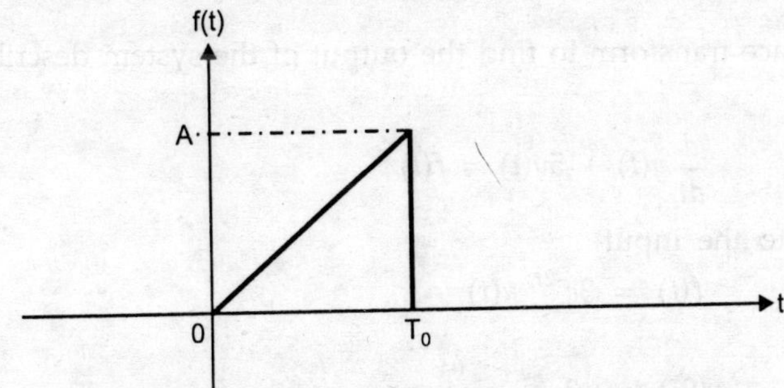

FIGURE 4.15 *Saw-tooth waveform of Problem (4).*

5. Find out the Laplace transform of the waveform shown in Figure 4.16.

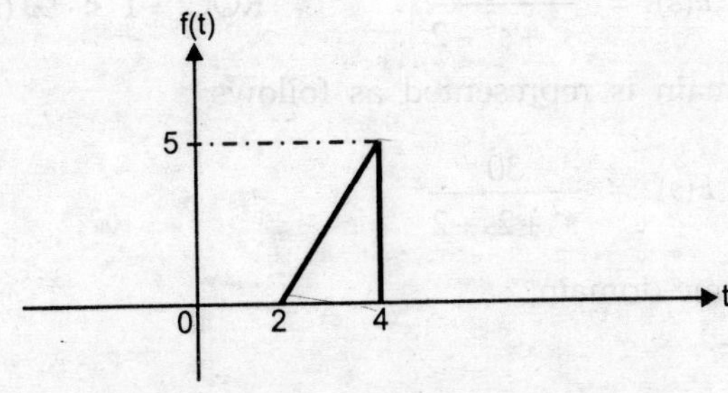

FIGURE 4.16 *Waveform of Problem (5).*

6. Obtain the Laplace transform of the square wave as depicted in Figure 4.17.

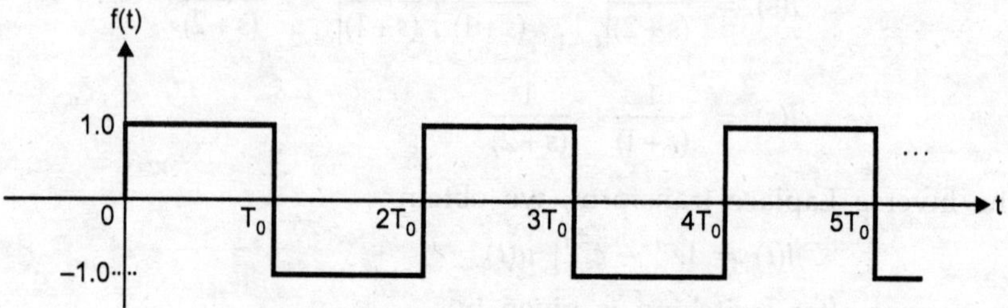

FIGURE 4.17 *Square waveform of Problem (6).*

7. A ramp signal occurs at $t = a$. Find its Laplace transform.

8. Obtain the Laplace transform of the square wave as depicted in Figure 4.18. Assume $T_0 = 1$ *Sec.*

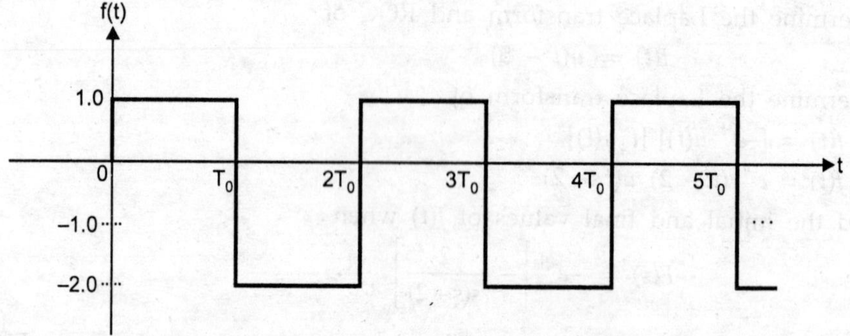

FIGURE 4.18 *Square waveform of Problem (8).*

9. Use the Laplace transform to find the output of the system described by the differential equation

$$\frac{d}{dt}y(t) + 5y(t) = f(t)$$

In response to the input

$$f(t) = 3e^{-2t}\,u(t)$$

Given that

$$y(0^-) = -2$$

10. Find the inverse Laplace transform of

$$F(s) = \frac{4s^2+6}{s^3+s^2-2} \qquad \text{ROC: } -1 < \mathcal{R}(s) < 1$$

11. A signal in *s*-domain is represented as follows

$$F(s) = \frac{30}{s^2+2s+2}$$

Represent it in time-domain.

12. Find the inverse Laplace transform of

$$F(s) = \frac{4s^2+6}{s^3+s^2-2}$$

13. A stable and causal system is described by the differential equation

$$\frac{d^2}{dt^2}y(t)+5\frac{d}{dt}y(t)+6y(t) = \frac{d^2}{dt^2}f(t)+8\frac{d}{dt}f(t)+13f(t)$$

Find the system impulse response.

14. A system has the transfer function

$$G(s) = \frac{2}{s+3}+\frac{1}{s-2}$$

Find the impulse response assuming

(*a*) the system is stable

(*b*) the system is causal

Can this system be both stable and causal?

15. A capacitor of 5μF being charged initially to 10 *V* is connected to a resistance of 10 *K*Ω and is allowed to discharge through it by switching a switch S. Obtain the expression for the discharging current $i(t)$.

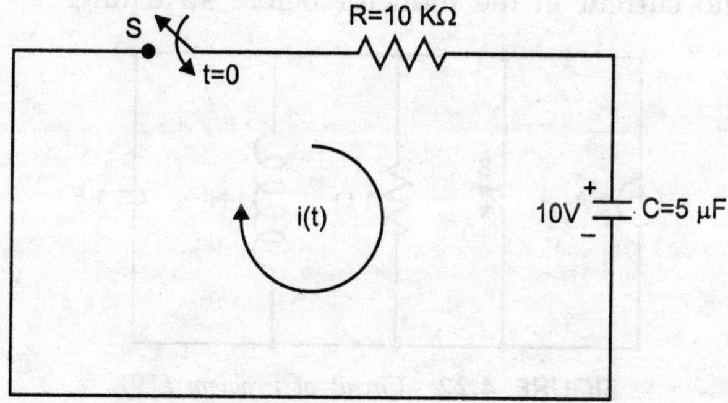

FIGURE 4.19 *Circuit of Problem (15).*

16. The circuit shown in Figure 4.20 is in steady-state condition when switch S is in position (1). The switch S is moved from position (1) to position (2) at $t = 0$. Find out the expression for the current $i(t)$ after switching.

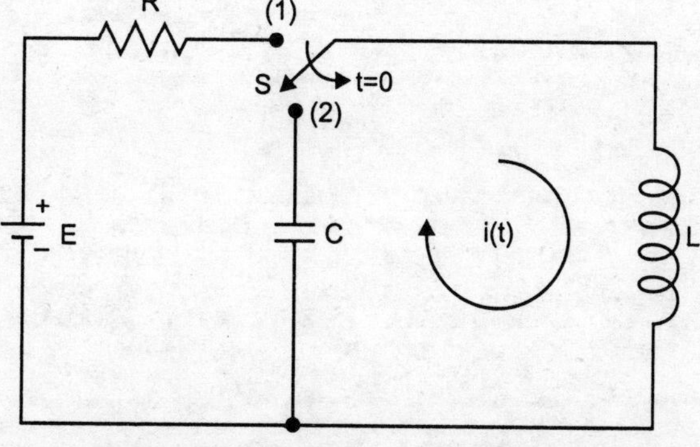

FIGURE 4.20 *Circuit of Problem (16).*

17. A unit step voltage u(t – 2) is applied to a series RL circuit at t = 0. Find the expression for the current through the inductor. Given that R = 2Ω and L = 1H. Assume zero initial conditions.

18. In the circuit of Figure 4.21, the switch S is closed at t = 0. Find the expression for the current $i(t)$. Assume that the initial charge on the capacitor is 250 μC as shown in the Figure. 4.21.

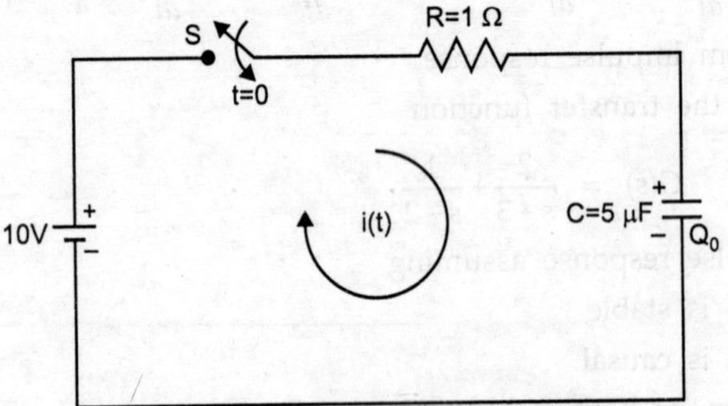

FIGURE 4.21 *Circuit of Problem (18).*

19. Consider a parallel RLC circuit shown in Figure 4.22, the switch is suddenly opened at t = 0. Find the voltage across the switch S assuming that there is no charge on capacitor and no current in the inductor before switching.

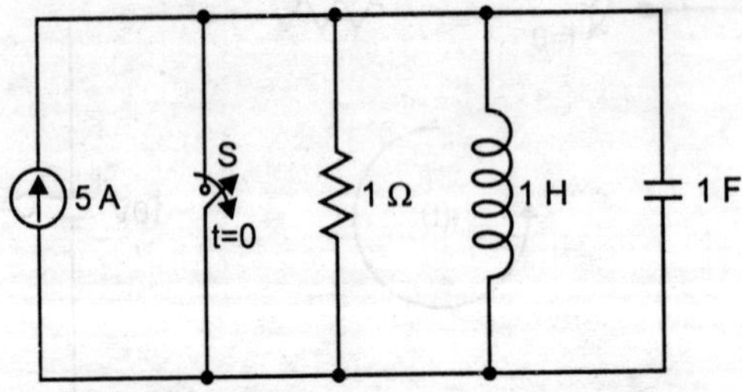

FIGURE 4.22 *Circuit of Problem (19).*

Chapter 5

z-Transform Analysis

5.1 INTRODUCTION

The advances made in microprocessors, micro-computers and digital signal processors have accelerated the growth of digital control systems theory. The terms **sampled-data control systems, discrete-time control systems and digital control systems** have all been used **interchangeably** in the control system literatures. Strictly speaking, sampled-data are pulse-amplitude modulated signals and are obtained by some means of sampling an analog signal. Digital signals are generated by means of digital transducers or digital computers, often in **digitally coded form.** The discrete-time systems, in broad sense, describe all systems having some form of digital or sampled signals.

Discrete-time systems differ from continuous-time systems in that the signals for a discrete-time system are in sampled data form. The discrete-time systems are dynamic systems in which the system variables are defined only at discrete instants of time. In contrast to continuous-time systems, the operation of discrete-time systems is described by a set of **difference-equations.** The analysis and design of discrete-time systems may effectively be carried out by use of the ***z*-transform** which was evolved from the Laplace transform as a special case.

5.2 THE *z*-TRANSFORM

The ***z*-transform** is a powerful mathematical tool for analysis of **discrete-time signals and systems** as **Laplace transform** is for **continuous-time signals and systems.** The ***z*-transform** $F(z)$ of a discrete-time signal $f(k)$ is defined as

$$F(z) = \mathcal{Z}[f(k)] = \sum_{k=-\infty}^{\infty} f(k)z^{-k} \quad \text{...(5.1 A)}$$

where, z is a **complex variable** with real and imaginary parts.

For **causal systems,** we have

$$f(k) = 0 \quad ; \quad k < 0$$

Therefore, the z-transform $F(z)$ of discrete-time signal $f(k)$, given by equation (5.1 A) reduces to

$$F(z) = \mathcal{Z}[f(k)] = \sum_{k=0}^{\infty} f(k) z^{-k} \qquad \text{...(5.1 B)}$$

The z-transform $F(z)$, defined by equation (5.1 A) is called the **bilateral** (or the **two-sided) z-transform** and the same defined by equation (5.1 B) is called the **unilateral** (or the **one-sided) z-transform.**

The difference between bilateral z-transform and unilateral z-transform is that the lower limit of summation is $-\infty$ for the bilateral z-transform and zero for the unilateral z-transform. Thus, the unilateral z-transform of discrete-time signal $f(k)$ can be thought of as the bilateral z-transform of $f(k)\, u(k)$, *i.e.*, $f(k)$ multiplied by a unit-step.

The portion of the z-plane for which the power series (5.1) converge, is called the **region of convergence (ROC).** For some values of *z*, the power series (5.1) may not converge to a finite value. Thus, the ROC depends upon the value of z. The possible configurations of the ROC for the bilateral z-transform may be

1. The entire z-plane [Figure 5.1 (*a*)]
2. Interior of a circle [Figure 5.1 (*b*)]
3. Exterior of a circle [Figure 5.1 (*c*)]
4. An annulus [Figure 5.1 (*d*)].

However, the ROC for a unilateral z-transform is **always** the exterior of a circle of radius equal to the largest magnitude of the poles of $F(z)$. The unilateral z-transform is particularly useful in analyzing **causal systems** specified by **linear constant coefficient difference equations** with **non-zero** initial conditions.

For most of the physical systems, $f(k) = 0$; $k < 0$, therefore, **unless otherwise stated,** we shall consider only the case of the **unilateral z-transform,** defined by equation (5.1 B).

5.2.1 Properties of the *z*-Transform

Let $\quad \mathcal{Z}[f(k)] = F(z)$

and $\quad \mathcal{Z}[g(k)] = G(z)$

Now, the properties of the z-transform are **summarized** in following sections one by one:

5.2.1.1 Linearity

$$\mathcal{Z}[af(k) \pm bg(k)] = aF(z) \pm bG(z) \qquad \text{...(5.2)}$$

Proof.

$$\mathcal{Z}[af(k) \pm bg(k)] = \sum_{k=0}^{\infty}[af(k) \pm bg(k)]z^{-k} = a\sum_{k=0}^{\infty}[f(k)]z^{-k} \pm b\sum_{k=0}^{\infty}[g(k)]z^{-k}$$

$$= aF(z) \pm bG(z)$$

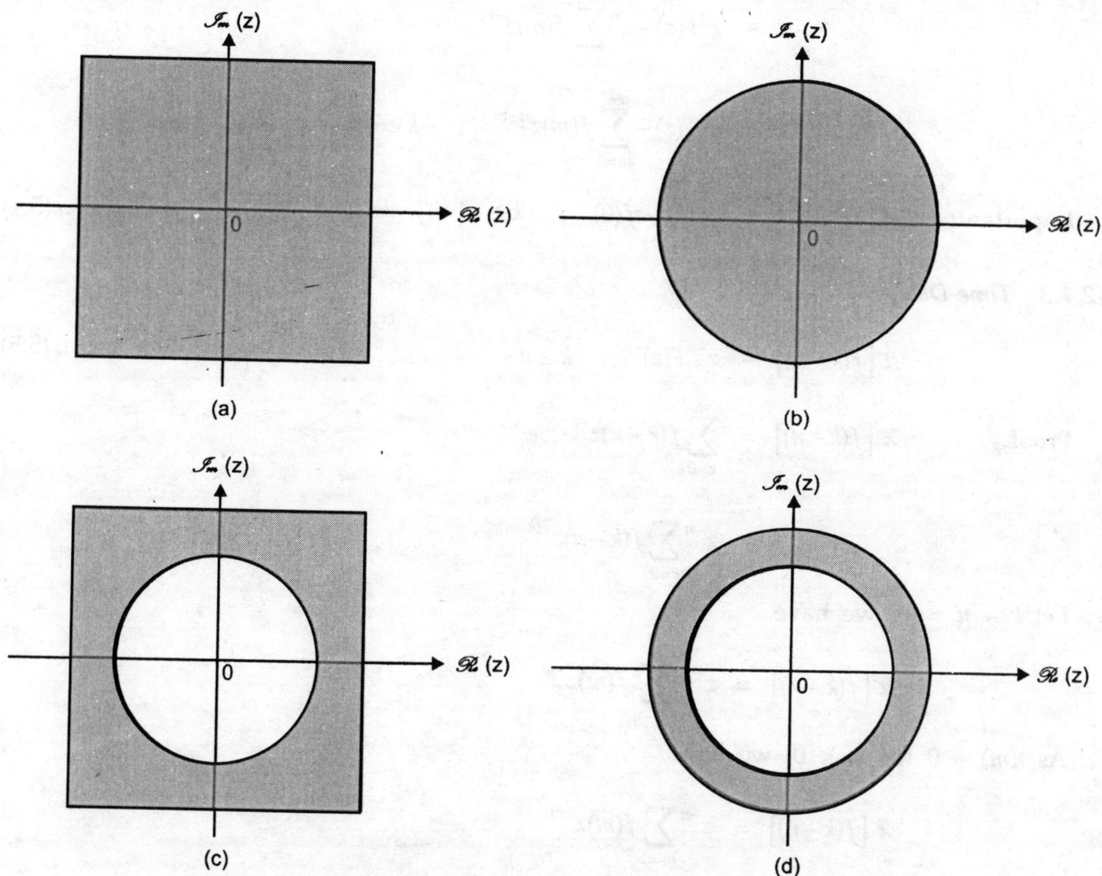

FIGURE 5.1 *Possible configurations of the ROC for the z-transform*
(a) The entire z-plane *(b) Interior of a circle*
(c) Exterior of a circle *(d) An annulus.*

5.2.1.2 Time-Advance

$$\mathscr{Z}[f(k+n)] = z^n F(z) - \sum_{m=0}^{n-1} f(m) z^{n-m} \quad ; \quad k \geq -n \qquad ...(5.3)$$

Proof.

$$\mathscr{Z}[f(k+n)] = \sum_{k=0}^{\infty} f(k+n) z^{-k}$$

$$= z^n \sum_{k=0}^{\infty} f(k+n) z^{-(k+n)}$$

Let $k + n = m$, we have

$$\mathscr{Z}[f(k+n)] = z^n \sum_{m=n}^{\infty} f(m) z^{-m}$$

$$= z^n \left[\sum_{m=0}^{\infty} f(m) z^{-m} - \sum_{m=0}^{n-1} f(m) z^{-m} \right]$$

$$= z^n F(z) - z^n \sum_{m=0}^{n-1} f(m) z^{-m}$$

$$= z^n F(z) - \sum_{m=0}^{n-1} f(m) z^{n-m} \quad ; \quad k \geq -n$$

Important: $$\mathcal{Z}[f(k+1)] = zF(z) - zf(0) \quad ; \quad k \geq -1 \qquad \text{...(5.4)}$$

5.2.1.3 *Time-Delay*

$$\mathcal{Z}[f(k-n)] = z^{-n} F(z) \quad ; \quad k \geq n \qquad \text{...(5.5)}$$

Proof. $$\mathcal{Z}[f(k-n)] = \sum_{k=0}^{\infty} f(k-n) z^{-k}$$

$$= z^{-n} \sum_{k=0}^{\infty} f(k-n) z^{-(k-n)}$$

Let $k - n = m$, we have

$$\mathcal{Z}[f(k-n)] = z^{-n} \sum_{m=-n}^{\infty} f(m) z^{-m}$$

As $f(m) = 0$ for $m < 0$, we have

$$\mathcal{Z}[f(k-n)] = z^{-n} \sum_{m=0}^{\infty} f(m) z^{-m}$$

$$= z^{-n} F(z) \quad ; \quad k \geq n$$

Important: $$\mathcal{Z}[f(k-1)] = z^{-1} F(z) \quad ; \quad k \geq 1 \qquad \text{...(5.6)}$$

5.2.1.4 *Scaling in z-Domain*

$$\mathcal{Z}[a^k f(k)] = F\left(\frac{z}{a}\right) \qquad \text{...(5.7)}$$

Proof. $$\mathcal{Z}[a^k f(k)] = \sum_{k=0}^{\infty} [a^k f(k)]\, z^{-k}$$

$$= \sum_{k=0}^{\infty} \left[\left(\frac{1}{a}\right)^{-k} f(k) \right] z^{-k}$$

$$= \sum_{k=0}^{\infty} f(k) \left(\frac{z}{a}\right)^{-k}$$

$$= F\left(\frac{z}{a}\right)$$

5.2.1.5 Differentiation in z-Domain

$$\mathcal{Z}[k\,f(k)] = -z\frac{d}{dz}F(z) \qquad ...(5.8)$$

Proof.

$$\mathcal{Z}[k\,f(k)] = \sum_{k=0}^{\infty}[k\,f(k)]\,z^{-k}$$

$$= -z\sum_{k=0}^{\infty}[-k\,f(k)]z^{-k-1}$$

$$= -z\sum_{k=0}^{\infty}\left[f(k)\frac{d}{dz}z^{-k}\right]$$

$$= -z\frac{d}{dz}\sum_{k=0}^{\infty}f(k)\,z^{-k}$$

$$= -z\frac{d}{dz}F(z)$$

Important: In general, we may write

$$\mathcal{Z}[k^n f(k)] = \left(-z\frac{d}{dz}\right)^n F(z) \qquad ...(5.9)$$

5.2.1.6 Convolution

$$\mathcal{Z}[f(k)*g(k)] = F(z)G(z) \qquad ...(5.10)$$

Proof.

$$\mathcal{Z}[f(k)*g(k)] = \sum_{k=0}^{\infty}[f(k)*g(k)]\,z^{-k}$$

$$= \sum_{k=0}^{\infty}\left[\sum_{m=0}^{\infty}f(m)g(k-m)\right]z^{-k}$$

Interchanging the order of summation, we get

$$\mathcal{Z}[f(k)*g(k)] = \sum_{m=0}^{\infty}f(m)\left[\sum_{k=0}^{\infty}g(k-m)\,z^{-k}\right]$$

$$= \sum_{m=0}^{\infty}f(m)\left[z^{-m}\sum_{k=0}^{\infty}g(k)\,z^{-k}\right] \qquad \text{[Refer Section 5.2.1.3]}$$

$$= \left[\sum_{m=0}^{\infty}f(m)\,z^{-m}\right]\left[\sum_{k=0}^{\infty}g(k)\,z^{-k}\right]$$

$$= F(z)G(z)$$

The **convolution property** is one of the most important properties of the z-transform because it is used to convert the **convolution** of two discrete-time signals in time-domain into **multiplication** of their z-transforms.

5.2.1.7 Initial Value Theorem

$$f(0) = \lim_{z \to \infty} F(z) \qquad \text{...(5.11)}$$

Proof.

$$\mathscr{Z}[f(k)] = \sum_{k=0}^{\infty} f(k)z^{-k}$$

$$= f(0) + f(1)z^{-k} + f(1)z^{-k} + f(1)z^{-k} + ...$$

Taking the limit $z \to \infty$ as $k \to 0$, we get

$$f(0) = \lim_{z \to \infty} F(z)$$

5.2.1.8 Final Value Theorem

$$f(\infty) = \lim_{z \to 1} [(z-1)F(z)] \qquad \text{...(5.12)}$$

Proof.

$$\mathscr{Z}[f(k+1)] = \sum_{k=0}^{\infty} f(k+1)z^{-k} \qquad \text{...(5.13 A)}$$

Also, $\mathscr{Z}[f(k+1)] = zF(z) - zf(0)$ [Refer Section 5.2.1.2] ...(5.13 B)

From equations (5.13), we may write

$$zF(z) - zf(0) = \sum_{k=0}^{\infty} f(k+1)z^{-k}$$

or,

$$zF(z) - zf(0) - F(z) = \sum_{k=0}^{\infty} f(k+1)z^{-k} - \sum_{k=0}^{\infty} f(k)z^{-k}$$

or,

$$(z-1)\ F(z) - zf(0) = \sum_{k=0}^{\infty} [f(k+1) - f(k)]\ z^{-k}$$

or,

$$(z-1)\ F(z) = zf(0) + \sum_{k=0}^{\infty} [f(k+1) - f(k)]\ z^{-k}$$

For final value theorem to be applicable, $F(z)$ must not have any pole in the region **outside a unit circle,** *i.e.,* $F(z)$ should be **analytic** for $|z| > 1$. Because of **assumed stability condition** stated above, taking the limit $z \to 1$ as $k \to \infty$, we obtain

$$\lim_{z \to 1} [(z-1)F(z)] = f(0) + f(\infty) - f(0)$$

or,

$$f(\infty) = \lim_{z \to 1} [(z-1)F(z)]$$

5.2.2 z-Transforms of Some Common Discrete-Time Signals

5.2.2.1 Unit-Impulse Signal

It is defined as

$$\delta(k) = \begin{cases} 1 & ; \ k = 0 \\ 0 & ; \ k \neq 0 \end{cases}$$

Using definition (5.1 B), we have

$$\mathscr{Z}[\delta(k)] = \sum_{k=0}^{\infty} \delta(k) z^{-k}$$

$$= \delta(0)z^{0} + \delta(1)z^{-1} + \delta(2)z^{-2} + \ldots\ldots$$

$$= 1 + 0 + 0 + \ldots\ldots$$

or, $\mathscr{Z}[\delta(k)] = 1$ **[ROC: Entire *z*-plane]** ...(5.14)

5.2.2.2 Unit-Step Signal

It is defined as

$$u(k) = \begin{cases} 1 & ; \quad k \geq 0 \\ 0 & ; \quad k < 0 \end{cases}$$

Using definition (5.1 B), we have

$$\mathscr{Z}[u(k)] = \sum_{k=0}^{\infty} u(k)\, z^{-k}$$

$$= 1 \cdot z^{0} + 1 \cdot z^{-1} + 1 \cdot z^{-2} + \ldots\ldots$$

$$= 1 + z^{-1} + z^{-2} + \ldots\ldots$$

$$= \frac{1}{1 - z^{-1}}$$

or, $\mathscr{Z}[u(k)] = \dfrac{z}{z-1}$ ROC: $|z| > 1$ ...(5.15)

The ROC is **exterior** of **unit-circle.**

5.2.2.3 Unit-Ramp Signal

A unit-ramp is obtained by multiplying a unit-step signal by k,

i.e., $r(k) = ku(k)$

$\therefore$ $\mathscr{Z}[r(k)] = \mathscr{Z}[k\, u(k)]$

$$= -z \frac{d}{dz} U(z)$$ [Property 5.8]

Since . $U(z) = \dfrac{z}{z-1}$

Therefore, $\mathscr{Z}[r(k)] = -z \dfrac{d}{dz}\left(\dfrac{z}{z-1}\right)$

or, $\mathscr{Z}[r(k)] = \dfrac{z}{(z-1)^2}$ ROC: $|z| > 1$ ...(5.16)

The ROC is **exterior** of **unit-circle.**

5.2.2.4 Sinusoidal Signals

Consider a continuous-time exponential signal

$$f(t) = e^{j\omega t} = \cos \omega t + j\sin \omega t \qquad ...(5.17)$$

A discrete-time signal can be generated from a continuous-time signal $f(t)$ by **mathematically sampling** it at time intervals T_0 seconds apart merely by substituting $t = kT_0$, *i.e.*,

$$f_s(t) = f(kT_0)$$

Therefore, we may write (5.17) as

$$f(kT_0) = e^{j\omega kT_0} = \cos\omega kT_0 + j\sin\omega kT_0$$

Let $\qquad e^{j\omega T_0} = a$

Then $\qquad f(kT_0) = a^k$

$$\therefore \qquad \mathcal{Z}[f(kT_0)] = \mathcal{Z}[a^k]$$

$$= \sum_{k=0}^{\infty} a^k z^{-k} = 1 + \frac{a}{z} + \frac{a^2}{z^2} + ...$$

$$= \frac{1}{1-\left(\frac{a}{z}\right)} = \frac{z}{z-a} = \frac{z}{z-e^{j\omega T_0}}$$

$$= \frac{z}{z-(\cos\omega T_0 + j\sin\omega T_0)}$$

$$= \frac{z[(z-\cos\omega T_0) + j\sin\omega T_0]}{(z-\cos\omega T_0)^2 + \sin^2\omega T_0}$$

$$= \frac{z^2 - z\cos\omega T_0 + jz\sin\omega T_0}{z^2 - 2z\cos\omega T_0 + (\cos^2\omega T_0 + \sin^2\omega T_0)}$$

$$= \frac{z^2 - z\cos\omega T_0}{z^2 - 2z\cos\omega T_0 + 1} + j\frac{z\sin\omega T_0}{z^2 - 2z\cos\omega T_0 + 1}$$

Expanding the L.H.S. of above equation, we get

$$\mathcal{Z}[\cos\omega kT_0 + j\sin\omega kT_0] = \frac{z^2 - z\cos\omega T_0}{z^2 - 2z\cos\omega T_0 + 1} + j\frac{z\sin\omega T_0}{z^2 - 2z\cos\omega T_0 + 1} \qquad ...(5.18)$$

Equating the real and imaginary parts of the equation (5.18), we obtain

$$\mathcal{Z}[\cos\omega kT_0] = \frac{z^2 - z\cos\omega T_0}{z^2 - 2z\cos\omega T_0 + 1} \qquad \text{ROC: } |z| > 1 \qquad ...(5.19)$$

The ROC is **exterior** of **unit-circle.**

and

$$\mathcal{Z}[\sin\omega kT_0] = \frac{z\sin\omega T_0}{z^2 - 2z\cos\omega T_0 + 1} \qquad \text{ROC: } |z| > 1 \qquad ...(5.20)$$

The ROC is **exterior** of **unit-circle.**

Example 5.1. *Determine the z-transforms of the following discrete-time signals.*

(a) $f(k) = u(k-1)$ (b) $f(k) = a^k u(k)$

(c) $f(k) = a^k u(-k-1)$ (d) $f(k) = a^k u(k) + b^k u(-k-1)$

Solution. (a) Given $f(k) = u(k-1)$

Using property (5.6), we have

$$\mathscr{Z}[f(k)] = \mathscr{Z}[u(k-1)]$$

$$= z^{-1}U(z)$$

But $$U(z) = \frac{z}{z-1}$$

Therefore, $$\mathscr{Z}[f(k)] = z^{-1}\left(\frac{z}{z-1}\right)$$

or, $$\mathscr{Z}[f(k)] = \frac{1}{z-1} \qquad \text{ROC: } |z| > 1$$

The ROC is **exterior** of **unit-circle.**

(b) Given $f(k) = a^k u(k)$

Using property (5.7), we have

$$\mathscr{Z}[f(k)] = \mathscr{Z}[a^k u(k)]$$

$$= U\left(\frac{z}{a}\right) = \frac{\left(\frac{z}{a}\right)}{\left(\frac{z}{a}-1\right)} = \left(\frac{z}{z-a}\right)$$

or, $$\mathscr{Z}[f(k)] = \left(\frac{z}{z-a}\right) \qquad \text{ROC: } |z| > |a|$$

The ROC is **exterior** of a **circle having radius** $|a|$.

(c) Given $f(k) = -a^k u(-k-1)$

Discrete-time unit-step signal is defined as

$$u(k) = \begin{cases} 1 & ; \; k \geq 0 \\ 0 & ; \; k < 0 \end{cases}$$

Therefore, $$u(-k-1) = \begin{cases} 1 & ; \; k \leq -1 \\ 0 & ; \; k > 0 \end{cases}$$

We find that $f(k) = -a^k u(-k-1)$ is a **non-causal signal**. Thus, we shall use the definition of **bilateral** *z***-transform** in this case.

Therefore, $$\mathscr{Z}[f(k)] = \mathscr{Z}[-a^k u(-k-1)]$$

$$= \sum_{k=-\infty}^{\infty} [-a^k u(-k-1)]\, z^{-k}$$

$$= \sum_{k=-\infty}^{-1} [-a^k] z^{-k} = -\sum_{k=-\infty}^{-1} (a^{-1}z)^{-k}$$

Let $m = -k$, we have

$$\mathcal{Z}[f(k)] = -\sum_{m=\infty}^{1} (a^{-1}z)^m$$

$$= -\sum_{m=1}^{\infty} (a^{-1}z)^m$$

$$= -\left[\left(\frac{z}{a}\right) + \left(\frac{z}{a}\right)^2 + \left(\frac{z}{a}\right)^3 + \ldots\ldots\right]$$

If $\left|\frac{z}{a}\right| < 1$ or equivalently $|z| < |a|$ the above series **converges.**

Therefore, $$\mathcal{Z}[f(k)] = -\left[\frac{\left(\frac{z}{a}\right)}{1-\left(\frac{z}{a}\right)}\right] = -\left(\frac{z}{a-z}\right)$$

or, $$\mathcal{Z}[f(k)] = \left(\frac{z}{z-a}\right) \qquad \text{ROC: } |z| < |a|$$

The ROC is **interior** of a **circle having radius** $|a|$.

(*d*) Given $$f(k) = a^k u(k) + b^k u(-k-1)$$

Discrete-time unit-step signal is defined as

$$u(k) = \begin{cases} 1 & ; \ k \geq 0 \\ 0 & ; \ k < 0 \end{cases}$$

Therefore, $$u(-k-1) = \begin{cases} 1 & ; \ k \leq -1 \\ 0 & ; \ k > 0 \end{cases}$$

Clearly, the first part, *i.e.*, $f(k) = a^k u(k)$ is a **causal signal** and the second part, *i.e.*, $f(k) = b^k u(-k-1)$ is a **non-causal signal.**

Using **linearity property** of the z-transform, we may write

$$\mathcal{Z}[f(k)] = \mathcal{Z}[a^k u(k)] + \mathcal{Z}[b^k u(-k-1)]$$

$$= \sum_{k=0}^{\infty} [a^k] z^{-k} + \sum_{k=-\infty}^{-1} [b^k] z^{-k}$$

$$= F_1(z) + F_2(z) \qquad \ldots(5.21)$$

Now, $$F_1(z) = \sum_{k=0}^{\infty} [a^k] z^{-k}$$

$$= \sum_{k=0}^{\infty}\left(\frac{a}{z}\right)^k$$

$$= 1+\frac{a}{z}+\frac{a^2}{z^2}+\ldots\ldots$$

If $\left|\frac{a}{z}\right|<1$ or equivalently $|z| > |a|$ the above series **converges.**

Therefore, $$F_1(z) = \frac{1}{1-\left(\frac{a}{z}\right)}$$

or, $$F_1(z) = \frac{z}{z-a}$$ ROC: of $F_1(z)$: $|z|>|a|$

Again, $$F_2(z) = \sum_{k=-\infty}^{-1}[b^k]\,z^{-k}$$

Let $m = -k$, we have

$$F_2(z) = \sum_{m=\infty}^{1}[b^{-m}]\,z^m$$

$$= \sum_{m=1}^{\infty}[b^{-m}]\,z^m = \sum_{m=1}^{\infty}\left(\frac{z}{b}\right)^m$$

$$= \frac{z}{b}+\frac{z^2}{b^2}+\frac{z^3}{b^3}+\ldots\ldots$$

If $\left|\frac{z}{b}\right|<1$ or equivalently $|z|<|b|$ the above series **converges.**

Therefore, $$F_2(z) = \frac{\left(\frac{z}{b}\right)}{1-\left(\frac{z}{b}\right)}$$

or, $$F_2(z) = \frac{z}{b-z}$$ ROC of $F_2(z)$: $|z|<|b|$

Now, from relation (5.21), we may write

$$\mathcal{Z}[f(k)] = F_1(z)+F_2(z)$$

$$= \frac{z}{z-a}+\frac{z}{b-z}$$

or $$\mathcal{Z}[f(k)] = \frac{b-a}{b+a-z-abz^{-1}}$$ ROC: $|a|<|z|<|b|$

The ROC is an **annulus.**

Example 5.2. *Consider a signal that is the sum of two real exponentials,*

$$f(k) = 7\left(\frac{1}{3}\right)^k u(k) - 6\left(\frac{1}{2}\right)^k u(k)$$

Determine the z-transform of the signal.

Solution. Using linearity property of the z-transform, we have

$$\mathscr{Z}[f(k)] = \mathscr{Z}\left[7\left(\frac{1}{3}\right)^k u(k)\right] - \mathscr{Z}\left[6\left(\frac{1}{2}\right)^k u(k)\right]$$

$$= \sum_{k=0}^{\infty}\left[7\left(\frac{1}{3}\right)^k\right] z^{-k} - \sum_{k=0}^{\infty}\left[6\left(\frac{1}{2}\right)^k\right] z^{-k}$$

$$= F_1(z) - F_2(z) \qquad \text{...(5.22)}$$

Now, $$F_1(z) = \sum_{k=0}^{\infty}\left[7\left(\frac{1}{3}\right)^k\right] z^{-k}$$

or, $$F_1(z) = 7\sum_{k=0}^{\infty}\left(\frac{1}{3}z^{-1}\right)^k$$

or, $$F_1(z) = 7\left[\frac{1}{1-\frac{1}{3}z^{-1}}\right] \qquad \text{ROC of } F_1(z) : |z| > \frac{1}{3}$$

Again $$F_2(z) = \sum_{k=0}^{\infty}\left[6\left(\frac{1}{2}\right)^k\right] z^{-k}$$

or, $$F_2(z) = 6\sum_{k=0}^{\infty}\left(\frac{1}{2}z^{-1}\right)^k$$

or, $$F_2(z) = 6\left[\frac{1}{1-\frac{1}{2}z^{-1}}\right] \qquad \text{ROC of } F_2(z) : |z| > \frac{1}{2}$$

Therefore, from relation (5.22), we may write

$$\mathscr{Z}[f(k)] = F_1(z) - F_2(z)$$

$$= 7\left[\frac{1}{1-\frac{1}{3}z^{-1}}\right] - 6\left[\frac{1}{1-\frac{1}{2}z^{-1}}\right]$$

$$= \frac{\left[1-\frac{3}{2}z^{-1}\right]}{\left[1-\frac{1}{3}z^{-1}\right]\left[1-\frac{1}{2}z^{-1}\right]}$$

or, $$\mathscr{Z}[f(k)] = \frac{z\left[z-\frac{3}{2}\right]}{\left[z-\frac{1}{3}\right]\left[z-\frac{1}{2}\right]} \qquad \text{ROC: } |z| > \frac{1}{2}$$

The ROC is **exterior of a circle** having **radius** 0.5.

Explanation : The ROC for a unilateral z-transform is **always** the exterior of a circle of radius equal to the **largest magnitude** of the poles of $F(z)$ [Refer Section 5.2], which in this case being 0.5 and, therefore, the ROC is **exterior of a circle** having **radius** 0.5.

Example 5.3. *Determine the z-transform of the signals defined as*

(a) $$f(k) = \begin{cases} \left(\frac{1}{a}\right)^k & \text{for } k \geq 0 \\ 0 & \text{otherwise} \end{cases}$$

(b) $$f(k) = \frac{a^k}{k\,!}$$

Solution. (a) Given

$$f(k) = \begin{cases} \left(\frac{1}{a}\right)^k & \text{for } k \geq 0 \\ 0 & \text{otherwise} \end{cases}$$

$$\therefore \quad \mathscr{Z}[f(k)] = \sum_{k=0}^{\infty} \left(\frac{1}{a}\right)^k z^{-k}$$

$$= 1 + \frac{1}{az} + \frac{1}{a^2 z^2} + \ldots\ldots$$

$$= \frac{1}{\left(1-\frac{1}{az}\right)}$$

or, $$\mathscr{Z}[f(k)] = \frac{az}{az-1} \qquad \text{ROC: } |z| > \left|\frac{1}{a}\right|$$

The ROC is **exterior** of a **circle having radius** $\left|\frac{1}{a}\right|$.

(b) Given $$f(k) = \frac{a^k}{k\,!}$$

$$\therefore \quad \mathscr{Z}[f(k)] = \sum_{k=0}^{\infty} \left(\frac{a^k}{k\,!}\right) z^{-k}$$

$$= \sum_{k=0}^{\infty} \frac{1}{k\,!}\left(\frac{a}{z}\right)^k$$

$$= 1+\left(\frac{a}{z}\right)+\frac{1}{2\,!}\left(\frac{a}{z}\right)^2+\ldots\ldots$$

or, $$\mathscr{Z}[f(k)] = \exp\left(\frac{a}{z}\right) \qquad \text{ROC: } |z| > 0$$

The ROC is **entire z-plane except the origin.**

Example 5.4. *Determine the initial and final values of the discrete-time signal f(k), whose z-transform is given by*

$$F(z) = 3+\frac{5}{z}+\frac{7}{z^2}$$

Solution. Using property (5.11), we have

$$f(0) = \lim_{z\to\infty} F(z)$$

$$= \lim_{z\to\infty}\left(3+\frac{5}{z}+\frac{7}{z^2}\right)$$

$$= 3$$

Now, using property (5.12), we have

$$f(\infty) = \lim_{z\to 1}\,[(z-1)F(z)]$$

$$= \lim_{z\to 1}\left[(z-1)\left(3+\frac{5}{z}+\frac{7}{z^2}\right)\right]$$

$$= 0$$

5.3 THE INVERSE *z*-TRANSFORM

The inverse z-transformation is a process of determining the discrete-time signal which generates a given z-transform. Thus, the mechanism for transforming $F(z)$ back to a discrete-time signal $f(k)$ is called the **inverse z-transformation.**

Three methods are often used for the evaluation of the inverse z-transform in practice. These are

1. Power series expansion method,
2. Partial-Fraction expansion method,
3. Contour-Integration Method.

We shall discuss here basically two methods. The third method of contour-integration is beyond the scope of the book.

5.3.1 Power-Series Expansion Method

This method of determining the inverse z-transform is **motivated** by the observation that the definition of the z-transform given in expression (5.1 A) can be interpreted as a **power-series** involving both positive and negative powers of z. The coefficients in this power series are, in fact, the sequence values of the discrete-time signal $f(k)$. To illustrate how a **power-series**

expansion can be used in evaluation of the inverse z-transform, let us consider the following examples:

Example 5.5. *Find the discrete-time sequence f(k) when F(z) is given by*

$$F(z) = \frac{36z^2 - 10z}{12z^2 - 7z + 1} \qquad ROC: |z| > \frac{1}{3}$$

Solution. Given,

$$F(z) = \frac{36z^2 - 10z}{12z^2 - 7z + 1}$$

or,

$$F(z) = \frac{36z^2 - 10z}{(3z-1)(4z-1)}$$

Here, $F(z)$ has two poles; one at $z = \frac{1}{3}$ and other at $z = \frac{1}{4}$. The ROC is $|z| > \frac{1}{3}$, *i.e.*, **exterior** of a circle having radius equal to the **largest magnitude** of the poles of $F(z)$, which in this case being $\frac{1}{3}$. This means $f(k)$ would be a **causal discrete-time sequence** and it, therefore, requires a **power-series expansion in negative powers of** z. The power series expansion can be carried out by **long division process** as follows:

$$\begin{array}{r|l}
 & 3 + \frac{11}{12}z^{-1} + \frac{41}{144}z^{-2} + \ldots\ldots \\
\hline
12z^2 - 7z + 1 & 36z^2 - 10z \\
 & +36z^2 - 21z + 3 \\
 & \;-\qquad\;\; +\qquad -\\
\hline
 & 11z - 3 \\
 & +11z - \frac{77}{12} - \frac{11}{12}z^{-1} \\
 & \;-\qquad\; + \qquad\; + \\
\hline
 & \frac{41}{12}z - \frac{11}{12}z^{-1} \\
 & +\frac{41}{12}z - \frac{287}{144}z^{-1} + \frac{41}{144}z^{-2} \\
 & \;-\qquad\quad + \qquad\quad - \\
\hline
 & \frac{155}{144}z^{-1} - \frac{41}{144}z^{-2} \\
 & \vdots \qquad\qquad \vdots
\end{array}$$

and so on.

Thus,

$$F(z) = 3 + \frac{11}{12}z^{-1} + \frac{41}{144}z^{-2} + \ldots\ldots \qquad \ldots(5.23)$$

But,

$$F(z) = \sum_{k=-\infty}^{\infty} f(k)z^{-k}$$

or,

$$F(z) = \ldots\ldots + f(-2)z^2 + f(-1)z^1 + f(0)z^0 + f(1)z^{-1} + f(2)z^{-2} + \ldots\ldots \qquad \ldots(5.24)$$

Comparing expression (5.23) and (5.24), we obtain

$$f(0) = 3$$

$$f(1) = \frac{11}{12}$$

$$f(2) = \frac{41}{144}$$

$$\vdots \quad \vdots$$

Therefore,
$$f(k) = \left\{ \underset{\uparrow}{3}, \frac{11}{12}, \frac{41}{144}, \ldots\ldots \right\}$$

Example 5.6. *Determine the inverse z-transform of*

$$F(z) = \frac{36z^2 - 10z}{12z^2 - 7z + 1} \qquad ROC: |z| < \frac{1}{4}$$

Solution. Given,
$$F(z) = \frac{36z^2 - 10z}{12z^2 - 7z + 1}$$

or,
$$F(z) = \frac{36z^2 - 10z}{(3z-1)(4z-1)}$$

Here, $F(z)$ has two poles; one at $z = \frac{1}{3}$ and other at $z = \frac{1}{4}$. The ROC is $|z| < \frac{1}{4}$, *i.e.*, **interior** of a circle having radius equal to the **smallest magnitude** of the poles of $F(z)$, which in this case being $\frac{1}{4}$. This means $f(k)$ would be a **non-causal discrete-time sequence** and it, therefore, requires a **power-series expansion in positive powers of** z. The power series expansion can be carried out by **long division process** as follows:

$$\begin{array}{r l}
 & -10z - 34z^2 - 118z^3 - \ldots\ldots \\
1 - 7z + 12z^2 \big) & -10z + 36z^2 \\
 & -10z + 70z^2 - 120z^3 \\
 & \;\; + \quad\;\; - \quad\;\; + \\
\hline
 & -34z^2 + 120z^3 \\
 & -34z^2 + 238z^3 - 408z^4 \\
 & \;\; + \quad\;\; - \quad\;\; + \\
\hline
 & -118z^3 + 408z^4 \\
 & -118z^3 + 826z^4 - 1416z^5 \\
 & \;\; + \quad\;\; - \quad\;\; + \\
\hline
 & -418z^4 + 1416z^5 \\
 & \vdots \qquad \vdots
\end{array}$$

and so on.

Thus, $$F(z) = -10z - 34z^2 - 118z^3 - \ldots\ldots\ldots \qquad \ldots(5.25)$$

But, $$F(z) = \sum_{k=-\infty}^{\infty} f(k)z^{-k}$$

or, $$F(z) = \ldots\ldots + f(-3)z^3 + f(-2)z^2 + f(-1)z^1 + f(0)z^0 + f(1)z^{-1} + f(2)z^{-2} + \ldots\ldots \qquad \ldots(5.26)$$

Comparing expressions (5.25) and (5.26), we obtain

$$f(0) = 0$$
$$f(-1) = 10$$
$$f(-2) = -34$$
$$f(-3) = -118$$
$$\vdots \qquad \vdots$$

Therefore, $$f(k) = \left\{ \ldots\ldots, \ -118, \ -34, \ -10, \ \underset{\uparrow}{0} \right\}$$

From above examples, it can be observed that the power-series expansion method gives $f(k)$ for any k, but it does not help us to get the **general form** of $f(k)$. It, therefore, is useful only in **numerical studies** and **not for analytical one.**

5.3.2 Partial-Fraction Expansion Method

The basic procedure of using partial-fraction expansion to determine the inverse z-transform relies on expressing the z-transform as a **linear combination of simpler terms**. The inverse z-transform of each term can then be obtained by inspection. In particular, suppose that the partial-fraction expansion of $F(z)$ is of the form

$$F(z) = \sum_{i=1}^{n} \frac{A_i z}{z - a_i} \qquad \ldots(5.27)$$

so that the inverse z-transform of $F(z)$ equals the sum of the inverse z-transforms of the individual terms in the equation (5.27).

Case 1: If ROC: $|z| > a_i$, then the inverse z-transform of the corresponding term in equation (5.27) is $A_i \, a_i^k \, u(k)$.

Case 2: If ROC: $|z| > a_i$, then the inverse z-transform of the corresponding term in equation (5.27) is $-A_i \, a_i^k \, u(-k-1)$.

In general, the partial-fraction expansion of a **rational** z-transform may include **higher-order terms in addition to the first-order terms** in equation. The inverse z-transform of each term can then be obtained by means of a list of **z-transform pairs** (given in **Appendix-D**) in conjunction with the **properties of the z-transform** (already discussed in **Section 5.2.1**). To illustrate how a **partial-fraction expansion** can be used to obtain the inverse z-transform, let us consider the following examples:

Example 5.7 *Find the inverse z-transform using partial-fraction expansion method.*

$$F(z) = \frac{36z^2 - 10z}{12z^2 - 7z + 1}$$

ROC:

(a) $|z| > \frac{1}{3}$ *(b)* $|z| < \frac{1}{4}$ *(c)* $\frac{1}{4} < |z| < \frac{1}{3}$

Solution. Given $F(z) = \dfrac{36z^2 - 10z}{12z^2 - 7z + 1}$

or, $F(z) = \dfrac{36z^2 - 10z}{(3z-1)(4z-1)}$

Here, $F(z)$ has two poles; one at $z = \frac{1}{3}$ and other at $z = \frac{1}{4}$.

Now,

$$\frac{F(z)}{z} = \frac{36z-10}{(3z-1)(4z-1)}$$

$$= \left.\frac{36z-10}{(4z-1)}\right|_{z=\frac{1}{3}} \frac{1}{(3z-1)} + \left.\frac{36z-10}{(3z-1)}\right|_{z=\frac{1}{4}} \frac{1}{(4z-1)}$$

$$= \frac{2}{\left(z-\frac{1}{3}\right)} + \frac{1}{\left(z-\frac{1}{4}\right)}$$

or,

$$F(z) = \frac{2z}{\left(z-\frac{1}{3}\right)} + \frac{z}{\left(z-\frac{1}{4}\right)} \quad \text{...(5.28)}$$

In order to determine the inverse z-transform of each of these individual terms, we **must specify the ROC** associated with each.

(*a*) ROC: $|z| > \frac{1}{3}$ (Given)

Here, the ROC for $F(z)$ is **outside** the **outermost pole** at $z = \frac{1}{3}$. Consequently, the ROC for each individual term in equation (5.28) **must also be outside** the pole associated with that term.

Thus, $f(k) = f_1(k) + f_2(k)$

where,

$$f_1(k) \stackrel{\mathcal{Z}}{\longleftrightarrow} \frac{2z}{\left(z-\frac{1}{3}\right)} \quad ; \quad |z| > \frac{1}{3}$$

$$f_2(k) \stackrel{\mathcal{Z}}{\longleftrightarrow} \frac{z}{\left(z-\frac{1}{4}\right)} \quad ; \quad |z| > \frac{1}{4}$$

By inspection, we can identify

$$f_1(k) = 2\left(\frac{1}{3}\right)^k u(k)$$

and

$$f_2(k) = \left(\frac{1}{4}\right)^k u(k)$$

so that $$f(k) = 2\left(\frac{1}{3}\right)^k u(k) + \left(\frac{1}{4}\right)^k u(k)$$

In order to check the result with $f(k)$ obtained in Example 5.5, put k = 0, 1, 2, ... in above expression, we obtain

$$f(0) = 2 + 1 = 3$$

$$f(1) = \frac{2}{3} + \frac{1}{4} = \frac{11}{12}$$

$$f(2) = \frac{2}{9} + \frac{1}{16} = \frac{41}{144}$$

$$\vdots \quad \vdots \quad \vdots$$

Which are consistent with Example 5.5.

(*b*) ROC: $|z| < \frac{1}{4}$ (Given)

Here, the ROC for $F(z)$ is **inside** both poles. Consequently, the ROC for each individual term in equation (5.28) **must also lie inside** the pole associated with that term.

Thus, $$f(k) = f_1(k) + f_2(k)$$

where, $$f_1(k) \overset{\mathcal{Z}}{\longleftrightarrow} \frac{2z}{\left(z - \frac{1}{3}\right)} \quad ; \quad |z| < \frac{1}{3}$$

$$f_2(k) \overset{\mathcal{Z}}{\longleftrightarrow} \frac{z}{\left(z - \frac{1}{4}\right)} \quad ; \quad |z| < \frac{1}{4}$$

By inspection, we can identify

$$f_1(k) = -2\left(\frac{1}{3}\right)^k u(-k-1)$$

and $$f_2(k) = -\left(\frac{1}{4}\right)^k u(-k-1)$$

so that $$f(k) = -2\left(\frac{1}{3}\right)^k u(-k-1) - \left(\frac{1}{4}\right)^k u(-k-1)$$

In order to check the result with $f(k)$ obtained in Example 5.6, put k = 0, – 1, –2, ... in above expression, we obtain

$$f(0) = -2 \times 0 - 1 \times 0 = 0$$

$$f(-1) = -6 \times 1 - 4 \times 1 = -10$$

$$f(-2) = -18 \times 1 - 16 \times 1 = -34$$

$$f(-3) = -54 \times 1 - 64 \times 1 = -118$$

$$\vdots \quad \vdots \quad \vdots$$

Which are **consistent** with Example 5.6.

(*c*) ROC: $\frac{1}{4} < |z| < \frac{1}{3}$ (Given)

Here, the ROC for $F(z)$ is **outside** the pole at $z = \frac{1}{4}$ and is **inside** the pole at $z = \frac{1}{3}$. Consequently, the ROC corresponding to first term in equation (5.28) **must also lie inside** the pole at $z = \frac{1}{3}$ and the ROC corresponding to second term in equation (5.28) **must be outside** the pole at $z = \frac{1}{4}$.

Thus,
$$f(k) = f_1(k) + f_2(k)$$

where,
$$f_1(k) \xleftrightarrow{\mathcal{Z}} \frac{2z}{\left(z - \frac{1}{3}\right)} \quad ; \quad |z| < \frac{1}{3}$$

$$f_2(k) \xleftrightarrow{\mathcal{Z}} \frac{z}{\left(z - \frac{1}{4}\right)} \quad ; \quad |z| > \frac{1}{4}$$

By inspection, we can identify

$$f_1(k) = -2\left(\frac{1}{3}\right)^k u(-k-1)$$

and
$$f_2(k) = \left(\frac{1}{4}\right)^k u(k)$$

so that
$$f(k) = -2\left(\frac{1}{3}\right)^k u(-k-1) + \left(\frac{1}{4}\right)^k u(k)$$

Example 5.8. *Find the inverse z-transform using partial-fraction expansion method. Given that*

$$F(z) = \frac{4z^2 - 2z}{z^3 - 5z^2 + 8z - 4}$$

Solution. Given
$$F(z) = \frac{4z^2 - 2z}{z^3 - 5z^2 + 8z - 4}$$

or,
$$F(z) = \frac{4z^2 - 2z}{(z-1)(z-2)^2}$$

Here, $F(z)$ is a proper rational function having one pole at $z = 1$ and two poles at $z = 2$.

Now,
$$\frac{F(z)}{z} = \frac{4z-2}{(z-1)(z-2)^2}$$

$$= \left.\frac{4z-2}{(z-2)^2}\right|_{z=1} \frac{1}{(z-1)} + \left.\frac{4z-2}{(z-1)}\right|_{z=2} \frac{1}{(z-2)^2} + \frac{d}{dz}\left[\frac{4z-2}{(z-1)}\right]_{z=2} \frac{1}{(z-2)}$$

$$= \frac{2}{(z-1)} + \frac{6}{(z-2)^2} - \frac{2}{(z-2)}$$

or, $$F(z) = \frac{2z}{(z-1)} + \frac{6z}{(z-2)^2} - \frac{2z}{(z-2)}$$

In order to determine the inverse z-transform of each of these individual terms, we **must specify the ROC** associated with each.

Thus,
$$f(k) = f_1(k) + f_2(k) - f_3(k)$$

where,
$$f_1(k) \overset{\mathscr{Z}}{\longleftrightarrow} \frac{2z}{(z-1)} \quad ; \quad |z| > 1$$

$$f_2(k) \overset{\mathscr{Z}}{\longleftrightarrow} \frac{6z}{(z-2)^2} \quad ; \quad |z| > 2$$

$$f_3(k) \overset{\mathscr{Z}}{\longleftrightarrow} \frac{2z}{(z-2)} \quad ; \quad |z| > 2$$

By inspection, we can identify
$$f_1(k) = 2(1)^k\, u(k)$$
$$f_2(k) = 6k(2)^{k-1}\, u(k) \qquad \text{[Refer Example 5.9]}$$
and $$f_3(k) = 2(2)^k\, u(k)$$
so that $$f(k) = 2(1)^k\, u(k) + 3k(2)^k\, u(k-1) - 2(2)^k\, u(k)$$

Alternative Method

$$F(z) = \frac{4z^2 - 2z}{z^3 - 5z^2 + 8z - 4}$$

$$= \frac{4z^2 - 2z}{(z-1)(z-2)^2}$$

$$= \left.\frac{4z^2 - 2z}{(z-2)^2}\right|_{z=1} \frac{1}{(z-1)} + \left.\frac{4z^2 - 2z}{(z-1)}\right|_{z=2} \frac{1}{(z-2)^2} + \frac{d}{dz}\left[\frac{4z^2 - 2z}{(z-1)}\right]_{z=2} \frac{1}{(z-2)}$$

or, $$F(z) = \frac{2}{(z-1)} + \frac{12}{(z-2)^2} + \frac{2}{(z-2)}$$

In order to determine the inverse z-transform of each of these individual terms, we **must specify the ROC** associated with each.

Thus, $$f(k) = f_1(k) + f_2(k) + f_3(k)$$

where,
$$f_1(k) \overset{\mathscr{Z}}{\longleftrightarrow} \frac{2}{(z-1)} \quad ; \quad |z| > 1$$

$$f_2(k) \overset{\mathscr{Z}}{\longleftrightarrow} \frac{12}{(z-2)^2} \quad ; \quad |z| > 2$$

$$f_3(k) \xleftrightarrow{\mathcal{Z}} \frac{2}{(z-2)} \quad ; \quad |z| > 2$$

By inspection, we can identify

$$f_1(k) = 2(1)^{k-1}\ u(k-1)$$

$$f_2(k) = 12(k-1)(2)^{k-2}\ u(k-2) \qquad \text{[Refer Example 5.9]}$$

and

$$f_3(k) = 2(2)^{k-1}\ u(k-1)$$

so that

$$f(k) = 2(1)^{k-1}\ u(k-1) + 6(k-1)(2)^{k-1}\ u(k-2) - 2(2)^{k-1}\ u(k-1)$$

Comment: In Example 5.8, for the same $F(z)$ we got **two different expressions** for $f(k)$ **by two different procedures.** It, however, can easily be verified that these two expressions of $f(k)$ are **equivalent.**

Example 5.9. *Determine the inverse z-transforms of*

(a) $$F(z) = \frac{1}{(z-2)} \quad ; \quad |z| > 2$$

(b) $$F(z) = \frac{1}{(z-2)^2} \quad ; \quad |z| > 2$$

(c) $$F(z) = \frac{z}{(z-2)^2} \quad ; \quad |z| > 2$$

Solution. (a)

$$f(k) = \mathcal{Z}^{-1}[F(z)]$$

$$= \mathcal{Z}^{-1}\left[\frac{1}{(z-2)}\right]$$

$$= \mathcal{Z}^{-1}\left[z^{-1}\left(\frac{z}{z-2}\right)\right]$$

$$= \mathcal{Z}^{-1}\left[z^{-1}\left\{\mathcal{Z}[(2)^k u(k)]\right\}\right]$$

$$= \mathcal{Z}^{-1}\left[\mathcal{Z}\left\{(2)^{k-1} u(k-1)\right\}\right] \qquad \text{[Property 5.5]}$$

or,

$$f(k) = (2)^{k-1}\ u(k-1)$$

(b)

$$f(k) = \mathcal{Z}^{-1}[F(z)]$$

$$= \mathcal{Z}^{-1}\left[\frac{1}{(z-2)^2}\right]$$

$$= \mathcal{Z}^{-1}\left[z^{-2}\left\{\frac{z^2}{(z-2)^2}\right\}\right]$$

$$= \mathcal{Z}^{-1}\left[z^{-2}\left\{\mathcal{Z}[(k+1)(2)^k u(k)]\right\}\right]$$

$$= \mathcal{Z}^{-1}\left[\mathcal{Z}\left\{(k-1)(2)^{k-2} u(k-2)\right\}\right] \qquad \text{[Property 5.5]}$$

or, $$f(k) = (k-1)(2)^{k-2}\, u(k-2)$$

(c) $$f(k) = \mathscr{Z}^{-1}[F(z)]$$

$$= \mathscr{Z}^{-1}\left[\frac{z}{(z-2)^2}\right]$$

$$= \mathscr{Z}^{-1}\left[z^{-1}\left\{\frac{z^2}{(z-2)^2}\right\}\right]$$

$$= \mathscr{Z}^{-1}\left[z^{-1}\left\{\mathscr{Z}\,[\,(k+1)(2)^k u(k)\,]\right\}\right]$$

$$= \mathscr{Z}^{-1}\left[\mathscr{Z}\left\{k\,(2)^{k-1}u(k-1)\right\}\right] \qquad \text{[Property 5.5]}$$

or, $$f(k) = k(2)^{k-1}\, u(k-1)$$

Example 5.10. *Determine the discrete-time signal for which the z-transform is given by*

$$f(z) = \log\,(1 + az^{-1}) \quad ; \quad |z| > |a|$$

Solution. Given

$$F(z) = \log\,(1 + az^{-1})$$

$$\therefore \quad \frac{d}{dz}F(z) = -\frac{az^{-2}}{1+az^{-1}}$$

or, $$-z\frac{d}{dz}F(z) = \frac{az^{-1}}{1+az^{-1}}$$

or, $$-z\frac{d}{dz}F(z) = \frac{a}{z+a} \qquad \text{...(5.29)}$$

Using property (5.8), we have

$$-z\frac{d}{dz}F(z) = \mathscr{Z}[k\,f(k)] \qquad \text{...(5.30)}$$

Comparing equations (5.29) and (5.30), we get

$$\mathscr{Z}[k\,f(k)] = \frac{a}{z+a}$$

Taking the inverse z-transform, we obtain

$$k\,f(k) = \mathscr{Z}^{-1}\left[\frac{a}{z+a}\right]$$

$$= a\,(-a)^{k-1}\,u(k-1)$$

$$= (-1)^{k-1}\,a^k\,u(k-1)$$

or, $$f(k) = (-1)^{k-1}\,\frac{a^k}{k}\,u(k-1)$$

5.4 SOLUTION OF DIFFERENCE EQUATIONS USING THE z-TRANSFORM

In contrast to the continuous-time systems, the operation of discrete-time systems is described by a set of difference equations. The **analysis** and **design** of **discrete-time systems** may effectively be carried out in terms of **difference equations**. Essentially, by using the z-transform method, we can transform difference equations into **algebraic equations** in z. In order to solve the difference equations we shall utilize the **time-advance property** (5.3) of the z-transform, written as

$$\mathcal{Z}[f(k+n)] = z^n F(z) - \sum_{m=0}^{n-1} f(m) z^{n-m} \quad ; \quad k \geq -n$$

Using above equation, we may write

$$\mathcal{Z}[f(k+1)] = zF(z) - zf(0)$$

$$\mathcal{Z}[f(k+2)] = z^2F(z) - z^2 f(0) - zf(1)$$

$$\mathcal{Z}[f(k+3)] = z^3F(z) - z^3 f(0) - z^2 f(1) - zf(2)$$

$$\vdots \qquad \vdots \qquad \vdots$$

and so on.

It is to be noted that when any difference equation is transformed into an algebraic equation in z by the z-transform method, the **initial data are automatically included** in the algebraic representation.

Example 5.11 *Find out the response of the system described by the difference equation*

$$f(k+2) - 5f(k+1) + 6f(k) = u(k)$$

Given that

$$f(0) = 0 \text{ and } f(1) = 1$$

Solution. Taking the z-transforms of both sides of the above difference equation, we get

$$z^2F(z) - z^2 f(0) - zf(1) - 5zF(z) + 5zf(0) + 6F(z) = \mathcal{Z}[u(k)]$$

or,
$$z^2F(z) - z - 5zf(z) + 6F(z) = \frac{z}{z-1}$$

or,
$$(z^2 - 5z + 6)\, F(z) = z + \frac{z}{z-1}$$

or,
$$(z-2)(z-3)\, F(z) = \frac{z^2}{z-1}$$

or,
$$F(z) = \frac{z^2}{(z-1)(z-2)(z-3)}$$

or,
$$\frac{F(z)}{z} = \frac{z}{(z-1)(z-2)(z-3)}$$

or, $$\frac{F(z)}{z} = \left.\frac{z}{(z-2)(z-3)}\right|_{z=1} \frac{1}{(z-1)} + \left.\frac{z}{(z-1)(z-3)}\right|_{z=2} \frac{1}{(z-2)} + \left.\frac{z}{(z-1)(z-2)}\right|_{z=3} \frac{1}{(z-3)}$$

or, $$\frac{F(z)}{z} = \frac{1}{2}\cdot\frac{1}{(z-1)} - \frac{2}{(z-2)} + \frac{3}{2}\cdot\frac{1}{(z-3)}$$

or, $$F(z) = \frac{1}{2}\cdot\frac{z}{(z-1)} - \frac{2z}{(z-2)} + \frac{3}{2}\cdot\frac{z}{(z-3)}$$

In order to determine the inverse z-transform of each of these individual terms, we **must specify the ROC** associated with each.

Thus, $$f(k) = f_1(k) - f_2(k) + f_3(k)$$

where, $$f_1(k) \overset{z}{\longleftrightarrow} \frac{1}{2}\cdot\frac{z}{(z-1)} \quad ; \quad |z| > 1$$

$$f_2(k) \overset{z}{\longleftrightarrow} \frac{2z}{(z-2)} \quad ; \quad |z| > 2$$

$$f_3(k) \overset{z}{\longleftrightarrow} \frac{3}{2}\cdot\frac{z}{(z-3)} \quad ; \quad |z| > 3$$

By inspection, we can identify

$$f_1(k) = \frac{1}{2}(1)^k\, u(k)$$

$$f_2(k) = 2(2)^k\, u(k)$$

and $$f_3(k) = \frac{3}{2}(3)^k\, u(k)$$

so that $$f(k) = \frac{1}{2}(1)^k\, u(k) - 2\,(2)^k\, u(k) + \frac{3}{2}\,(3)^k\, u(k).$$

5.5 THE z-TRANSFER FUNCTION (PULSE TRANSFER FUNCTION)

Figure 5.2 shows a **linear discrete-time system** wherein the output signal $y(k)$ of the system is given by the **convolution sum** between the input signal $x(k)$ and the **impulse response function** $g(k)$. We may write

$$y(k) = x(k) * g(k) \qquad \text{...(5.31)}$$

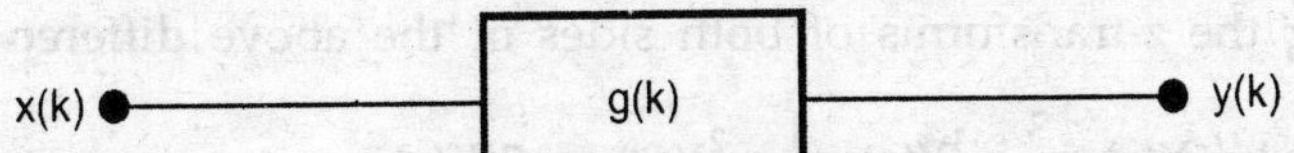

x(k) → g(k) → y(k)

FIGURE 5.2 *Linear discrete-time system.*

Using **convolution property** of z-transform, we have

$$Y(z) = X(z)\ G(z) \qquad ...(5.32)$$

This is an important result wherein $G(z)$ is interpreted as the **z-transfer function** or **pulse transfer function** of the **linear discrete-time system** as shown in Figure (5.3).

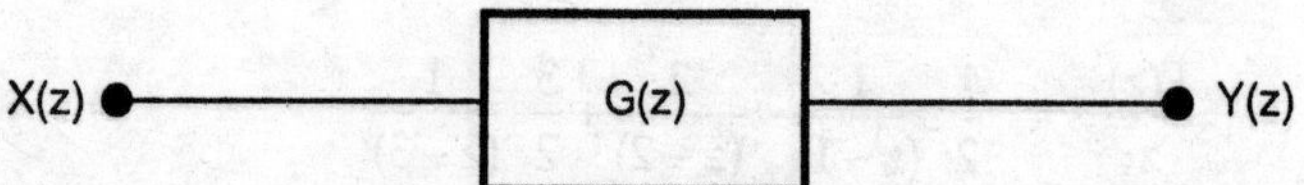

FIGURE 5.3 *Block-diagram for a pulse transfer function system.*

Thus, from equation (5.32), the **pulse transfer function** is expressed as

$$G(z) = \frac{Y(z)}{X(z)} \qquad ...(5.33)$$

It is to be noted that $G(z)$ is the z-transform of the impulse response function $g(k)$ of an **initially relaxed system,** *i.e.,* a system with **zero initial conditions.**

As seen from equation (5.33), we can define the z-transfer function or pulse transfer function as the ratio of the z-transform of the output to the z-transform of the input, keeping all initial condition zero, *i.e.,*

$$G(z) = \left.\frac{\text{z-transform of output}}{\text{z-transform of input}}\right|_{\text{Initial conditions}=0}$$

or,

$$G(z) = \left.\frac{Y(z)}{X(z)}\right|_{\text{Initial conditions}=0} \qquad ...(5.34)$$

Hence, the z-transfer function is a **mathematical model of an initially relaxed system.** From the knowledge of this model the output of the system to any input can be obtained by using the relation (5.34) and then taking the inverse z-transform to obtain $y(k)$.

The **z-transfer function** has the same **important role** in **z-transform analysis** of **discrete-time systems** as the **transfer function** in the **Laplace transform analysis** of **continuous-time systems.**

Example 5.12 *Find the impulse response function g(k) of an LTI system described by the following difference equation.*

$$y(k) - \frac{3}{4}y(k-1) + \frac{1}{8}y(k-2) = 2\,x(k)$$

What will be the response y(k) to the input

$$x(k) = \left(\frac{1}{4}\right)^k u(k)$$

Solution. Taking the z-transforms of both sides of the above difference equation, we get

$$Y(z) - \frac{3}{4}z^{-1}Y(z) + \frac{1}{8}z^{-2}Y(z) = 2\,X(z) \qquad \text{[Property 5.5]}$$

or,

$$z^2Y(z) - \frac{3}{4}zY(z) + \frac{1}{8}Y(z) = 2z^2X(z)$$

or,
$$\frac{Y(z)}{X(z)} = \frac{2z^2}{z^2 - \frac{3}{4}z + \frac{1}{8}}$$

or,
$$G(z) = \frac{2z^2}{\left(z-\frac{1}{2}\right)\left(z-\frac{1}{4}\right)}$$

or,
$$\frac{G(z)}{z} = \frac{2z}{\left(z-\frac{1}{2}\right)\left(z-\frac{1}{4}\right)}$$

or,
$$\frac{G(z)}{z} = \left.\frac{2z}{\left(z-\frac{1}{4}\right)}\right|_{z=\frac{1}{2}} \frac{1}{\left(z-\frac{1}{2}\right)} + \left.\frac{2z}{\left(z-\frac{1}{2}\right)}\right|_{z=\frac{1}{4}} \frac{1}{\left(z-\frac{1}{4}\right)}$$

or,
$$\frac{G(z)}{z} = \frac{4}{\left(z-\frac{1}{2}\right)} - \frac{2}{\left(z-\frac{1}{4}\right)}$$

or,
$$G(z) = \frac{4z}{\left(z-\frac{1}{2}\right)} - \frac{2z}{\left(z-\frac{1}{4}\right)}$$

In order to determine the inverse z-transform of each of these individual terms, we **must specify the ROC** associated with each.

Thus,
$$g(k) = g_1(k) - g_2(k)$$

where,
$$g_1(k) \overset{\mathcal{Z}}{\longleftrightarrow} \frac{4z}{\left(z-\frac{1}{2}\right)} \quad ; \quad |z| > \frac{1}{2}$$

$$g_2(k) \overset{\mathcal{Z}}{\longleftrightarrow} \frac{2z}{\left(z-\frac{1}{4}\right)} \quad ; \quad |z| > \frac{1}{4}$$

By inspection, we can identify

$$g_1(k) = 4\left(\frac{1}{2}\right)^k u(k)$$

and
$$g_2(k) = 2\left(\frac{1}{4}\right)^k u(k)$$

so that
$$g(k) = 4\left(\frac{1}{2}\right)^k u(k) - 2\left(\frac{1}{4}\right)^k u(k)$$

Now, the input is given as

$$x(k) = \left(\frac{1}{4}\right)^k u(k)$$

Taking the z-transforms of both sides, we have

$$X(z) = \frac{z}{\left(z-\frac{1}{4}\right)}$$

Pulse transfer function is defined as

$$G(z) = \left.\frac{Y(z)}{X(z)}\right|_{\text{Initial conditions}=0}$$

or,
$$Y(z) = G(z)\ X(z)$$

or,
$$Y(z) = \left[\frac{4z}{\left(z-\frac{1}{2}\right)} - \frac{2z}{\left(z-\frac{1}{4}\right)}\right]\frac{z}{\left(z-\frac{1}{4}\right)}$$

or,
$$\frac{Y(z)}{z} = \frac{4z}{\left(z-\frac{1}{2}\right)\left(z-\frac{1}{4}\right)} - \frac{2z}{\left(z-\frac{1}{4}\right)^2}$$

or,
$$\frac{Y(z)}{z} = \left.\frac{4z}{\left(z-\frac{1}{4}\right)}\right|_{z=\frac{1}{2}}\frac{1}{\left(z-\frac{1}{2}\right)} + \left.\frac{4z}{\left(z-\frac{1}{2}\right)}\right|_{z=\frac{1}{4}}\frac{1}{\left(z-\frac{1}{4}\right)} - \frac{2z}{\left(z-\frac{1}{4}\right)^2}$$

or,
$$\frac{Y(z)}{z} = \frac{8}{\left(z-\frac{1}{2}\right)} - \frac{4}{\left(z-\frac{1}{4}\right)} - \frac{2z}{\left(z-\frac{1}{4}\right)^2}$$

or
$$Y(z) = \frac{8z}{\left(z-\frac{1}{2}\right)} - \frac{4z}{\left(z-\frac{1}{4}\right)} - \frac{2z^2}{\left(z-\frac{1}{4}\right)^2}$$

In order to determine the inverse z-transform of each of these individual terms, we **must specify the ROC** associated with each.

Thus,
$$y(k) = y_1(k) - y_2(k) - y_3(k)$$

where,
$$y_1(k) \overset{\mathcal{Z}}{\longleftrightarrow} \frac{8z}{\left(z-\frac{1}{2}\right)} \quad ; \quad |z| > \frac{1}{2}$$

$$y_2(k) \overset{\mathcal{Z}}{\longleftrightarrow} \frac{4z}{\left(z-\frac{1}{4}\right)} \quad ; \quad |z| > \frac{1}{4}$$

$$y_3(k) \overset{\mathcal{Z}}{\longleftrightarrow} \frac{2z^2}{\left(z-\frac{1}{4}\right)^2} \quad ; \quad |z| > \frac{1}{4}$$

By inspection, we can identify

$$y_1(k) = 8\left(\frac{1}{2}\right)^k u(k)$$

$$y_2(k) = 4\left(\frac{1}{4}\right)^k u(k)$$

and

$$y_3(k) = 2\,(k+1)\left(\frac{1}{4}\right)^k u(k)$$

so that

$$y(k) = 8\left(\frac{1}{2}\right)^k u(k) - 4\left(\frac{1}{4}\right)^k u(k) - 2\,(k+1)\left(\frac{1}{4}\right)^k u(k)$$

or,

$$y(k) = 8\left(\frac{1}{2}\right)^k u(k) - 6\left(\frac{1}{4}\right)^k u(k) - 2\,k\left(\frac{1}{4}\right)^k u(k).$$

PROBLEMS

1. Determine the z-transform of the following
   (*a*) $f(k) = k^2$
   (*b*) $f(k) = k^2 a^{k-1}$
   (*c*) $f(k) = \sinh \beta k$
   (*d*) $f(k) = \cosh \beta k.$
2. Find the z-transform of the following continuous-time signals
   (*a*) $f(t) = e^{at}$
   (*b*) $f(t) = te^{-at}$
   (*c*) $f(t) = e^{-at} \sin \omega t$
   (*d*) $f(t) = 2u(t) + t.$
3. Find the inverse z-transform of the following
   (*a*) $F(z) = \dfrac{z}{(z-1)(z+0.85)}$
   (*b*) $F(z) = \dfrac{z^2}{(z-1)(z^2-5z+6)}$
   (*c*) $F(z) = \dfrac{z^2}{(z-1)(z^2-1.6z+1.64)}.$
4. Find the z-domain transfer function of the following s-domain transfer functions
   (*a*) $\dfrac{\omega}{s^2+\omega^2}$
   (*b*) $\dfrac{\omega}{(s+\alpha)^2+\omega^2}$
   (*c*) $\dfrac{s}{s^2+\omega^2}$
   (*d*) $\dfrac{s+\alpha}{(s+\alpha)^2+\omega^2}.$
5. Determine the z-transfer function of two cascaded systems each described by the difference equation

$$y(k) = 0.5\,y(k-1) + u(k).$$

6. Solve the following difference equation by means of the z-transform

$$f(k+2) - f(k) = 0; f(0) = 1, f(1) = 1.$$

7. Solve the difference equation

$$y(k+2) + 3y(k+1) + 2y(k) = u(k) \; ; \; y(0) = 1$$
$$y(k) = 0 \quad for \quad k < 0.$$

[**Hint:** $y(1)$ needed in the solution which can be obtained by putting $k = -1$ in the above difference equation, *viz.*

$$y(1) + 3y(0) + 2y(-1) = u(-1)$$

or, $$y(1) = -3.$$

8. Find the unit-step time response for the pulse transfer function

$$\frac{Y(z)}{U(z)} = \frac{z}{z^2 - z + 0.5}.$$

Chapter 6

State-Variable Analysis

6.1 INTRODUCTION

The methods of analysis and design of physical systems require the system to be modelled in the form of a transfer function. Though, the transfer function approach of analysis is simple and powerful technique, it suffers from certain drawbacks :

(*a*) Transfer function model is only applicable to linear time-invariant systems.

(*b*) A transfer function is only defined under zero initial conditions; meaning that the system is initially **at rest** and the time-solution is in a general form due to input only. However, for multi-input-multi-output; the systems initially do not at rest. Hence, as a conclusion, the transfer function model is generally restricted to single-input-single-output (SISO) systems.

(*c*) Transfer function technique only reveals the system output for a given input and do not provide any information regarding the **internal state** of the system.

Thus, the methods of analysis and design of physical systems, based on the transfer function approach are **inadequate** and not convenient.

The limitations listed above made us to feel the need of a more general mathematical representation of a physical system which along with the output, yields information about the state of the system-variables at some predetermined points along the flow of signals. This lead to the development of the **state variable approach** which has the following advantages over **transfer function approach:**

(*a*) It is a direct **time-domain approach.** Thus, this approach is suitable for digital computer computations.

(*b*) It is a very powerful technique for the design and analysis of linear or non-linear, time-variant or time-invariant and **SISO** or **MIMO** systems.

(c) In this technique the nth order differential equations can be expressed as 'n' equations of first order. Thus, making the solutions easier.

(d) Using this approach, the systems can be designed for optimal conditions with respect to given **performance indices.**

6.2 STATE-VARIABLE TERMINOLOGY

6.2.1 State

The **state** of a dynamical system is a minimal set of variables $x_1(t)$, $x_2(t)$, $x_3(t)$ ... $x_n(t)$ such that the knowledge of these variables at $t = t_0$ (initial condition) together with the knowledge of inputs $u_1(t)$, $u_2(t)$, $u_3(t)$ .... $u_m(t)$ for $t \geq t_0$ completely determines the behaviour of the system for $t > t_0$.

6.2.2 State-Variables

The variables $x_1(t)$, $x_2(t)$, $x_3(t)$ ... $x_n(t)$ such that the knowledge of these variables at $t = t_0$ (initial condition) together with the knowledge of inputs $u_1(t)$, $u_2(t)$, $u_3(t)$ ..... $u_m(t)$ for $t \geq t_0$ completely determines the behaviour of the system for $t > t_0$ are called **state-variables.**

In other words—the variables which determine the **state** of a dynamical system, are called **state-variables.**

6.2.3 State-Vector

If n state-variables $x_1(t)$, $x_2(t)$, $x_3(t)$ ... $x_n(t)$ are necessary to determine the behaviour of a dynamical system, then these n state-variables can be considered as n components of a vector $\mathbf{x}(t)$, called **state-vector.**

6.2.4 State-Space

The n dimensional space, whose elements are the n state-variables, is called **state-space.** Any state can be represented by a **point** in the state-space.

6.3 BLOCK-DIAGRAM REPRESENTATION OF A GENERAL CONTROL SYSTEM

In state-variable formulation of a system, the state-variables are usually represented by $x_1(t)$, $x_2(t)$, $x_3(t)$ ...; the inputs by $u_1(t)$, $u_2(t)$, $u_3(t)$...; and the outputs by $y_1(t)$, $y_2(t)$, $y_3(t)$.... . The state-space representation of a system which has m inputs, p outputs and n state-variables may be visualized in block-diagram form as shown in Figure 6.1. For notational economy, the different variables may be represented by the input vector $\mathbf{u}(t)$, output vector $\mathbf{y}(t)$ and the state-vector $\mathbf{x}(t)$, where

$$\mathbf{u}(t) = \begin{bmatrix} u_1(t) \\ u_2(t) \\ \vdots \\ u_m(t) \end{bmatrix}_{m \times 1} ; \ \mathbf{y}(t) = \begin{bmatrix} y_1(t) \\ y_2(t) \\ \vdots \\ y_p(t) \end{bmatrix}_{p \times 1} ; \ \mathbf{x}(t) = \begin{bmatrix} x_1(t) \\ x_2(t) \\ \vdots \\ x_n(t) \end{bmatrix}_{n \times 1}$$

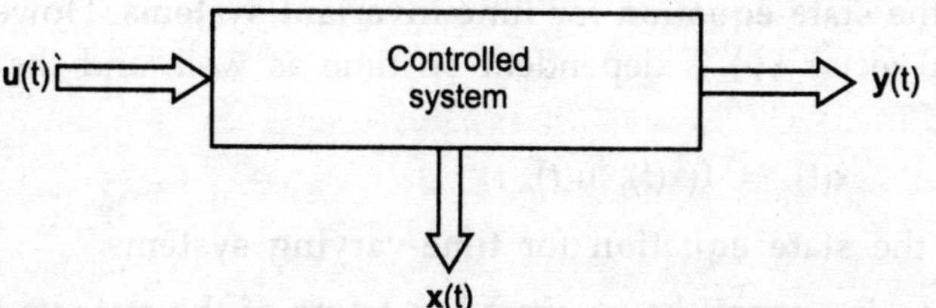

FIGURE 6.1 *Block-diagram representation of a General Control System.*

6.4 STATE MODEL OF GENERAL CONTROL SYSTEM

For a general system of Figure 6.1, the state representation can be arranged in the form of n first order differential equation as

$$\left.\begin{aligned} \dot{x}_1(t) &= f_1(x_1(t), x_2(t)...x_n(t); u_1(t), u_2(t)...u_m(t)) \\ \dot{x}_2(t) &= f_2(x_1(t), x_2(t)...x_n(t); u_1(t), u_2(t)...u_m(t)) \\ &\vdots \qquad\qquad \vdots \\ \dot{x}_n(t) &= f_n(x_1(t), x_2(t)...x_n(t); u_1(t), u_2(t)...u_m(t)) \end{aligned}\right\} \qquad ...(6.1)$$

Integrating (6.1), we have

$$x_i(t) = x_i(t_0) + \int_{t_0}^{t} f_i(x_1(\tau), x_2(\tau), ...x_n(\tau); u_1(\tau), u_2(\tau)...u_m(\tau))d\tau$$

$$; i = 1, 2, 3, ... n.$$

Thus, the n state-variables and hence, the state of the system can uniquely be determined at any $t > t_0$, provided each state-variable is known at $t = t_0$ and all the m control forces are known throughout the interval t_0 to t.

The n differential equation of (6.1) may be written in vector form as

$$\dot{\mathbf{x}}(t) = \mathbf{f}(\mathbf{x}(t), \mathbf{u}(t)) \qquad ...(6.2)$$

where, $$\mathbf{x}(t) = \begin{bmatrix} x_1(t) \\ x_2(t) \\ \vdots \\ x_n(t) \end{bmatrix}_{n\times 1} \quad ; \text{State-Vector}$$

$$\mathbf{u}(t) = \begin{bmatrix} u_1(t) \\ u_2(t) \\ \vdots \\ u_m(t) \end{bmatrix}_{m\times 1} \quad ; \text{Input Vector}$$

and, $$\mathbf{f}(\cdot) = \begin{bmatrix} f_1(\cdot) \\ f_2(\cdot) \\ \vdots \\ f_n(\cdot) \end{bmatrix}_{n\times 1} \quad ; \text{Function Vector}$$

Equation (6.2) is the **state equation** for **time-invariant systems.** However, for **time-varying systems,** the function vector $\mathbf{f}(\cdot)$ is dependent on time as well, and the vector equation may be given as

$$\dot{\mathbf{x}}(t) = \mathbf{f}(\mathbf{x}(t), \mathbf{u}(t), t) \qquad ...(6.3)$$

Equation (6.3) is the **state equation** for **time-varying systems.**

The output $\mathbf{y}(t)$ can in general be expressed in terms of the state-vector $\mathbf{x}(t)$ and the input vector $\mathbf{u}(t)$ as :

For **time-invariant systems:**

$$\mathbf{y}(t) = \mathbf{g}(\mathbf{x}(t), \mathbf{u}(t)) \qquad ...(6.4)$$

For **time-varying systems:**

$$\mathbf{y}(t) = \mathbf{g}(\mathbf{x}(t), \mathbf{u}(t), t) \qquad ...(6.5)$$

where, $$\mathbf{y}(t) = \begin{bmatrix} y_1(t) \\ y_2(t) \\ \vdots \\ y_p(t) \end{bmatrix}_{p\times 1} \quad ; \quad \text{Output Vector}$$

$$\mathbf{u}(t) = \begin{bmatrix} u_1(t) \\ u_2(t) \\ \vdots \\ u_m(t) \end{bmatrix}_{m\times 1} \quad ; \quad \text{Input Vector}$$

and, $$\mathbf{g}(\cdot) = \begin{bmatrix} g_1(\cdot) \\ g_2(\cdot) \\ \vdots \\ g_p(\cdot) \end{bmatrix}_{p\times 1} \quad ; \quad \text{Function Vector}$$

Equations (6.4) and (6.5) are the **output equations** for **time-invariant** and **time-varying systems** respectively.

The state equations and output equations together constitute the **state-model** of the system. Thus, the state-model of a general control system (shown in Figure 6.1) is given by the following equations :

For **time-invariant systems:**

$$\dot{\mathbf{x}}(t) = \mathbf{f}(\mathbf{x}(t), \mathbf{u}(t)) \quad ; \quad \text{State Equation} \qquad ...(6.6\ A)$$

$$\mathbf{y}(t) = \mathbf{g}(\mathbf{x}(t), \mathbf{u}(t)) \quad ; \quad \text{Output Equation} \qquad ...(6.6\ B)$$

For **time-varying systems:**

$$\dot{\mathbf{x}}(t) = \mathbf{f}(\mathbf{x}(t), \mathbf{u}(t), t) \quad ; \quad \text{State Equation} \qquad ...\ (6.7\ A)$$

$$\mathbf{y}(t) = \mathbf{g}(\mathbf{x}(t), \mathbf{u}(t), t) \quad ; \quad \text{Output Equation} \qquad ...\ (6.7\ B)$$

6.5 STATE-MODEL OF A LINEAR MULTI-INPUT-MULTI-OUTPUT SYSTEM

The state model of a linear time-invariant MIMO system is a special case of the general time-invariant state-model of equation (6.6). In state representation the derivative of each state-variables can be written as a linear combination of system states and inputs, *i.e.*,

$$\left.\begin{aligned}\dot{x}_1(t) &= a_{11}x_1(t)+a_{12}x_2(t)+\ldots+a_{1n}x_n(t)+b_{11}u_1(t)+b_{12}u_2(t)+\ldots+b_{1m}u_m(t)\\ \dot{x}_2(t) &= a_{21}x_1(t)+a_{22}x_2(t)+\ldots+a_{2n}x_n(t)+b_{21}u_1(t)+b_{22}u_2(t)+\ldots+b_{2m}u_m(t)\\ &\vdots\\ \dot{x}_n(t) &= a_{n1}x_1(t)+a_{n2}x_2(t)+\ldots+a_{nn}x_n(t)+b_{n1}u_1(t)+b_{n2}u_2(t)+\ldots+b_{nm}u_m(t)\end{aligned}\right\} \quad \ldots(6.8)$$

where coefficients $a_{ij}(i = 1, 2, \ldots n;\ j = 1, 2, \ldots n)$ and $b_{ik}(i = 1, 2, \ldots n;\ k = 1, 2, \ldots m)$ are constants. Equations (6.8) may be written in vector-matrix form as

$$\dot{\mathbf{x}}(t) = \mathbf{A}\mathbf{x}(t) + \mathbf{B}\mathbf{u}(t) \quad ; \text{ State Equation}$$

where,

$$\mathbf{x}(t) = \begin{bmatrix} x_1(t) \\ x_2(t) \\ \vdots \\ x_n(t) \end{bmatrix}_{n\times 1} \quad ; \text{ State Vector}$$

$$\mathbf{u}(t) = \begin{bmatrix} u_1(t) \\ u_2(t) \\ \vdots \\ u_m(t) \end{bmatrix}_{m\times 1} \quad ; \text{ Input Vector}$$

$$\mathbf{A} = \begin{bmatrix} a_{11} & a_{12} & \cdots & a_{1n} \\ a_{21} & a_{22} & \cdots & a_{2n} \\ \vdots & \vdots & & \vdots \\ a_{n1} & a_{n2} & \cdots & a_{nn} \end{bmatrix}_{n\times n} \quad ; \text{ System Matrix}$$

and,

$$\mathbf{B} = \begin{bmatrix} b_{11} & b_{12} & \cdots & b_{1m} \\ b_{21} & b_{22} & \cdots & b_{2m} \\ \vdots & \vdots & & \vdots \\ b_{n1} & b_{n2} & \cdots & b_{nm} \end{bmatrix}_{n\times m} \quad ; \text{ Input Matrix}$$

Similarly, the output variables can be written as a linear combination of system states and inputs, *i.e.*,

$$\left.\begin{aligned} y_1(t) &= c_{11}x_1(t)+c_{12}x_2(t)+\ldots+c_{1n}x_n(t)+d_{11}u_1(t)+d_{12}u_2(t)+\ldots+d_{1m}u_m(t) \\ y_2(t) &= c_{21}x_1(t)+c_{22}x_2(t)+\ldots+c_{2n}x_n(t)+d_{21}u_1(t)+d_{22}u_2(t)+\ldots+d_{2m}u_m(t) \\ &\vdots \\ y_p(t) &= c_{p1}x_1(t)+c_{p2}x_2(t)+\ldots+c_{pn}x_n(t)+d_{p1}u_1(t)+d_{p2}u_2(t)+\ldots+d_{pm}u_m(t) \end{aligned}\right\} \quad \ldots(6.9)$$

where, coefficients $c_{ij}(i = 1, 2, \ldots p;\ j = 1, 2, \ldots n)$ and $d_{ik}(i = 1, 2, \ldots p;\ k = 1, 2, \ldots, m)$ are constants. The equation (6.9) may be written in vector-matrix form as

$$\mathbf{y}(t) = \mathbf{Cx}(t) + \mathbf{Du}(t) \quad ; \quad \text{Output Equation}$$

where,

$$\mathbf{y}(t) = \begin{bmatrix} y_1(t) \\ y_2(t) \\ \vdots \\ y_p(t) \end{bmatrix}_{p\times 1} \quad ; \quad \text{Output Vector}$$

$$\mathbf{u}(t) = \begin{bmatrix} u_1(t) \\ u_2(t) \\ \vdots \\ u_m(t) \end{bmatrix}_{m\times 1} \quad ; \quad \text{Input Vector}$$

$$\mathbf{C} = \begin{bmatrix} c_{11} & c_{12} & \cdots & c_{1n} \\ c_{21} & c_{22} & \cdots & c_{2n} \\ \vdots & \vdots & & \vdots \\ c_{p1} & c_{p2} & \cdots & c_{pn} \end{bmatrix}_{p\times n} \quad ; \quad \text{Output Matrix}$$

and

$$\mathbf{D} = \begin{bmatrix} d_{11} & d_{12} & \cdots & d_{1m} \\ d_{21} & d_{22} & \cdots & d_{2m} \\ \vdots & \vdots & & \vdots \\ d_{p1} & d_{p2} & \cdots & d_{pm} \end{bmatrix}_{p\times m} \quad ; \quad \text{Transmission Matrix}$$

As the state equation and output equation together constitute the state model of the system. Thus, the state-model of a linear time-invariant MIMO system is given as

$$\dot{\mathbf{x}}(t) = \mathbf{Ax}(t) + \mathbf{Bu}(t) \quad ; \quad \text{State Equation} \quad \ldots(6.10\text{ A})$$

$$\mathbf{y}(t) = \mathbf{Cx}(t) + \mathbf{Du}(t) \quad ; \quad \text{Output Equation} \quad \ldots(6.10\text{ B})$$

The block-diagram representation of the state-model of linear multi-input-multi-output system is shown in Figure 6.2.

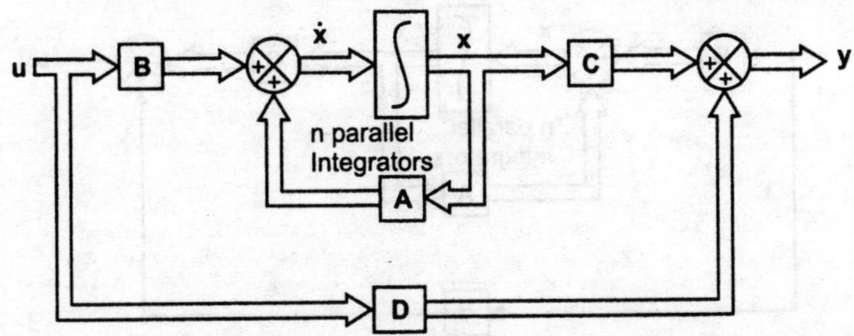

FIGURE 6.2 *Block-diagram representation of the state-model of linear multi-input-multi-output system.*

6.6 STATE-MODEL OF A LINEAR SINGLE-INPUT-SINGLE-OUTPUT SYSTEM

The state-model of a linear single-input-single-output system can be obtained by putting $m = 1$ and $p = 1$ in the state-model of a linear multi-input-multi-output system as

$$\dot{\mathbf{x}}(t) = \mathbf{A}\mathbf{x}(t) + \mathbf{B}u(t) \quad ; \quad \text{State Equation} \quad \text{...(6.11 A)}$$

$$y(t) = \mathbf{C}\mathbf{x}(t) + du(t) \quad ; \quad \text{Output Equation} \quad \text{...(6.11 B)}$$

where,
$$\mathbf{x}(t) = \begin{bmatrix} x_1(t) \\ x_2(t) \\ \vdots \\ x_n(t) \end{bmatrix}_{n\times 1} \quad ; \quad \text{State Vector}$$

$$\mathbf{A} = \begin{bmatrix} a_{11} & a_{12} & \cdots & a_{1n} \\ a_{21} & a_{22} & \cdots & a_{2n} \\ \vdots & \vdots & \ddots & \vdots \\ a_{n1} & a_{n2} & \cdots & a_{nn} \end{bmatrix}_{n\times n} \quad ; \quad \text{System Matrix}$$

$$\mathbf{B} = \begin{bmatrix} b_1 \\ b_2 \\ \vdots \\ b_n \end{bmatrix}_{n\times 1} \quad ; \quad \text{Input Matrix}$$

$$\mathbf{C} = [c_1 \quad c_2 \quad c_3 \quad \dots \quad c_n]_{1\times n} \quad ; \quad \text{Output Matrix}$$

$$d = \text{Transmission Constant}$$

$$u(t) = \text{Input or Control Variable (Scalar)}$$

and,
$$y(t) = \text{Output Variable (Scalar)}$$

The block-diagram representation of the state-model of linear single-input-single-output system is shown in Figure 6.3.

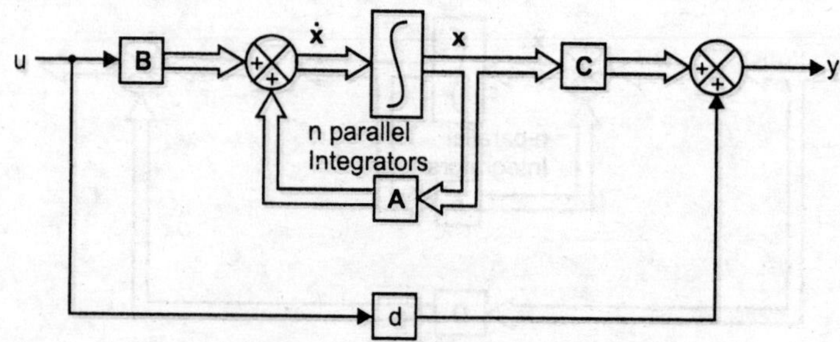

FIGURE 6.3 *Block-diagram representation of the state-model of linear single-input-single-output system.*

6.7 HOMOGENEOUS AND NON-HOMOGENEOUS LINEAR TIME-INVARIANT SYSTEMS

The state equation of a linear time-invariant system is given by

$$\dot{\mathbf{x}}(t) = \mathbf{A}\mathbf{x}(t) + \mathbf{B}\mathbf{u}(t) \qquad \text{...(6.12)}$$

Case 1: If **A** is a constant matrix and **u**(*t*) is a **zero vector,** *i.e.,* no control forces are applied to the system, then the equation (6.12) represents a **homogeneous** linear time-invariant system.

Case 2: If **A** is a constant matrix and **u**(*t*) is **non-zero vector,** *i.e.,* control forces are applied to the system, then the equation (6.12) represents a **non-homogeneous** linear time-invariant system.

6.8 SOLUTION OF STATE EQUATION FOR LINEAR TIME-INVARIANT SYSTEM

6.8.1 State Transition Matrix

The state equation of a linear time-invariant system is given by

$$\dot{\mathbf{x}}(t) = \mathbf{A}\mathbf{x}(t) + \mathbf{B}\mathbf{u}(t)$$

For homogeneous (unforced) system

$$\mathbf{u}(t) = 0$$

We have,

$$\dot{\mathbf{x}}(t) = \mathbf{A}\mathbf{x}(t) \qquad \text{...(6.13)}$$

Take the Laplace Transform on both sides

$$s\mathbf{X}(s) - \mathbf{x}(0) = \mathbf{A}\mathbf{X}(s) \qquad \text{...(6.14)}$$

Equation (6.14) may also be written as

$$s\mathbf{I}\mathbf{X}(s) - \mathbf{x}(0) = \mathbf{A}\mathbf{X}(s)$$

or,

$$[s\mathbf{I} - \mathbf{A}]\mathbf{X}(s) = \mathbf{x}(0)$$

or,

$$\mathbf{X}(s) = [s\mathbf{I} - \mathbf{A}]^{-1}\mathbf{x}(0)$$

Take Inverse Laplace transform

$$\mathbf{x}(t) = e^{\mathbf{A}t}\mathbf{x}(0) \qquad ...(6.15)$$

Equation (6.15) gives the **solution** of the **LTI homogeneous state equation (6.13).** from equation (6.15) it is observed that the initial state $\mathbf{x}(0)$ at $t = 0$, is driven to a state $\mathbf{x}(t)$ at time t. This transition in state is carried out by the matrix exponential $e^{\mathbf{A}t}$. Because of this property, $e^{\mathbf{A}t}$ is termed as **State Transition Matrix** and is denoted by $\phi(t)$.

Thus, $$\phi(t) = e^{\mathbf{A}t} = \mathscr{L}^{-1}\{[s\mathbf{I} - \mathbf{A}]^{-1}\} = \mathscr{L}^{-1}\Phi(s)$$

where, $\Phi(s) = [s\mathbf{I} - \mathbf{A}]^{-1}$ is called **Resolvant Matrix.**

As $e^{\mathbf{A}t}$ represents a power series of the matrix $\mathbf{A}t$, thus,

$$\phi(t) = e^{\mathbf{A}t} = \mathbf{I} + \mathbf{A}t + \frac{\mathbf{A}^2t^2}{2!} + ...$$

6.8.1.1 Properties of the State Transition Matrix

1. $$\phi(0) = \mathbf{I}$$

Proof. $\phi(t) = e^{\mathbf{A}t}$

Put $t = 0$

We have, $\phi(0) = \mathbf{I}$.

2. $$\phi^{-1}(t) = \phi(-t)$$

Proof. $\phi(t) = e^{\mathbf{A}t}$

Post multiply both the sides by $e^{-\mathbf{A}t}$

We have, $\phi(t)e^{-\mathbf{A}t} = e^{\mathbf{A}t}\cdot e^{-\mathbf{A}t}$

or, $\phi(t)\,\phi(-t) = \mathbf{I}$

Premultiply both the sides by $\phi^{-1}(t)$

We have,

$$\phi^{-1}(t)\,\phi(t)\,\phi(-t) = \phi^{-1}(t)$$

or, $\phi(-t) = \phi^{-1}(t)$

or, $\phi^{-1}(t) = \phi(-t)$

An interesting result from this property of STM is that Equation (6.15) can be rearranged as,

$$\mathbf{x}(0) = \phi(-t)\,\mathbf{x}(t) \qquad ...(6.16)$$

which clearly means that the transition in time can take place in either direction **(bilateral in time).**

3. $$\phi(t_1 + t_2) = \phi(t_1)\,\phi(t_2) = \phi(t_2)\,\phi(t_1)$$

Proof. $$\phi(t_1 + t_2) = e^{\mathbf{A}(t_1+t_2)}$$

$$= e^{\mathbf{A}t_1}\cdot e^{\mathbf{A}t_2}$$

$$= \phi(t_1)\phi(t_2)$$

Again, $\phi(t_1 + t_2) = e^{\mathbf{A}(t_1+t_2)}$

$= e^{\mathbf{A}(t_2+t_1)}$

$= e^{\mathbf{A}t_2} \cdot e^{\mathbf{A}t_1}$

$= \phi(t_2)\,\phi(t_1).$

4. $[\phi(t)]^n = \phi(nt);\ n = \text{integer}$

Proof. $[\phi(t)]^n = e^{\mathbf{A}t}\,e^{\mathbf{A}t}\,e^{\mathbf{A}t}$ ...upto n terms

$= e^{\mathbf{A}nt}$

$= \phi(nt).$

5. $\dot{\phi}(t) = \mathbf{A}\phi(t)$

Proof. $\dot{\phi}(t) = \dfrac{d}{dt}e^{\mathbf{A}t}$

$= \mathbf{A}e^{\mathbf{A}t} = \mathbf{A}\phi(t).$

6. $\phi(t_2 - t_1)\,\phi(t_1 - t_0) = \phi(t_2 - t_0) = \phi(t_1 - t_0)\,\phi(t_2 - t_1)$

Proof.

$\phi(t_2 - t_1)\,\phi(t_1 - t_0) = e^{\mathbf{A}(t_2-t_1)}\,e^{\mathbf{A}(t_1-t_0)}$

$= e^{\mathbf{A}(t_2-t_1+t_1-t_0)}$

$= e^{\mathbf{A}(t_2-t_0)}$

$= \phi(t_2 - t_0)$

Similarly,

$\phi(t_1 - t_0)\,\phi(t_2 - t_1) = \phi(t_2 - t_0)$

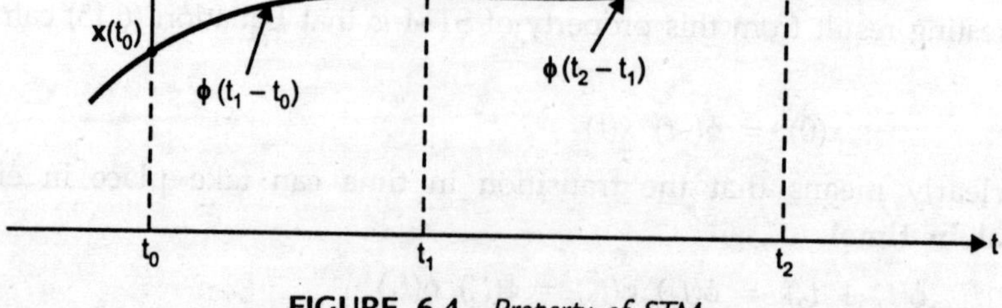

FIGURE 6.4 *Property of STM.*

This is a very important property of State Transition Matrix (STM) as it implies that a **state transition process** can be divided into a number of **sequential transitions.**

Figure 6.4 illustrates that the transition from $t = t_0$ to $t = t_2$ is equal to the transition from t_0 to t_1 and then from t_1 to t_2. In general, of course, the transition process can be broken up into any number of parts.

6.8.2 State Transition Equation

The state transition equation is defined as the **solution** of the linear non-homogeneous state equation (forced system).

Consider the state equation of a linear time-invariant system

$$\dot{\mathbf{x}}(t) = \mathbf{A}\mathbf{x}(t) + \mathbf{B}\mathbf{u}(t) \qquad ...(6.17)$$

For non-homogeneous (forced) system $\mathbf{u}(t)$ is non-zero vector, *i.e.*, control forces are applied to the system. Thus, equation (6.17) represents a non-homogeneous linear time-invariant system.

The solution of equation (6.17) can be obtained by considering two cases as follows:

Case 1: When initial state is known at $t = 0$

Rewrite equation (6.17) in the form,

$$\dot{\mathbf{x}}(t) - \mathbf{A}\mathbf{x}(t) = \mathbf{B}\mathbf{u}(t)$$

Multiplying both sides by $e^{-\mathbf{A}t}$, we have

$$e^{-\mathbf{A}t}[\dot{\mathbf{x}}(t) - \mathbf{A}\mathbf{x}(t)] = e^{-\mathbf{A}t}\mathbf{B}\mathbf{u}(t)$$

or,
$$\frac{d}{dt}[e^{-\mathbf{A}t}\mathbf{x}(t)] = e^{-\mathbf{A}t}\mathbf{B}\mathbf{u}(t)$$

Integrate both sides with respect to t between the limits 0 and t,

We have
$$e^{-\mathbf{A}t}\mathbf{x}(t)\Big|_0^t = \int_0^t e^{-\mathbf{A}\tau}\,\mathbf{B}\mathbf{u}(\tau)d\tau$$

or,
$$e^{-\mathbf{A}t}\mathbf{x}(t) - \mathbf{x}(0) = \int_0^t e^{-\mathbf{A}\tau}\,\mathbf{B}\mathbf{u}(\tau)d\tau$$

Premultiply both sides by $e^{\mathbf{A}t}$, we have

$$\mathbf{x}(t) = e^{\mathbf{A}t}\,\mathbf{x}(0) + \int_0^t e^{\mathbf{A}(t-\tau)}\,\mathbf{B}\mathbf{u}(\tau)d\tau$$

or,
$$\mathbf{x}(t) = \underbrace{\phi(t)\,\mathbf{x}(0)}_{\text{Homogeneous solution}} + \underbrace{\int_0^t \phi(t-\tau)\,\mathbf{B}\mathbf{u}(\tau)d\tau}_{\text{Forced solution}} \qquad ...(6.18)$$

Case 2: When initial state is known at $t = t_0$

Put $t = t_0$ in equation (6.18) and solve for $\mathbf{x}(0)$

$$\mathbf{x}(t_0) = \phi(t_0)\,\mathbf{x}(0) + \int_0^{t_0} \phi(t_0-\tau)\,\mathbf{B}\mathbf{u}(\tau)d\tau$$

Multiplying both sides by $\phi^{-1}(t_0)$, we have

$$\phi^{-1}(t_0)\;\mathbf{x}(t_0) = \mathbf{x}(0) + \int_0^{t_0} \phi(-t_0)\,\phi(t_0-\tau)\,\mathbf{B}\mathbf{u}(\tau)d\tau \qquad \because\; [\phi^{-1}(t_0) = \phi(-t_0)]$$

or, $$\mathbf{x}(0) = \phi(-t_0)\,\mathbf{x}(t_0) - \int_0^{t_0} \phi(-t_0 + t_0 - \tau)\,\mathbf{Bu}(\tau)d\tau$$

or, $$\mathbf{x}(0) = \phi(-t_0)\,\mathbf{x}(t_0) - \int_0^{t_0} \phi(-\tau)\,\mathbf{Bu}(\tau)d\tau \qquad ...(6.19)$$

Put the value of $\mathbf{x}(0)$ from equation (6.19) in equation (6.18), we have

$$\mathbf{x}(t) = \phi(t)\left[\phi(-t_0)\mathbf{x}(t_0) - \int_0^{t_0} \phi(-\tau)\,\mathbf{Bu}(\tau)d\tau\right] + \int_0^{t} \phi(t-\tau)\,\mathbf{Bu}(\tau)d\tau$$

or, $$\mathbf{x}(t) = \phi(t-t_0)\,\mathbf{x}(t_0) - \int_0^{t_0} \phi(t-\tau)\,\mathbf{Bu}(\tau)d\tau + \int_0^{t} \phi(t-\tau)\,\mathbf{Bu}(\tau)d\tau$$

or, $$\mathbf{x}(t) = \phi(t-t_0)\,\mathbf{x}(t_0) + \int_{t_0}^{t} \phi(t-\tau)\,\mathbf{Bu}(\tau)d\tau \qquad ...(6.20)$$

Once the state transition equation (state response) is determined, the output vector can easily by expressed as

$$\mathbf{y}(t) = \mathbf{C}\phi(t-t_0)\,\mathbf{x}(t_0) + \int_{t_0}^{t} \mathbf{C}\phi(t-\tau)\,\mathbf{Bu}(\tau)d\tau + \mathbf{Du}(t) \qquad ...(6.21)$$

Example 6.1. *Compute STM, when*

$$\mathbf{A} = \begin{bmatrix} 0 & 1 \\ -2 & -3 \end{bmatrix}$$

Solution. $$\mathbf{A} = \begin{bmatrix} 0 & 1 \\ -2 & -3 \end{bmatrix}$$

$\therefore$ $$[s\mathbf{I} - \mathbf{A}] = \begin{bmatrix} s & 0 \\ 0 & s \end{bmatrix} - \begin{bmatrix} 0 & 1 \\ -2 & -3 \end{bmatrix}$$

$$= \begin{bmatrix} s & -1 \\ 2 & s+3 \end{bmatrix}$$

Now, $$\text{Adj } \mathbf{A} = \begin{bmatrix} s+3 & -2 \\ 1 & s \end{bmatrix}^T = \begin{bmatrix} s+3 & 1 \\ -2 & s \end{bmatrix}$$

$\therefore$ $$[s\mathbf{I} - \mathbf{A}]^{-1} = \frac{\text{Adj } \mathbf{A}}{|s\mathbf{I} - \mathbf{A}|}$$

$$= \frac{\begin{bmatrix} s+3 & 1 \\ -2 & s \end{bmatrix}}{\begin{vmatrix} s & -1 \\ 2 & s+3 \end{vmatrix}}$$

$$= \begin{bmatrix} \frac{(s+3)}{(s+1)(s+2)} & \frac{1}{(s+1)(s+2)} \\ \frac{-2}{(s+1)(s+2)} & \frac{s}{(s+1)(s+2)} \end{bmatrix}$$

$$= \begin{bmatrix} \left(\frac{2}{s+1} - \frac{1}{s+2}\right) & \left(\frac{1}{s+1} - \frac{1}{s+2}\right) \\ \left(\frac{-2}{s+1} + \frac{2}{s+2}\right) & \left(\frac{-1}{s+1} + \frac{2}{s+2}\right) \end{bmatrix}$$

$\therefore$ STM is given by

$$\boldsymbol{\phi}(t) = \mathscr{L}^{-1}[s\mathbf{I} - \mathbf{A}]^{-1}$$

or,

$$\boldsymbol{\phi}(t) = \begin{bmatrix} (2e^{-t} - e^{-2t}) & (e^{-t} - e^{-2t}) \\ (-2e^{-t} + 2e^{-2t}) & (-e^{-t} + 2e^{-2t}) \end{bmatrix}$$

Example 6.2. *A Linear time-invariant system is characterised by the homogeneous state equation.*

$$\begin{bmatrix} \dot{x}_1 \\ \dot{x}_2 \end{bmatrix} = \begin{bmatrix} 1 & 0 \\ 1 & 1 \end{bmatrix} \begin{bmatrix} x_1 \\ x_2 \end{bmatrix}$$

Find the time response of the system. Assume the initial state vector

$$\mathbf{x}(0) = \begin{bmatrix} 1 \\ 0 \end{bmatrix}$$

Solution. We have,

$$\mathbf{A} = \begin{bmatrix} 1 & 0 \\ 1 & 1 \end{bmatrix}$$

$$\therefore \quad [s\mathbf{I} - \mathbf{A}] = \begin{bmatrix} s & 0 \\ 0 & s \end{bmatrix} - \begin{bmatrix} 1 & 0 \\ 1 & 1 \end{bmatrix} = \begin{bmatrix} s-1 & 0 \\ -1 & s-1 \end{bmatrix}$$

Now,

$$\text{Adj } \mathbf{A} = \begin{bmatrix} s-1 & 1 \\ 0 & s-1 \end{bmatrix}^T$$

$$= \begin{bmatrix} s-1 & 0 \\ 1 & s-1 \end{bmatrix}$$

$$\therefore \quad [s\mathbf{I} - \mathbf{A}]^{-1} = \frac{\text{Adj } \mathbf{A}}{|s\mathbf{I} - \mathbf{A}|}$$

$$= \frac{\begin{bmatrix} s-1 & 0 \\ 1 & s-1 \end{bmatrix}}{\begin{vmatrix} s-1 & 0 \\ -1 & s-1 \end{vmatrix}} = \begin{bmatrix} \frac{1}{(s-1)} & 0 \\ \frac{1}{(s-1)^2} & \frac{1}{(s-1)} \end{bmatrix}$$

∴ STM is given by

$$\phi(t) = \mathscr{L}^{-1}[s\mathbf{I} - \mathbf{A}]^{-1}$$

$$= \begin{bmatrix} e^t & 0 \\ te^t & e^t \end{bmatrix}$$

Now, the time response of a linear time-invariant, homogeneous system is given by

$$\mathbf{x}(t) = \phi(t)\ \mathbf{x}(0)$$

$$= \begin{bmatrix} e^t & 0 \\ te^t & e^t \end{bmatrix}\begin{bmatrix} 1 \\ 0 \end{bmatrix} = \begin{bmatrix} e^t \\ te^t \end{bmatrix}$$

Example 6.3. *Obtain the time response of the following system,*

$$\begin{bmatrix} \dot{x}_1 \\ \dot{x}_2 \end{bmatrix} = \begin{bmatrix} 1 & 0 \\ 1 & 1 \end{bmatrix}\begin{bmatrix} x_1 \\ x_2 \end{bmatrix} + \begin{bmatrix} 1 \\ 1 \end{bmatrix} u$$

Where, $u(t)$ is a unit step occurring at $t = 0$ and $\mathbf{x}^T(0) = [1 \quad 0]$.

Solution. We have,

$$\mathbf{A} = \begin{bmatrix} 1 & 0 \\ 1 & 1 \end{bmatrix}, \quad \mathbf{B} = \begin{bmatrix} 1 \\ 1 \end{bmatrix}$$

The STM can be calculated as in Example 6.2.

We have, $\qquad \phi(t) = \begin{bmatrix} e^t & 0 \\ te^t & e^t \end{bmatrix}$

The time response of the linear time-invariant, non-homogeneous system is given by,

$$\mathbf{x}(t) = \phi(t)\,\mathbf{x}(0) + \int_0^t \phi(t-\tau)\,\mathbf{B}u(\tau)d\tau$$

$$= \phi(t)\left\{x(0) + \int_0^t \phi(-\tau)\,\mathbf{B}u(\tau)d\tau\right\} \qquad \text{...(6.22)}$$

Now, with $u(t) = 1;\ t > 0$

We may write

$$\phi(-\tau)\ \mathbf{B}u(\tau) = \begin{bmatrix} e^{-\tau} & 0 \\ -\tau e^{-\tau} & e^{-\tau} \end{bmatrix}\begin{bmatrix} 1 \\ 1 \end{bmatrix}$$

$$= \begin{bmatrix} e^{-\tau} \\ e^{-\tau}(1-\tau) \end{bmatrix}$$

$$\therefore \qquad \int_0^t \phi(-\tau)\,\mathbf{B}u(\tau)d\tau = \begin{bmatrix} -e^{-\tau} \\ \tau e^{-\tau} \end{bmatrix}_0^t = \begin{bmatrix} 1-e^{-t} \\ te^{-t} \end{bmatrix}$$

Now, from equation (6.22) we get

$$\mathbf{x}(t) = \begin{bmatrix} e^t & 0 \\ te^t & e^t \end{bmatrix} \left\{ \begin{bmatrix} 1 \\ 0 \end{bmatrix} + \begin{bmatrix} 1-e^{-t} \\ te^{-t} \end{bmatrix} \right\}$$

$$= \begin{bmatrix} e^t \\ te^t \end{bmatrix} + \begin{bmatrix} e^{-t}-1 \\ te^t \end{bmatrix} = \begin{bmatrix} 2e^t - 1 \\ 2te^t \end{bmatrix}$$

6.9 TRANSFER MATRIX FROM STATE-MODEL

The concept of the transfer matrix is an extension of that of the transfer function. Here we shall first obtain transfer function of linear single-input-single output control system and then transfer matrix of linear multiple-input-multiple-output control system using respective state and output equations.

6.9.1 Transfer Function

Consider the state-model of a linear single-input-single-output system

$$\dot{\mathbf{x}}(t) = \mathbf{Ax}(t) + \mathbf{B}u(t) \quad ; \quad \text{State Equation} \qquad \text{...(6.23 A)}$$

$$y(t) = \mathbf{Cx}(t) + du(t) \quad ; \quad \text{Output Equation} \qquad \text{...(6.23 B)}$$

Taking the Lapales transform of equations (6.23), we have

$$s\mathbf{X}(s) - \mathbf{x}(0) = \mathbf{AX}(s) + \mathbf{B}U(s) \qquad \text{...(6.24 A)}$$

$$Y(s) = \mathbf{CX}(s) + dU(s) \qquad \text{...(6.24 B)}$$

Consider equation (6.24 A)

$$[s\mathbf{I} - \mathbf{A}]\mathbf{X}(s) = \mathbf{B}U(s) + \mathbf{x}(0)$$

Assume zero initial conditions, *i.e.*, $\mathbf{x}(0) = 0$, we get,

$$[s\mathbf{I} - \mathbf{A}]\mathbf{X}(s) = \mathbf{B}U(s)$$

or,

$$\mathbf{X}(s) = [s\mathbf{I} - \mathbf{A}]^{-1}\,\mathbf{B}U(s) \qquad \text{...(6.25)}$$

Put $\mathbf{X}(s)$ from equation (6.25) in (6.24 B), we get,

$$Y(s) = \mathbf{C}[s\mathbf{I} - \mathbf{A}]^{-1}\,\mathbf{B}U(s) + dU(s)$$

or,

$$\frac{Y(s)}{U(s)} = \mathbf{C}[s\mathbf{I} - \mathbf{A}]^{-1}\,\mathbf{B} + d \qquad \text{...(6.26)}$$

or,

$$G(s) = \mathbf{C}[s\mathbf{I} - \mathbf{A}]^{-1}\mathbf{B} + d \qquad \text{...(6.27)}$$

Equation (6.27) gives the expression of **transfer function** of linear single-input-single-output system.

6.9.2 Transfer Matrix

Consider state-model of a linear multi-input-multi-output system,

$$\dot{\mathbf{x}}(t) = \mathbf{Ax}(t) + \mathbf{Bu}(t) \quad ; \quad \text{State Equation} \qquad \text{...(6.28 A)}$$

$$\mathbf{y}(t) = \mathbf{Cx}(t) + \mathbf{Du}(t) \quad ; \quad \text{Output Equation} \qquad \text{...(6.28 B)}$$

Take Laplace transform of equations (6.28), we have,

$$s\mathbf{X}(s) - \mathbf{x}(0) = \mathbf{AX}(s) + \mathbf{BU}(s) \qquad \text{...(6.29 A)}$$

$$\mathbf{Y}(s) = \mathbf{CX}(s) + \mathbf{DU}(s) \qquad \text{...(6.29 B)}$$

Consider equation (6.29 A) assuming zero initial conditions, *i.e.*, $\mathbf{x}(0) = 0$, we get

$$[s\mathbf{I} - \mathbf{A}]\mathbf{X}(s) = \mathbf{BU}(s)$$

or,
$$\mathbf{X}(s) = [s\mathbf{I} - \mathbf{A}]^{-1}\, \mathbf{BU}(s) \qquad \text{...(6.30)}$$

Put $\mathbf{X}(s)$ from equation (6.30) in (6.29 *B*), we get

$$\mathbf{Y}(s) = \mathbf{C}[s\mathbf{I} - \mathbf{A}]^{-1}\, \mathbf{BU}(s) + \mathbf{DU}(s)$$

or,
$$\frac{\mathbf{Y}(s)}{\mathbf{U}(s)} = \mathbf{C}[s\mathbf{I} - \mathbf{A}]^{-1}\, \mathbf{B} + \mathbf{D} \qquad \text{...(6.31)}$$

or,
$$\mathbf{G}(s) = \mathbf{C}[s\mathbf{I} - \mathbf{A}]^{-1}\mathbf{B} + \mathbf{D} \qquad \text{...(6.32)}$$

Equation (6.32) gives the expression of **transfer matrix** of linear multi-input-multi-output system, where transfer matrix $\mathbf{G}(s)$ relates the output $\mathbf{Y}(s)$ to the input $\mathbf{U}(s)$ as

$$\mathbf{Y}(s) = \mathbf{G}(s)\ \mathbf{U}(s)$$

or, in an expanded form

$$\begin{bmatrix} Y_1(s) \\ Y_2(s) \\ \vdots \\ Y_p(s) \end{bmatrix}_{p\times 1} = \begin{bmatrix} G_{11}(s) & G_{12}(s) & \cdots & G_{1m}(s) \\ G_{21}(s) & G_{22}(s) & \cdots & G_{2m}(s) \\ \vdots & \vdots & & \vdots \\ G_{p1}(s) & G_{p2}(s) & \cdots & G_{pm}(s) \end{bmatrix}_{p\times m} \begin{bmatrix} U_1(s) \\ U_2(s) \\ \vdots \\ U_m(s) \end{bmatrix}_{m\times 1}$$

The $(i, j)^{th}$ element $G_{ij}(s)$ ($i = 1, 2, .., p$; $j = 1, 2, ..., m$) of $\mathbf{G}(s)$ is the transfer function relating the ith output to the jth input.

It is clear that the transfer function expression given by equation (6.27) is a **special case** of the transfer matrix expression given by equation (6.32).

Example 6.4. *Determine the transfer function for the system given below*

$$\begin{bmatrix} \dot{x}_1 \\ \dot{x}_2 \end{bmatrix} = \begin{bmatrix} 1 & -2 \\ 4 & -5 \end{bmatrix} \begin{bmatrix} x_1 \\ x_2 \end{bmatrix} + \begin{bmatrix} 2 \\ 1 \end{bmatrix} u$$

and,
$$y = [1 \ \ 1] \begin{bmatrix} x_1 \\ x_2 \end{bmatrix}$$

Solution. We have,

$$\mathbf{A} = \begin{bmatrix} 1 & -2 \\ 4 & -5 \end{bmatrix} \quad ; \quad \mathbf{B} = \begin{bmatrix} 2 \\ 1 \end{bmatrix}$$

$$\mathbf{C} = [1 \ \ 1] \quad ; \quad d = 0$$

$$[s\mathbf{I} - \mathbf{A}] = \begin{bmatrix} s & 0 \\ 0 & s \end{bmatrix} - \begin{bmatrix} 1 & -2 \\ 4 & -5 \end{bmatrix} = \begin{bmatrix} (s-1) & 2 \\ -4 & (s+5) \end{bmatrix}$$

Now, $$\text{Adj }[s\mathbf{I} - \mathbf{A}] = \begin{bmatrix} (s+5) & -2 \\ 4 & (s-1) \end{bmatrix}$$

$\therefore$ $$[s\mathbf{I} - \mathbf{A}]^{-1} = \frac{\text{Adj }[s\mathbf{I}-\mathbf{A}]}{|s\mathbf{I}-\mathbf{A}|}$$

$$= \frac{1}{(s^2+4s+3)}\begin{bmatrix} (s+5) & -2 \\ 4 & (s-1) \end{bmatrix}$$

$$= \begin{bmatrix} \dfrac{(s+5)}{(s+1)(s+3)} & \dfrac{-2}{(s+1)(s+3)} \\ \dfrac{4}{(s+1)(s+3)} & \dfrac{(s-1)}{(s+1)(s+3)} \end{bmatrix}$$

The transfer function is given by

$$G(s) = \mathbf{C}[s\mathbf{I} - \mathbf{A}]^{-1}\mathbf{B}$$

$$= [1 \;\; 1]\begin{bmatrix} \dfrac{(s+5)}{(s+1)(s+3)} & \dfrac{-2}{(s+1)(s+3)} \\ \dfrac{4}{(s+1)(s+3)} & \dfrac{(s-1)}{(s+1)(s+3)} \end{bmatrix}\begin{bmatrix} 2 \\ 1 \end{bmatrix} = [1 \;\; 1]\begin{bmatrix} \dfrac{2(s+5)-2}{(s+1)(s+3)} \\ \dfrac{8+(s-1)}{(s+1)(s+3)} \end{bmatrix}$$

$$= [1 \;\; 1]\begin{bmatrix} \dfrac{(2s+8)}{(s+1)(s+3)} \\ \dfrac{(s+7)}{(s+1)(s+3)} \end{bmatrix} = \frac{(2s+8)}{(s+1)(s+3)} + \frac{(s+7)}{(s+1)(s+3)}$$

or, $$G(s) = \frac{3(s+5)}{(s+1)(s+3)}$$

Example 6.5. *Determine the transfer matrix for the system given below*

$$\begin{bmatrix} \dot{x}_1 \\ \dot{x}_2 \end{bmatrix} = \begin{bmatrix} 0 & 3 \\ -2 & -5 \end{bmatrix}\begin{bmatrix} x_1 \\ x_2 \end{bmatrix} + \begin{bmatrix} 1 & 1 \\ 1 & 1 \end{bmatrix}\mathbf{u}$$

and, $$\mathbf{y} = \begin{bmatrix} 2 & 1 \\ 1 & 0 \end{bmatrix}\begin{bmatrix} x_1 \\ x_2 \end{bmatrix}$$

Solution. We have, $$\mathbf{A} = \begin{bmatrix} 0 & 3 \\ -2 & -5 \end{bmatrix} \quad ; \quad \mathbf{B} = \begin{bmatrix} 1 & 1 \\ 1 & 1 \end{bmatrix}$$

$$\mathbf{C} = \begin{bmatrix} 2 & 1 \\ 1 & 0 \end{bmatrix} \quad ; \quad \mathbf{D} = 0$$

$$[s\mathbf{I} - \mathbf{A}] = \begin{bmatrix} s & 0 \\ 0 & s \end{bmatrix} - \begin{bmatrix} 0 & 3 \\ -2 & -5 \end{bmatrix}$$

$$= \begin{bmatrix} s & -3 \\ 2 & (s+5) \end{bmatrix}$$

Now, $\quad \text{Adj}\,[s\mathbf{I} - \mathbf{A}] = \begin{bmatrix} (s+5) & -2 \\ 3 & s \end{bmatrix}^T$

$$= \begin{bmatrix} (s+5) & 3 \\ -2 & s \end{bmatrix}$$

$$\therefore \quad [s\mathbf{I} - \mathbf{A}]^{-1} = \frac{\text{Adj}\,[s\mathbf{I}-\mathbf{A}]}{|s\mathbf{I}-\mathbf{A}|}$$

$$= \frac{1}{(s^2+5s+6)}\begin{bmatrix} (s+5) & 3 \\ -2 & s \end{bmatrix}$$

$$= \begin{bmatrix} \dfrac{(s+5)}{(s+2)(s+3)} & \dfrac{3}{(s+2)(s+3)} \\ \dfrac{-2}{(s+2)(s+3)} & \dfrac{s}{(s+2)(s+3)} \end{bmatrix}$$

$$\therefore \quad \mathbf{G}(s) = \mathbf{C}[s\mathbf{I} - \mathbf{A}]^{-1}\mathbf{B}$$

or, $\quad \mathbf{G}(s) = \begin{bmatrix} 2 & 1 \\ 1 & 0 \end{bmatrix}\begin{bmatrix} \dfrac{(s+5)}{(s+2)(s+3)} & \dfrac{3}{(s+2)(s+3)} \\ \dfrac{(s-2)}{(s+2)(s+3)} & \dfrac{s}{(s+2)(s+3)} \end{bmatrix}\begin{bmatrix} 1 & 1 \\ 1 & 1 \end{bmatrix}$

$$= \begin{bmatrix} 2 & 1 \\ 1 & 0 \end{bmatrix}\begin{bmatrix} \dfrac{(s+8)}{(s+2)(s+3)} & \dfrac{(s+8)}{(s+2)(s+3)} \\ \dfrac{(s-2)}{(s+2)(s+3)} & \dfrac{(s-2)}{(s+2)(s+3)} \end{bmatrix}$$

or, $\quad \mathbf{G}(s) = \begin{bmatrix} \dfrac{(3s+14)}{(s+2)(s+3)} & \dfrac{(3s+14)}{(s+2)(s+3)} \\ \dfrac{(s+8)}{(s+2)(s+3)} & \dfrac{(s+8)}{(s+2)(s+3)} \end{bmatrix}$

6.10 CHARACTERISTIC EQUATION, EIGEN-VALUES AND EIGEN-VECTORS

The **eigen-values** of an $n \times n$ matrix **A** are the roots of **characteristic equation** given by

$$|\lambda\mathbf{I} - \mathbf{A}| = 0 \quad \text{...(6.33)}$$

The eigen values are sometimes called the **characteristic roots.**

Any non-zero vector $\mathbf{m}_i$; $i = 1, 2,\ldots n$ that satisfies the matrix equation

$$(\lambda_i\mathbf{I} - \mathbf{A})\mathbf{m}_i = \mathbf{0} \quad \text{...(6.34)}$$

is known as **eigen-vector** of **A** associated with eigen value λ_i; $i = 1, 2, \ldots n$.

6.11 STATE-SPACE REPRESENTATION OF PHYSICAL SYSTEMS

6.11.1 State-Space Representation Using Physical Variables

A simple electrical system which is an **RLC network** is shown in Figure 6.5. If the initial conditions $v_c(0)$, $i_1(0)$, $i_2(0)$ and the input signal $e(t)$ for $t \geq 0$ are known; then the behaviour of electrical network is completely specified for $t \geq 0$. Therefore, initial conditions $v_c(0)$, $i_1(0)$, $i_2(0)$ together with the input signal $e(t)$ for $t \geq 0$ constitute the minimal information needed. It then follows that a natural selection of the state-variables would be

$$x_1(t) = v_c(t) \quad \text{...(6.35 A)}$$

$$x_2(t) = i_1(t) \quad \text{...(6.35 B)}$$

$$x_3(t) = i_2(t) \quad \text{...(6.35 C)}$$

However, the choice of the state-variables for a given system is **not unique.**

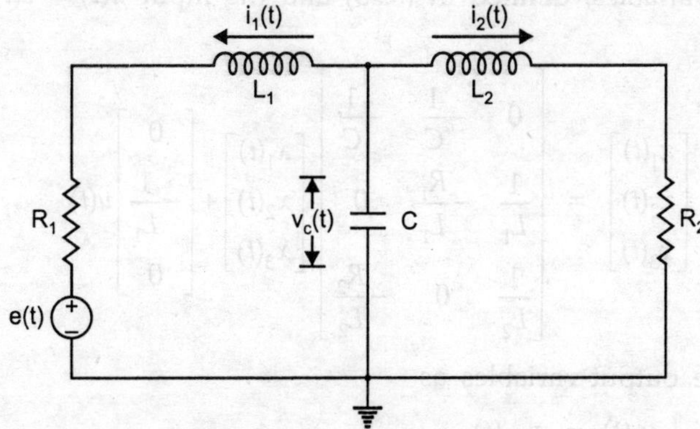

FIGURE 6.5 *An RLC Network.*

Now, the differential equations, governing the behaviour of the RLC network are

$$i_1(t) + i_2(t) + C\frac{dv_c(t)}{dt} = 0 \quad \text{...(6.36 A)}$$

$$L_1\frac{di_1(t)}{dt} + R_1 i_1(t) + e(t) - v_c(t) = 0 \quad \text{...(6.36 B)}$$

$$L_2\frac{di_2(t)}{dt} + R_2 i_2(t) - v_c(t) = 0 \quad \text{...(6.36 C)}$$

We are interested in expressing the variables $\frac{dv_c(t)}{dt}, \frac{di_1(t)}{dt}$ and $\frac{di_2(t)}{dt}$ as linear combinations of the variables $v_c(t)$, $i_1(t)$, $i_2(t)$ and $e(t)$ as required in representation (6.10). For this purpose equations (6.36) may be rewritten as

$$\frac{dv_c(t)}{dt} = -\frac{1}{C}i_1(t) - \frac{1}{C}i_2(t)$$

$$\frac{di_1(t)}{dt} = \frac{1}{L_1}v_c(t) - \frac{R_1}{L_1}i_1(t) - \frac{1}{L_1}e(t)$$

$$\frac{di_2(t)}{dt} = \frac{1}{L_2}v_c(t) - \frac{R_2}{L_2}i_2(t)$$

or, In matrix form,

$$\begin{bmatrix} \frac{dv_c(t)}{dt} \\ \frac{di_1(t)}{dt} \\ \frac{di_2(t)}{dt} \end{bmatrix} = \begin{bmatrix} 0 & -\frac{1}{C} & -\frac{1}{C} \\ \frac{1}{L_1} & -\frac{R_1}{L_1} & 0 \\ \frac{1}{L_2} & 0 & -\frac{R_2}{L_2} \end{bmatrix} \begin{bmatrix} v_c(t) \\ i_1(t) \\ i_2(t) \end{bmatrix} + \begin{bmatrix} 0 \\ -\frac{1}{L_1} \\ 0 \end{bmatrix} e(t)$$

In terms of state-variables, defined in (6.35) and the input $u(t) = e(t)$, we have the state equations as

$$\begin{bmatrix} \dot{x}_1(t) \\ \dot{x}_2(t) \\ \dot{x}_3(t) \end{bmatrix} = \begin{bmatrix} 0 & -\frac{1}{C} & -\frac{1}{C} \\ \frac{1}{L_1} & -\frac{R_1}{L_1} & 0 \\ \frac{1}{L_2} & 0 & -\frac{R_2}{L_2} \end{bmatrix} \begin{bmatrix} x_1(t) \\ x_2(t) \\ x_3(t) \end{bmatrix} + \begin{bmatrix} 0 \\ -\frac{1}{L_1} \\ 0 \end{bmatrix} u(t) \qquad ...(6.37)$$

Let us assume the output-variables as

$$y_1(t) = v_{R_2}(t) \qquad ...(6.38\ A)$$

$$y_2(t) = i_{R_2}(t) \qquad ...(6.38\ B)$$

But

$$v_{R_2}(t) = R_2 i_2(t) \qquad ...(6.39\ A)$$

$$i_{R_2}(t) = i_2(t) \qquad ...(6.39\ B)$$

Writing equations (6.39) in matrix form, we get

$$\begin{bmatrix} v_{R_2}(t) \\ i_{R_2}(t) \end{bmatrix} = \begin{bmatrix} 0 & 0 & R_2 \\ 0 & 0 & 1 \end{bmatrix} \begin{bmatrix} v_c(t) \\ i_1(t) \\ i_2(t) \end{bmatrix}$$

Thus, in terms of output-variables, defined in (6.38) and state-variables defined in (6.35), we get the output equations as

$$\begin{bmatrix} y_1(t) \\ y_2(t) \end{bmatrix} = \begin{bmatrix} 0 & 0 & R_2 \\ 0 & 0 & 1 \end{bmatrix} \begin{bmatrix} x_1(t) \\ x_2(t) \\ x_3(t) \end{bmatrix} \qquad ...(6.40)$$

Equations (6.37) and (6.40) provide the state-model of the RLC network.

In the above example, the selected state-variables are the physical quantities of the system, which can be measured. The choice of physical-variable of a system as state-variable therefore, helps in implementation of system-design.

Other advantage of selecting physical-variables for state-space representation is that the solution of state-equations gives time-variation of variables, which have direct relevance to the physical system. However, with the choice of physical-variables, the solution of state equations may become a difficult task.

6.11.2 State-Space Representation Using Phase-Variables

The **Phase-variables** are defined as those particular state variables, which are obtained from one of the system-variables and its derivatives. Often the variable used is the system output and remaining state-variables are then derivatives of the output.

The phase-variable state-model is easily determined if the system-model is already known in the form of differential equation/transfer function.

The transfer function of a general control system may be expressed as

$$G(s) = \frac{Y(s)}{U(s)} = \frac{b_0 s^m + b_1 s^{m-1} + \ldots + b_{m-1}s + b_m}{s^n + a_1 s^{n-1} + \ldots + a_{n-1}s + a_n} \qquad \text{...(6.41)}$$

Most of the practical control schemes are realized with $m < n$. However, for the sake of generality, we consider the case that $m = n$. Thus,

$$G(s) = \frac{Y(s)}{U(s)} = \frac{b_0 s^n + b_1 s^{n-1} + \ldots + b_{n-1}s + b_n}{s^n + a_1 s^{n-1} + \ldots + a_{n-1}s + a_n} \qquad \text{...(6.42)}$$

We may divide the transfer function $G(s)$ given by (6.42) into two parts as

$$G(s) = \frac{Y(s)}{U(s)} = \frac{Q(s)}{U(s)} \cdot \frac{Y(s)}{Q(s)} \qquad \text{...(6.43)}$$

where, $$\frac{Q(s)}{U(s)} = \frac{1}{s^n + a_1 s^{n-1} + \ldots + a_{n-1}s + a_n} \qquad \text{...(6.44)}$$

and, $$\frac{Y(s)}{Q(s)} = b_0 s^n + b_1 s^{n-1} + \ldots + b_{n-1}s + b_n \qquad \text{...(6.45)}$$

Equation (6.44) can be rewritten as

$$s^n Q(s) = -a_1 s^{n-1} Q(s) - \ldots - a_{n-1} s Q(s) - a_n Q(s) + U(s) \qquad \text{...(6.46)}$$

Define state-variables as follows

$$\begin{aligned} X_1(s) &= Q(s) \\ X_2(s) &= sQ(s) \\ &\vdots \\ X_{n-1}(s) &= s^{n-2}Q(s) \\ X_n(s) &= s^{n-1}Q(s) \end{aligned}$$

Then clearly,

$$\left.\begin{aligned} sX_1(s) &= sQ(s) = X_2(s) \\ sX_2(s) &= s^2Q(s) = X_3(s) \\ &\vdots \\ sX_{n-1}(s) &= s^{n-1}Q(s) = X_n(s) \\ sX_n(s) &= s^nQ(s) = -a_1s^{n-1}Q(s) - \ldots -a_{n-1}sQ(s) - a_nQ(s) + U(s) \\ &= -a_1X_n(s) - \ldots - a_{n-1}X_2(s) - a_nX_1(s) + U(s) \end{aligned}\right\} \quad \ldots(6.47)$$

Taking inverse Laplace transform of set of equations (6.47); assuming all initial conditions to be zero, we have

$$\left.\begin{aligned} \dot{x}_1(t) &= x_2(t) \\ \dot{x}_2(t) &= x_3(t) \\ &\vdots \\ \dot{x}_{n-1}(t) &= x_n(t) \\ \dot{x}_n(t) &= -a_1x_n(t) - \ldots -a_{n-1}x_2(t) - a_nx_1(t) + u(t) \end{aligned}\right\} \quad \ldots(6.48)$$

The set of equations (6.48) may be expressed in matrix form as

$$\begin{bmatrix} \dot{x}_1(t) \\ \dot{x}_2(t) \\ \vdots \\ \dot{x}_{n-1}(t) \\ \dot{x}_n(t) \end{bmatrix}_{n\times 1} = \begin{bmatrix} 0 & 1 & 0 & \ldots & 0 \\ 0 & 0 & 1 & \ldots & 0 \\ \vdots & \vdots & \vdots & \ldots & \vdots \\ 0 & 0 & 0 & \ldots & 1 \\ -a_n & -a_{n-1} & -a_{n-2} & \ldots & -a_1 \end{bmatrix}_{n\times n} \begin{bmatrix} x_1(t) \\ x_2(t) \\ \vdots \\ x_{n-1}(t) \\ x_n(t) \end{bmatrix}_{n\times 1} + \begin{bmatrix} 0 \\ 0 \\ \vdots \\ 0 \\ 1 \end{bmatrix}_{n\times 1} u(t) \quad \ldots(6.49)$$

or, $$\dot{\mathbf{x}}(t) = \mathbf{Ax}(t) + \mathbf{B}u(t) \quad \ldots(6.50\text{ A})$$

Now, the equation (6.45) can be rewritten as

$$Y(s) = b_0s^nQ(s) + b_1s^{n-1}\,Q(s) + \ldots + b_{n-1}sQ(s) + b_nQ(s)$$

Putting the values from (6.47), we get

$$Y(s) = b_0\{-a_1\,X_n(s) - \ldots - a_{n-1}X_2(s) - a_nX_1(s) + U(s)\} + b_1X_n(s) + \ldots + b_{n-1}X_2(s) + b_nX_1(s)$$

or, $$Y(s) = (b_n - a_nb_0)X_1(s) + (b_{n-1} - a_{n-1}b_0)X_2(s) + \ldots + (b_2 - a_2b_0)\,X_{n-1}(s) + (b_1 - a_1b_0)X_n(s) + b_0U(s)$$

Taking inverse Laplace transform; assuming all initial conditions to be zero; we get

$$y(t) = (b_n - a_nb_0)x_1(t) + (b_{n-1} - a_{n-1}b_0)x_2(t) + \ldots + (b_2 - a_2b_0)x_{n-1}(t) + (b_1 - a_1b_0)x_n(t) + b_0u(t)$$

or, in matrix form,

$$y(t) = \mathbf{Cx}(t) + du(t) \quad \ldots(6.50\text{ B})$$

where, $\mathbf{C} = [(b_n - a_n b_0) \;\; (b_{n-1} - a_{n-1} b_0) \;\; \cdots \;\; (b_1 - a_1 b_0)]_{1 \times n}$

$$\mathbf{x}(t) = \begin{bmatrix} x_1(t) \\ x_2(t) \\ \vdots \\ x_n(t) \end{bmatrix} \;;\; \text{State Vector}$$

$$d = b_0 \;;\; \text{Transmission Constant}$$

Equations (6.50) give the phase-variable state-model of the system. The block-diagram representation of the phase-variable state-model is shown in Figure 6.6.

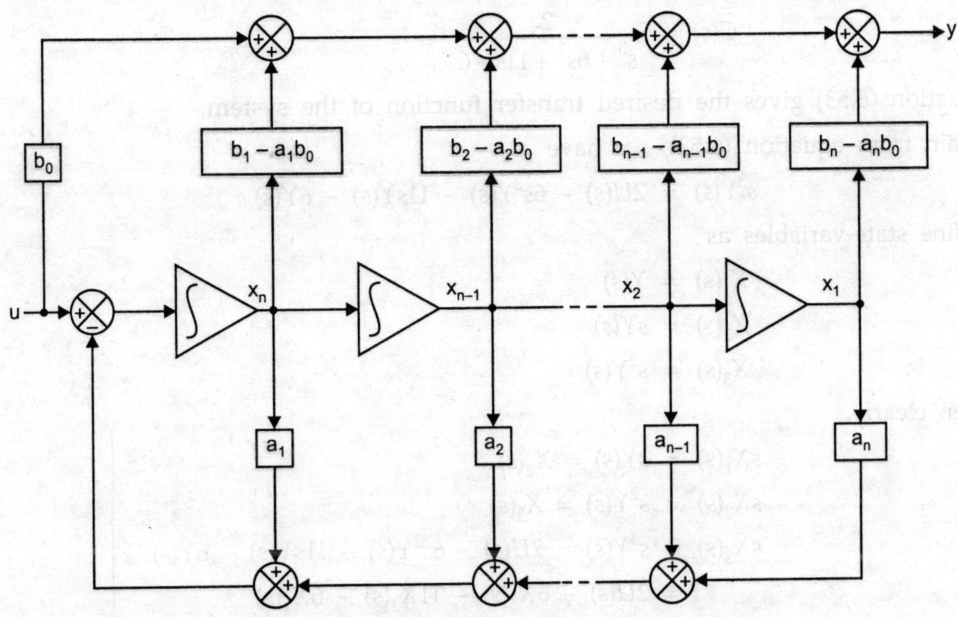

FIGURE 6.6 *Block-diagram representation of the state-model given by equation (6.50).*

The matrix **A** in equation (6.49) has a very special form having all 1's in the upper off-diagonal. Its last row comprises of the negative of the coefficients a_n, a_{n-1}, ..., a_1 and remaining all elements are zero. This very special form of matrix **A** is known as **Bush Form** or **Companion Form.**

We observed that the phase-variable formulation can be obtained by inspection from the transfer function and **vice versa.** A disadvantage of phase-variable formulation is that the phase variables, in general, are not physical (real) variables of the system and therefore, are not available for **measurement** and **control purposes;** though phase-variables are simple to realize mathematically.

In spite of these disadvantages, phase-variables provide a powerful method of state-variable formulation. A link between the transfer function design approach and time-domain design approach is established through phase-variables.

Example 6.6. *A system is described by following differential equation*

$$\frac{d^3y}{dt^3}+6\frac{d^2y}{dt^2}+11\frac{dy}{dt}+6y = 2u \qquad ...(6.51)$$

Obtain the state-space representation of the system. Also find the transfer function of the system.

Solution. Writing equation (6.51) in s-domain, assuming all initial conditions to be zero.

$$s^3Y(s) + 6s^2Y(s) + 11sY(s) + 6Y(s) = 2U(s) \qquad ...(6.52)$$

or,
$$\frac{Y(s)}{U(s)} = \frac{2}{s^3+6s^2+11s+6}$$

or,
$$G(s) = \frac{2}{s^3+6s^2+11s+6} \qquad ...(6.53)$$

Equation (6.53) gives the desired transfer function of the system.

Again from equation (6.52), we have

$$s^3Y(s) = 2U(s) - 6s^2Y(s) - 11sY(s) - 6Y(s) \qquad ...(6.54)$$

Define state-variables as

$$X_1(s) = Y(s)$$
$$X_2(s) = sY(s)$$
$$X_3(s) = s^2Y(s)$$

Then clearly,

$$\left.\begin{aligned} sX_1(s) &= sY(s) = X_2(s) \\ sX_2(s) &= s^2Y(s) = X_3(s) \\ sX_3(s) &= s^3Y(s) = 2U(s) - 6s^2Y(s) - 11sY(s) - 6Y(s) \\ &= 2U(s) - 6X_3(s) - 11X_2(s) - 6X_1(s) \end{aligned}\right\} \qquad ...(6.55)$$

Taking inverse Laplace transform of set of equations (6.55); assuming all initial conditions to be zero, we get

$$\dot{x}_1(t) = x_2(t)$$
$$\dot{x}_2(t) = x_3(t)$$
$$\dot{x}_3(t) = 2u(t) - 6x_3(t) - 11x_2(t) - 6x_1(t)$$

or, In matrix form

$$\begin{bmatrix}\dot{x}_1(t)\\ \dot{x}_2(t)\\ \dot{x}_3(t)\end{bmatrix} = \begin{bmatrix}0 & 1 & 0\\ 0 & 0 & 1\\ -6 & -11 & -6\end{bmatrix}\begin{bmatrix}x_1(t)\\ x_2(t)\\ x_3(t)\end{bmatrix} + \begin{bmatrix}0\\ 0\\ 2\end{bmatrix}u(t) \qquad ...(6.56\text{ A})$$

Also, $Y(s) = X_1(s)$

Taking inverse Laplace transform, we have

$$y(t) = x_1(t)$$

or, $$y(t) = \begin{bmatrix} 1 & 0 & 0 \end{bmatrix} \begin{bmatrix} x_1(t) \\ x_2(t) \\ x_3(t) \end{bmatrix} \quad \text{...(6.56 B)}$$

Equations (6.56) give the desired state-model of the system.

Example 6.7. *Obtain the state-model for the given transfer function*

$$G(s) = \frac{Y(s)}{U(s)} = \frac{K(C_2 s + C_1)}{s^3 + a_3 s^2 + a_2 s + a_1} \quad \text{...(6.57)}$$

Also show it in block-diagram.

Solution. Dividing the transfer function $G(s)$ given by (6.57) into two parts as

$$G(s) = \frac{Y(s)}{U(s)} = \frac{Q(s)}{U(s)} \cdot \frac{Y(s)}{Q(s)}$$

where, $$\frac{Q(s)}{U(s)} = \frac{K}{s^3 + a_3 s^2 + a_2 s + a_1} \quad \text{...(6.58)}$$

and $$\frac{Y(s)}{Q(s)} = C_2 s + C_1 \quad \text{...(6.59)}$$

Equation (6.58) can be rewritten as

$$s^3 Q(s) = -a_3 s^2 Q(s) - a_2 s Q(s) - a_1 Q(s) + KU(s)$$

Define state-variables as

$$X_1(s) = Q(s)$$
$$X_2(s) = sQ(s)$$
$$X_3(s) = s^2 Q(s)$$

Then, clearly,

$$\left.\begin{aligned} sX_1(s) &= sQ(s) = X_2(s) \\ sX_2(s) &= s^2Q(s) = X_3(s) \\ sX_3(s) &= s^3Q(s) = -a_3 s^2 Q(s) - a_2 sQ(s) - a_1 Q(s) + KU(s) \\ &= -a_3 X_3(s) - a_2 X_2(s) - a_1 X_1(s) + KU(s) \end{aligned}\right\} \quad \text{...(6.60)}$$

Taking inverse Laplace transform of set of equations (6.60); assuming all initial conditions to be zero, we have

$$\dot{x}_1(t) = x_2(t)$$
$$\dot{x}_2(t) = x_3(t)$$
$$\dot{x}_3(t) = -a_3 x_3(t) - a_2 x_2(t) - a_1 x_1(t) + Ku(t)$$

or, In matrix form,

$$\begin{bmatrix} \dot{x}_1(t) \\ \dot{x}_2(t) \\ \dot{x}_3(t) \end{bmatrix} = \begin{bmatrix} 0 & 1 & 0 \\ 0 & 0 & 1 \\ -a_1 & -a_2 & -a_3 \end{bmatrix} \begin{bmatrix} x_1(t) \\ x_2(t) \\ x_3(t) \end{bmatrix} + \begin{bmatrix} 0 \\ 0 \\ K \end{bmatrix} u(t) \quad \text{...(6.61 A)}$$

Consider equation (6.59), we have

$$Y(s) = C_2 sQ(s) + C_1 Q(s)$$

or

$$Y(s) = C_2 X_2(s) + C_1 X_1(s)$$

Taking inverse Laplace transform, we obtain

$$y(t) = C_2 x_2(t) + C_1 x_1(t)$$

or, in matrix form,

$$y(t) = \begin{bmatrix} C_1 & C_2 & 0 \end{bmatrix} \begin{bmatrix} x_1(t) \\ x_2(t) \\ x_3(t) \end{bmatrix} \quad \text{...(6.61 B)}$$

Equations (6.61) give the desired state-model of the system. Block-diagram representation of this state-model is shown in Figure (6.7).

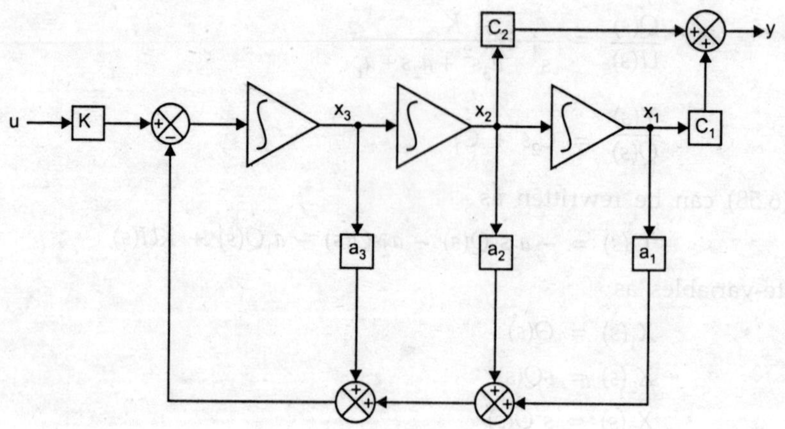

FIGURE 6.7 *Block-diagram representation of state-model given by equations (6.61).*

6.11.3 State-Space Representation in Diagonal Canonical Form (Normal Form)

Consider the transfer function system, defined in equation (6.42). Here we consider the case where the denominator polynomial involves only **distinct roots.** For the **distinct root case,** equation (6.42) can be rewritten as

$$G(s) = \frac{Y(s)}{U(s)} = \frac{b_0 s^n + b_1 s^{n-1} + \ldots + b_{n-1}s + b_n}{(s-\lambda_1)(s-\lambda_2)\ldots(s-\lambda_n)}$$

$$= b_0 + \frac{K_1}{s-\lambda_1} + \frac{K_2}{s-\lambda_2} + \ldots + \frac{K_n}{s-\lambda_n} \quad \text{...(6.62)}$$

where, K_i $(i = 1, 2, \ldots n)$ are the residues of the poles at $s = \lambda_i$ $(i = 1, 2, \ldots, n)$.

Equation (6.62) may be rewritten as

$$Y(s) = b_0 U(s) + \frac{K_1}{s-\lambda_1} U(s) + \frac{K_2}{s-\lambda_2} U(s) + \ldots + \frac{K_n}{s-\lambda_n} U(s) \quad \text{...(6.63)}$$

Define state-variables as follows

$$\left.\begin{aligned} X_1(s) &= \frac{K_1}{s-\lambda_1}U(s) \\ X_2(s) &= \frac{K_2}{s-\lambda_2}U(s) \\ &\vdots \qquad \vdots \\ X_n(s) &= \frac{K_n}{s-\lambda_n}U(s) \end{aligned}\right\} \quad \text{...(6.64)}$$

Then, clearly

$$\begin{aligned} sX_1(s) &= \lambda_1 X_1(s) + K_1 U(s) \\ sX_2(s) &= \lambda_2 X_2(s) + K_2 U(s) \\ \vdots \quad & \quad \vdots \qquad\quad \vdots \\ sX_n(s) &= \lambda_n X_n(s) + K_n U(s) \end{aligned}$$

Taking inverse Laplace transforms of above equations; assuming all initial conditions to be zero, we have

$$\begin{aligned} \dot{x}_1(t) &= \lambda_1 x_1(t) + K_1 u(t) \\ \dot{x}_2(t) &= \lambda_2 x_2(t) + K_2 u(t) \\ \vdots \quad & \quad \vdots \qquad\quad \vdots \\ \dot{x}_n(t) &= \lambda_n x_n(t) + K_n u(t) \end{aligned}$$

or, In matrix form,

$$\begin{bmatrix} \dot{x}_1(t) \\ \dot{x}_2(t) \\ \vdots \\ \dot{x}_n(t) \end{bmatrix}_{n\times 1} = \begin{bmatrix} \lambda_1 & 0 & 0 & \cdots & 0 \\ 0 & \lambda_2 & 0 & \cdots & 0 \\ \vdots & \vdots & \vdots & \cdots & \vdots \\ 0 & 0 & 0 & \cdots & \lambda_n \end{bmatrix}_{n\times n} \begin{bmatrix} x_1(t) \\ x_2(t) \\ \vdots \\ x_n(t) \end{bmatrix}_{n\times 1} + \begin{bmatrix} K_1 \\ K_2 \\ \vdots \\ K_n \end{bmatrix}_{n\times 1} u(t) \quad \text{...(6.65 A)}$$

Now, in terms of state-variables defined in (6.64), the equation (6.63) can be rewritten as

$$Y(s) = b_0 U(s) + X_1(s) + X_2(s) + \ldots + X_n(s)$$

Taking inverse Laplace transform, we get

$$y(t) = b_0 u(t) + x_1(t) + x_2(t) + \ldots + x_n(t)$$

or, in matrix form

$$y(t) = \begin{bmatrix} 1 & 1 & \cdots & 1 \end{bmatrix}_{1\times n} \begin{bmatrix} x_1(t) \\ x_2(t) \\ \vdots \\ x_n(t) \end{bmatrix}_{n\times 1} + b_0 u(t) \quad \text{...(6.65 B)}$$

Equations (6.65) give the state-model of the system in **diagonal canonical form**. The block-diagram representation is shown in Figure 6.8.

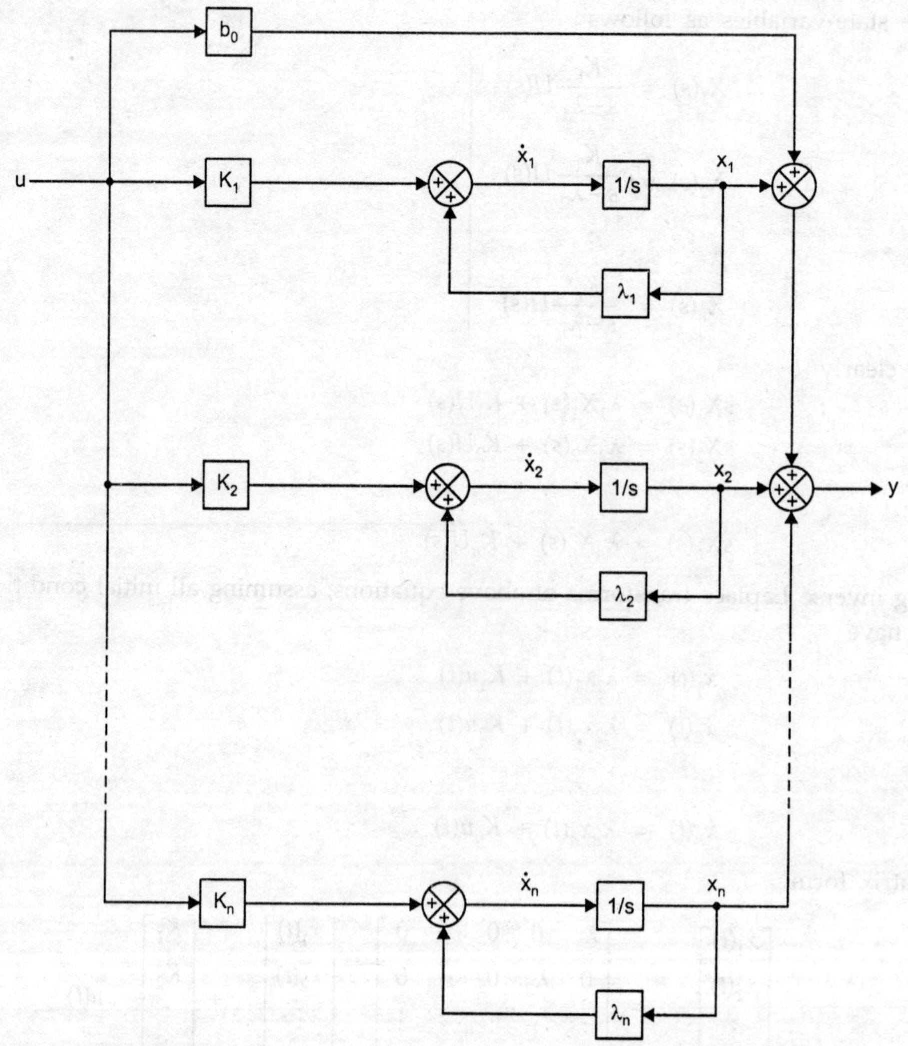

FIGURE 6.8 *Block-diagram representation of system defined by equations (6.65).*

In normal form or diagonal canonical form of state-model; the n first order differential equations are completely independent of each other. This unique **decoupling feature** of normal form or diagonal canonical form greatly helps in the analysis of the system. Furthermore, the diagonal canonical form is sometimes also referred as **Canonical form** only.

6.11.4 State-Space Representation in Jordan Canonical Form

Consider the transfer function system, defined in equation (6.42). Here, we consider the case where the denominator polynomial involves **multiple roots.** Let us assume that the denominator polynomial involves a **triple pole** at $s = \lambda_1$ and a **double pole** at $s = \lambda_4$. All other roots are **distinct** from each other. For this case, the equation (6.42) can be rewritten as

$$G(s) = \frac{Y(s)}{U(s)} = \frac{b_0 s^n + b_1 s^{n-1} + \ldots + b_{n-1}s + b_n}{(s-\lambda_1)^3 (s-\lambda_4)^2 (s-\lambda_6)(s-\lambda_7)\ldots(s-\lambda_n)}$$

The partial fraction expansion becomes

$$\frac{Y(s)}{U(s)} = b_0 + \frac{K_1}{(s-\lambda_1)^3} + \frac{K_2}{(s-\lambda_1)^2} + \frac{K_3}{(s-\lambda_1)} + \frac{K_4}{(s-\lambda_4)^2} + \frac{K_5}{(s-\lambda_4)} + \frac{K_6}{(s-\lambda_6)} + \frac{K_7}{(s-\lambda_7)} + \ldots + \frac{K_n}{(s-\lambda_n)}$$

or,

$$Y(s) = b_0 U(s) + \frac{K_1}{(s-\lambda_1)^3}U(s) + \frac{K_2}{(s-\lambda_1)^2}U(s) + \frac{K_3}{(s-\lambda_1)}U(s) + \frac{K_4}{(s-\lambda_4)^2}U(s) + \frac{K_5}{(s-\lambda_4)}U(s) + \frac{K_6}{(s-\lambda_6)}U(s) + \frac{K_7}{(s-\lambda_7)}U(s) + \ldots + \frac{K_n}{(s-\lambda_n)}U(s) \quad \ldots(6.66)$$

Define state-variables as

$$X_1(s) = \frac{1}{(s-\lambda_1)^3}U(s)$$

$$X_2(s) = \frac{1}{(s-\lambda_1)^2}U(s)$$

$$X_3(s) = \frac{1}{(s-\lambda_1)}U(s)$$

$$X_4(s) = \frac{1}{(s-\lambda_4)^2}U(s)$$

$$X_5(s) = \frac{1}{(s-\lambda_4)}U(s)$$

$$X_6(s) = \frac{1}{(s-\lambda_6)}U(s)$$

$$X_7(s) = \frac{1}{(s-\lambda_7)}U(s)$$

$$\vdots \qquad \vdots$$

$$X_n(s) = \frac{1}{(s-\lambda_n)}U(s)$$

Also, note that the following relationships exist among $X_1(s)$, $X_2(s)$, $X_3(s)$ and in between $X_4(s)$, $X_5(s)$

$$\frac{X_1(s)}{X_2(s)} = \frac{1}{(s-\lambda_1)}$$

$$\frac{X_2(s)}{X_3(s)} = \frac{1}{(s-\lambda_1)}$$

and,

$$\frac{X_4(s)}{X_5(s)} = \frac{1}{(s-\lambda_4)}$$

Then, from the preceding definition of state-variables and the preceding relationships, we have

$$
\begin{aligned}
sX_1(s) &= \lambda_1 X_1(s) + X_2(s) \\
sX_2(s) &= \lambda_1 X_2(s) + X_3(s) \\
sX_3(s) &= \lambda_1 X_3(s) + U(s) \\
sX_4(s) &= \lambda_4 X_4(s) + X_5(s) \\
sX_5(s) &= \lambda_4 X_5(s) + U(s) \\
sX_6(s) &= \lambda_6 X_6(s) + U(s) \\
sX_7(s) &= \lambda_7 X_7(s) + U(s) \\
&\vdots \\
sX_n(s) &= \lambda_n X_n(s) + U(s)
\end{aligned}
$$

Taking inverse Laplace transforms of preceding equations; assuming all initial conditions to be zero, we have

$$
\begin{aligned}
\dot{x}_1(t) &= \lambda_1 x_1(t) + x_2(t) \\
\dot{x}_2(t) &= \lambda_1 x_2(t) + x_3(t) \\
\dot{x}_3(t) &= \lambda_1 x_3(t) + u(t) \\
\dot{x}_4(t) &= \lambda_4 x_4(t) + x_5(t) \\
\dot{x}_5(t) &= \lambda_4 x_5(t) + u(t) \\
\dot{x}_6(t) &= \lambda_6 x_6(t) + u(t) \\
\dot{x}_7(t) &= \lambda_7 x_7(t) + u(t) \\
&\vdots \\
\dot{x}_n(t) &= \lambda_n x_n(t) + u(t)
\end{aligned}
$$

or, In matrix form

Jordan Blocks

$$
\begin{bmatrix} \dot{x}_1(t) \\ \dot{x}_2(t) \\ \dot{x}_3(t) \\ \dot{x}_4(t) \\ \dot{x}_5(t) \\ \dot{x}_6(t) \\ \dot{x}_7(t) \\ \vdots \\ \dot{x}_n(t) \end{bmatrix}_{n\times 1}
=
\begin{bmatrix}
\lambda_1 & 1 & 0 & 0 & 0 & 0 & 0 & \cdots & 0 \\
0 & \lambda_1 & 1 & 0 & 0 & 0 & 0 & \cdots & 0 \\
0 & 0 & \lambda_1 & 0 & 0 & 0 & 0 & \cdots & 0 \\
0 & 0 & 0 & \lambda_4 & 1 & 0 & 0 & \cdots & 0 \\
0 & 0 & 0 & 0 & \lambda_4 & 0 & 0 & \cdots & 0 \\
0 & 0 & 0 & 0 & 0 & \lambda_6 & 0 & \cdots & 0 \\
0 & 0 & 0 & 0 & 0 & 0 & \lambda_7 & \cdots & 0 \\
\vdots & \vdots & \vdots & \vdots & \vdots & \vdots & \vdots & & \vdots \\
0 & 0 & 0 & 0 & 0 & 0 & 0 & \cdots & \lambda_n
\end{bmatrix}_{n\times n}
\begin{bmatrix} \dot{x}_1(t) \\ \dot{x}_2(t) \\ \dot{x}_3(t) \\ \dot{x}_4(t) \\ \dot{x}_5(t) \\ \dot{x}_6(t) \\ \dot{x}_7(t) \\ \vdots \\ \dot{x}_n(t) \end{bmatrix}_{n\times 1}
+
\begin{bmatrix} 0 \\ 0 \\ 1 \\ 0 \\ 1 \\ 1 \\ 1 \\ \vdots \\ 1 \end{bmatrix}_{n\times 1} u(t)
\qquad \text{...(6.67 A)}
$$

Now in terms of state-variables, the equation (6.66) can be rewritten as

$$Y(s) = b_0U(s) + K_1X_1(s) + K_2X_2(s) + K_3X_3(s) + K_4X_4(s) + \dots + K_nX_n(s)$$

Taking inverse Laplace transform, we get

$$y(t) = b_0u(t) + K_1x_1(t) + K_2x_2(t) + K_3x_3(t) + K_4x_4(t) + \dots + K_nx_n(t)$$

or, In matrix form,

$$y(t) = \begin{bmatrix} K_1 & K_2 & K_3 & K_4 & \cdots & K_n \end{bmatrix}_{1\times n} \begin{bmatrix} x_1(t) \\ x_2(t) \\ x_3(t) \\ x_4(t) \\ \vdots \\ x_n(t) \end{bmatrix}_{n\times 1} + b_0u(t) \qquad \text{...(6.67 B)}$$

The state-space representation in the form given by equations (6.67) is said to be in the **Jordan Canonical Form.** Figure (6.9) shows a block-diagram representation of the system given by equations (6.67). The dotted sections in equation (6.67 A) are called **Jordan Blocks.**

6.12 SIMILARITY TRANSFORMATION

The state equation of a system is not unique *i.e.*, there exist more than one set of state-variables in terms of which the system behaviour can be completely described. In fact, there are infinitely many state-models for a given system and any two models are uniquely related. The transformation of one set of dynamic equations to another set of dynamic equations is termed as **'similarity transformation'**.

Consider a multi-input-multi-output state-model

$$\dot{\mathbf{x}}(t) = \mathbf{Ax}(t) + \mathbf{Bu}(t) \qquad \text{...(6.68 A)}$$

$$\mathbf{y}(t) = \mathbf{Cx}(t) + \mathbf{Du}(t) \qquad \text{...(6.68 B)}$$

Let us define a new state-vector $\mathbf{z}(t)$ such that

$$\mathbf{x}(t) = \mathbf{Tz}(t)$$

where $\mathbf{T}$ is an $n \times n$ non-singular, constant matrix.

Since $\mathbf{T}$ is a constant matrix, it follows that

$$\dot{\mathbf{x}}(t) = \mathbf{T}\dot{\mathbf{z}}(t)$$

Substituting $\mathbf{x}(t)$ and $\dot{\mathbf{x}}(t)$ in equations (6.68), we obtain

$$\mathbf{T}\dot{\mathbf{z}}(t) = \mathbf{ATz}(t) + \mathbf{Bu}(t) \qquad \text{...(6.69 A)}$$

$$\mathbf{y}(t) = \mathbf{CTz}(t) + \mathbf{Du}(t) \qquad \text{...(6.69 B)}$$

Equations (6.69) may also be rewritten as

$$\dot{\mathbf{z}}(t) = \mathbf{T}^{-1}\mathbf{ATz}(t) + \mathbf{T}^{-1}\mathbf{Bu}(t) \qquad \text{...(6.70 A)}$$

$$\mathbf{y}(t) = \mathbf{CTz}(t) + \mathbf{Du}(t) \qquad \text{...(6.70 B)}$$

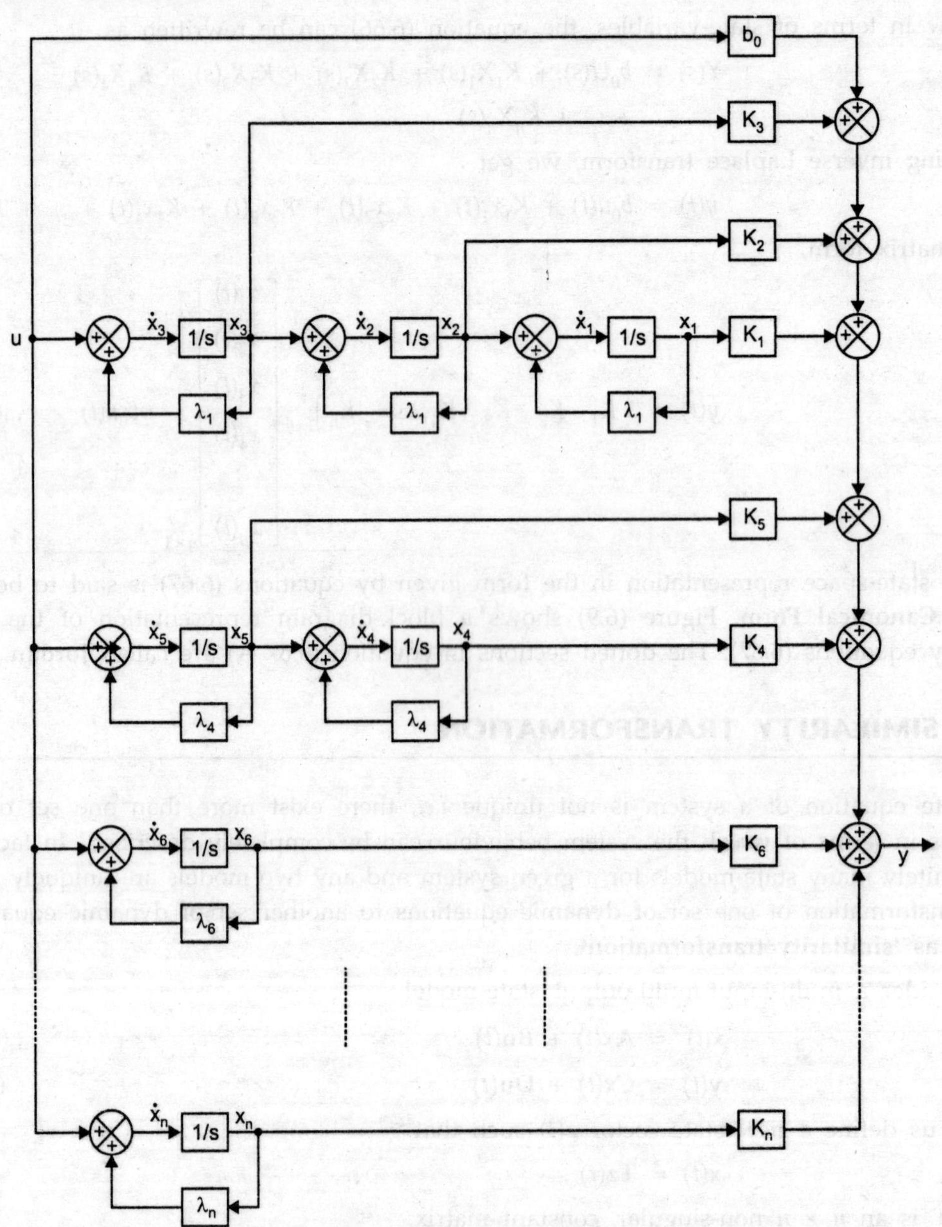

FIGURE 6.9 *Block-diagram representation of system defined by equation (6.67).*

Thus, the state-model given by equation (6.68) modifies to a new state-model given by.

$$\dot{\mathbf{z}}(t) = \tilde{\mathbf{A}}\,\mathbf{z}(t) + \tilde{\mathbf{B}}\,\mathbf{u}(t) \qquad \text{...(6.71 A)}$$

$$\mathbf{y}(t) = \tilde{\mathbf{C}}\,\mathbf{z}(t) + \mathbf{D}\mathbf{u}(t) \qquad \text{...(6.71 B)}$$

where,

$$\tilde{\mathbf{A}} = \mathbf{T}^{-1}\mathbf{A}\mathbf{T}$$

$$\tilde{\mathbf{B}} = \mathbf{T}^{-1}\,\mathbf{B}$$

and,

$$\tilde{\mathbf{C}} = \mathbf{C}\mathbf{T}$$

Equation (6.71) give another state-model for a given system. Since **T** is assumed to be non-unique, non singular, constant matrix; the state-model is also **non-unique**.

It is important to note that the transformation matrix **T** must be non-singular, *i.e.*,

$$|\mathbf{T}| \neq 0$$

If this was not the case it is obvious that the inverse transformation would not exist. The above discussion, therefore, demonstrates that for a given system, infinitely many state-models are possible and any two models are **uniquely related.**

6.12.1 Diagonalization

The physical-variable state-model, in general, is not convenient for investigation of system properties and evaluation of time-response. The diagonal canonical form or normal form of state-model, wherein the matrix **A** turns out to be a diagonal matrix, is most suitable for this purpose. Thus, it is useful to study the techniques by means of which, a general state-model can be transformed into a diagonal canonical form. These techniques are often referred to as **'diagonalization techniques'.**

Consider a multi-input-multi-output state-model

$$\dot{\mathbf{x}}(t) = \mathbf{A}\mathbf{x}(t) + \mathbf{B}\mathbf{u}(t) \quad \text{...(6.72 A)}$$

$$\mathbf{y}(t) = \mathbf{C}\mathbf{x}(t) + \mathbf{D}\mathbf{u}(t) \quad \text{...(6.72 B)}$$

In the state-model given by equation (6.72), the matrix **A** is non-diagonal.

Let us define a new state-vector $\mathbf{z}(t)$ such that

$$\mathbf{x}(t)= \mathbf{M}\mathbf{z}(t)$$

where, **M** is an n×n non-singular, constant matrix.

Under this similarity transformation, the original state-model given by equation (6.72) modifies to

$$\dot{\mathbf{z}}(t) = \mathbf{M}^{-1}\mathbf{A}\mathbf{M}\mathbf{z}(t) + \mathbf{M}^{-1}\mathbf{B}\mathbf{u}(t) \quad \text{...(6.73 A)}$$

$$\mathbf{y}(t) = \mathbf{C}\mathbf{M}\mathbf{z}(t) + \mathbf{D}\mathbf{u}(t) \quad \text{...(6.73 B)}$$

If we select the matrix **M** such that $\mathbf{M}^{-1}\mathbf{A}\mathbf{M}$ is a diagonalized matrix of matrix **A**; then the model given by equation (6.73) is a state-model in diagonal canonical form. Under this condition, the matrix **M** is called the **diagonalizing matrix** or the **modal matrix,** and is constructed by placing the eigen-vectors of matrix **A** together *i.e.,*

If $\mathbf{m}_1, \mathbf{m}_2, \mathbf{m}_3, ..., \mathbf{m}_n$ be the eigen-vectors of matrix A corresponding to the eigen-values $\lambda_1, \lambda_2, \lambda_3 ... \lambda_n$ respectively; then the modal matrix is given by

$$\mathbf{M} = [\mathbf{m}_1 : \mathbf{m}_2 : \mathbf{m}_3 : \cdots : \mathbf{m}_n]_{n\times n}$$

Thus, the general state-model given by equation (6.72) modifies to a new state-model (in diagonal canonical form), given by

$$\dot{\mathbf{z}}(t) = \Lambda\mathbf{z}(t) + \tilde{\mathbf{B}}\mathbf{u}(t) \quad \text{...(6.74 A)}$$

$$\mathbf{y}(t) = \tilde{\mathbf{C}}\mathbf{z}(t) + \mathbf{D}\mathbf{u}(t) \quad \text{...(6.74 B)}$$

where, $\Lambda = \mathbf{M}^{-1}\mathbf{AM}$; A Diagonal Matrix

$\tilde{\mathbf{B}} = \mathbf{M}^{-1}\mathbf{B}$

and, $\tilde{\mathbf{C}} = \mathbf{CM}$

Infact the matrix Λ could be obtained directly without the need to compute $\mathbf{M}^{-1}\mathbf{AM}$, since the diagonal elements of matrix Λ are given by distinct eigen-values $\lambda_1, \lambda_2, \lambda_3, \ldots \lambda_n$ of matrix **A**, *i.e.*,

$$\Lambda = \begin{bmatrix} \lambda_1 & 0 & 0 & \cdots & 0 \\ 0 & \lambda_2 & 0 & \cdots & 0 \\ 0 & 0 & \lambda_3 & \cdots & 0 \\ \vdots & \vdots & \vdots & & \vdots \\ 0 & 0 & 0 & \ldots & \lambda_n \end{bmatrix}_{n\times n} = \mathbf{M}^{-1}\mathbf{AM}$$

Note that **A** and Λ matrices have the same characteristic equation ; therefore, the eigen-values are **invariant under** the transformation.

If matrix **A** is given in **Bush's or phase–variable or companion form,** *i.e.*,

$$\mathbf{A} = \begin{bmatrix} 0 & 1 & 0 & \cdots & 0 \\ 0 & 0 & 1 & \cdots & 0 \\ \vdots & \vdots & \vdots & & \vdots \\ 0 & 0 & 0 & \cdots & 1 \\ -a_n & -a_{n-1} & -a_{n-2} & \cdots & -a_1 \end{bmatrix}_{n\times n}$$

then the modal matrix with reference to Bush's or phase-variable or companion form of matrix **A** can be shown to be a special matrix, called the **Vander Monde Matrix** and is given by

$$\mathbf{V} = \begin{bmatrix} 1 & 1 & \cdots & 1 \\ \lambda_1 & \lambda_2 & \cdots & \lambda_n \\ \lambda_1^2 & \lambda_2^2 & \cdots & \lambda_n^2 \\ \vdots & \vdots & & \vdots \\ \lambda_1^{n-1} & \lambda_2^{n-1} & \cdots & \lambda_n^{n-1} \end{bmatrix}_{n\times n}$$

where, $\lambda_1, \lambda_2, \ldots \lambda_n$ are distinct **eigen-values** of matrix **A**.

An advantage of diagonal matrix is that the inverse of such a matrix can be obtained merely by inspection *e.g.*,

$$\begin{bmatrix} \alpha & 0 & 0 \\ 0 & \beta & 0 \\ 0 & 0 & \gamma \end{bmatrix}^{-1} = \begin{bmatrix} 1/\alpha & 0 & 0 \\ 0 & 1/\beta & 0 \\ 0 & 0 & 1/\gamma \end{bmatrix}$$

Now consider the case, in which the matrix **A** involves multiple eigen-values, it is impossible to be diagonalized. However, there exists a similarity transformation $\mathbf{x}(t) = \mathbf{Sz}(t)$

such that the matrix $\mathbf{J} = \mathbf{S}^{-1}\mathbf{AS}$ is **almost a diagonal matrix.** The matrix **J** is called to be in **Jordan Canonical Form.** (Refer section 6.11.4). Under this similarity transformation the state-model in Jordan canonical form, thus, can be expressed as

$$\dot{\mathbf{z}}(t) = \mathbf{Jz}(t) + \tilde{\mathbf{B}}\mathbf{u}(t) \quad \text{...(6.75 A)}$$

$$\mathbf{y}(t) = \tilde{\mathbf{C}}\mathbf{z}(t) + \mathbf{Du}(t) \quad \text{...(6.75 B)}$$

where, $\mathbf{J} = \mathbf{S}^{-1}\mathbf{AS}$; *A* Jordan matrix (Almost Diagonal)

$$\tilde{\mathbf{B}} = \mathbf{S}^{-1}\mathbf{B}$$

and, $\tilde{\mathbf{C}} = \mathbf{CS}$

Note that, for each λ_i of **multiplicity** q, the Jordan matrix $\mathbf{J} = \mathbf{S}^{-1}\mathbf{AS}$ will have a $q \times q$ **Jordan Block** corresponding to eigen-value λ_i.

6.12.2 Computation of State Transition Matrix by Use of Diagonal

In previous section, we have seen that under the similarity transformation $\mathbf{x}(t) = \mathbf{Mz}(t)$, the matrix $\Lambda = \mathbf{M}^{-1}\mathbf{AM}$ is a diagonalized matrix of system matrix **A** with eigen-values on its main diagonal.

Thus, $\Lambda = \mathbf{M}^{-1}\mathbf{AM}$

or, $e^{\Lambda t} = \mathbf{M}^{-1}\, e^{\mathbf{A}t}\, \mathbf{M}$

or, $e^{\mathbf{A}t} = \mathbf{M}\, e^{\Lambda t}\, \mathbf{M}^{-1}$

or, $\phi(t) = \mathbf{M}\, e^{\Lambda t}\, \mathbf{M}^{-1}$ ...(6.76)

where, $\phi(t) = e^{\mathbf{A}t}$; State Transition Matrix

Note that, if diagonalized matrix of **A** is given by

$$\Lambda = \begin{bmatrix} \lambda_1 & 0 & 0 \\ 0 & \lambda_2 & 0 \\ 0 & 0 & \lambda_3 \end{bmatrix}$$

then,
$$e^{\Lambda t} = \begin{bmatrix} e^{\lambda_1 t} & 0 & 0 \\ 0 & e^{\lambda_2 t} & 0 \\ 0 & 0 & e^{\lambda_3 t} \end{bmatrix}$$

In case of **Jordan matrix J**, given by

$$\mathbf{J} = \left[\begin{array}{ccc|cc|cc} \lambda_1 & 1 & 0 & 0 & 0 & 0 & 0 \\ 0 & \lambda_1 & 1 & 0 & 0 & 0 & 0 \\ 0 & 0 & \lambda_1 & 0 & 0 & 0 & 0 \\ \hline 0 & 0 & 0 & \lambda_4 & 1 & 0 & 0 \\ 0 & 0 & 0 & 0 & \lambda_4 & 0 & 0 \\ \hline 0 & 0 & 0 & 0 & 0 & \lambda_6 & 0 \\ 0 & 0 & 0 & 0 & 0 & 0 & \lambda_7 \end{array}\right]$$

$$e^{Jt} = \left[\begin{array}{ccc|cc|cc} e^{\lambda_1 t} & te^{\lambda_1 t} & \frac{1}{2}t^2 e^{\lambda_1 t} & 0 & 0 & 0 & 0 \\ 0 & e^{\lambda_1 t} & te^{\lambda_1 t} & 0 & 0 & 0 & 0 \\ 0 & 0 & e^{\lambda_1 t} & 0 & 0 & 0 & 0 \\ \hline 0 & 0 & 0 & e^{\lambda_4 t} & te^{\lambda_4 t} & 0 & 0 \\ 0 & 0 & 0 & 0 & e^{\lambda_4 t} & 0 & 0 \\ \hline 0 & 0 & 0 & 0 & 0 & e^{\lambda_6 t} & 0 \\ 0 & 0 & 0 & 0 & 0 & 0 & e^{\lambda_7 t} \end{array}\right]$$

Example 6.8. *Consider a matrix* **A** *given by*

$$\mathbf{A} = \begin{bmatrix} 0 & 1 & 0 \\ 3 & 0 & 2 \\ -12 & -7 & -6 \end{bmatrix}$$

Compute diagonal matrix $\mathbf{\Lambda}$ *and STM of* **A** .

Solution. Characteristic equation of matrix **A** is given by

$$|\lambda\mathbf{I} - \mathbf{A}| = 0$$

Now,

$$\lambda\mathbf{I} - \mathbf{A} = \begin{bmatrix} \lambda & -1 & 0 \\ -3 & \lambda & -2 \\ 12 & 7 & \lambda+6 \end{bmatrix}$$

$\therefore$

$$|\lambda\ \mathbf{I} - \mathbf{A}| = \begin{vmatrix} \lambda & -1 & 0 \\ -3 & \lambda & -2 \\ 12 & 7 & \lambda+6 \end{vmatrix} = 0$$

or,

$$(\lambda + 1)\ (\lambda + 2)\ (\lambda + 3) = 0$$

Therefore, the eigen–values of matrix **A** are

$$\lambda_1 = -1$$

$$\lambda_2 = -2$$

$$\lambda_3 = -3$$

Now, the eigen-vector $\mathbf{m}_1$ associated with eigen-value $\lambda_1 = -1$ is obtained by solving the equation

$$(\lambda_1\mathbf{I} - \mathbf{A})\mathbf{m}_1 = 0$$

or,

$$\begin{bmatrix} -1 & -1 & 0 \\ -3 & -1 & -2 \\ 12 & 7 & 5 \end{bmatrix}\begin{bmatrix} m_{11} \\ m_{21} \\ m_{31} \end{bmatrix} = 0$$

or,

$$-m_{11} - m_{21} = 0$$

$$-3m_{11} - m_{21} - 2m_{31} = 0$$

$$12m_{11} + 7m_{21} + 5m_{31} = 0$$

Select $m_{11} = 1$, we get

$$m_{21} = -1$$

and, $m_{31} = -1$

thus,

$$\mathbf{m}_1 = \begin{bmatrix} m_{11} \\ m_{21} \\ m_{31} \end{bmatrix} = \begin{bmatrix} 1 \\ -1 \\ -1 \end{bmatrix}$$

Again, the eigen-vector $\mathbf{m}_2$ associated with eigen-value $\lambda_2 = -2$ is obtained by

$$(\lambda\mathbf{I} - \mathbf{A})\mathbf{m}_2 = 0$$

or,

$$\begin{bmatrix} -2 & -1 & 0 \\ -3 & -2 & -2 \\ 12 & 7 & 4 \end{bmatrix}\begin{bmatrix} m_{12} \\ m_{22} \\ m_{32} \end{bmatrix} = 0$$

or,

$$-2m_{12} - m_{22} = 0$$

$$-3m_{12} - 2m_{22} - 2m_{32} = 0$$

$$12m_{12} + 7m_{22} + 4m_{32} = 0$$

Select, $m_{12} = 2$, we get

$$m_{22} = -4$$

and, $m_{32} = 1$

thus,

$$\mathbf{m}_2 = \begin{bmatrix} m_{12} \\ m_{22} \\ m_{32} \end{bmatrix} = \begin{bmatrix} 2 \\ -4 \\ 1 \end{bmatrix}$$

Similarly, we can compute the eigen–vector $\mathbf{m}_3$ associated with $\lambda_3 = -3$, as

$$\mathbf{m}_3 = \begin{bmatrix} m_{13} \\ m_{23} \\ m_{33} \end{bmatrix} = \begin{bmatrix} 1 \\ -3 \\ 3 \end{bmatrix}$$

The modal matrix $\mathbf{M}$ is obtained by

$$\mathbf{M} = [\mathbf{m}_1 \vdots \mathbf{m}_2 \vdots \mathbf{m}_3]$$

or,

$$\mathbf{M} = \begin{bmatrix} 1 & 2 & 1 \\ -1 & -4 & -3 \\ -1 & 1 & 3 \end{bmatrix}$$

Now,

$$\text{Adj } \mathbf{M} = \begin{bmatrix} +\begin{vmatrix} -4 & -3 \\ 1 & 3 \end{vmatrix} & -\begin{vmatrix} -1 & -3 \\ -1 & 3 \end{vmatrix} & +\begin{vmatrix} -1 & -4 \\ -1 & 1 \end{vmatrix} \\ -\begin{vmatrix} 2 & 1 \\ 1 & 3 \end{vmatrix} & +\begin{vmatrix} 1 & 1 \\ -1 & 3 \end{vmatrix} & -\begin{vmatrix} 1 & 2 \\ -1 & 1 \end{vmatrix} \\ +\begin{vmatrix} 2 & 1 \\ -4 & -3 \end{vmatrix} & -\begin{vmatrix} 1 & 1 \\ -1 & -3 \end{vmatrix} & +\begin{vmatrix} 1 & 2 \\ -1 & -4 \end{vmatrix} \end{bmatrix}^T$$

$$= \begin{bmatrix} -9 & 6 & -5 \\ -5 & 4 & -3 \\ -2 & 2 & -2 \end{bmatrix}^T$$

$$= \begin{bmatrix} -9 & -5 & -2 \\ 6 & 4 & 2 \\ -5 & -3 & -2 \end{bmatrix}$$

Again,

$$|\mathbf{M}| = \begin{bmatrix} 1 & 2 & 1 \\ -1 & -4 & -3 \\ -1 & 1 & 3 \end{bmatrix} = -2$$

$\therefore$

$$\mathbf{M}^{-1} = \frac{\text{Adj } \mathbf{M}}{|\mathbf{M}|} = -\frac{1}{2}\begin{bmatrix} -9 & -5 & -2 \\ 6 & 4 & 2 \\ -5 & -3 & -2 \end{bmatrix}$$

or,

$$\mathbf{M}^{-1} = \begin{bmatrix} 4.5 & 2.5 & 1 \\ -3 & -2 & -1 \\ 2.5 & 1.5 & 1 \end{bmatrix}$$

$\therefore$

$$\Lambda = \mathbf{M}^{-1}\mathbf{A}\mathbf{M}$$

$$= \begin{bmatrix} 4.5 & 2.5 & 1 \\ -3 & -2 & -1 \\ 2.5 & 1.5 & 1 \end{bmatrix}\begin{bmatrix} 0 & 1 & 0 \\ 3 & 0 & 2 \\ -12 & -7 & -6 \end{bmatrix}\begin{bmatrix} 1 & 2 & 1 \\ -1 & -4 & -3 \\ -1 & 1 & 3 \end{bmatrix}$$

$$= \begin{bmatrix} 4.5 & 2.5 & 1 \\ -3 & -2 & -1 \\ 2.5 & 1.5 & 1 \end{bmatrix}\begin{bmatrix} -1 & -4 & -3 \\ 1 & 8 & 9 \\ 1 & -2 & -9 \end{bmatrix} = \begin{bmatrix} -1 & 0 & 0 \\ 0 & -2 & 0 \\ 0 & 0 & -3 \end{bmatrix}$$

$\therefore$

$$e^{\Lambda t} = \begin{bmatrix} e^{-t} & 0 & 0 \\ 0 & e^{-2t} & 0 \\ 0 & 0 & e^{-3t} \end{bmatrix}$$

The STM is given by

$$\phi(t) = e^{\mathbf{A}t} = \mathbf{M}e^{\Lambda t}\,\mathbf{M}^{-1}$$

or,
$$\phi(t) = \begin{bmatrix} 1 & 2 & 1 \\ -1 & -4 & -3 \\ -1 & 1 & 3 \end{bmatrix} \begin{bmatrix} e^{-t} & 0 & 0 \\ 0 & e^{-2t} & 0 \\ 0 & 0 & e^{-3t} \end{bmatrix} \begin{bmatrix} 4.5 & 2.5 & 1 \\ -3 & -2 & -1 \\ 2.5 & 1.5 & 1 \end{bmatrix}$$

or,
$$\phi(t) = \begin{bmatrix} 1 & 2 & 1 \\ -1 & -4 & -3 \\ -1 & 1 & 3 \end{bmatrix} \begin{bmatrix} 4.5e^{-t} & 2.5e^{-t} & e^{-t} \\ -3e^{-2t} & -2e^{-2t} & -e^{-2t} \\ 2.5e^{-3t} & 1.5e^{-3t} & e^{-3t} \end{bmatrix}$$

or,
$$\phi(t) = \begin{bmatrix} (4.5e^{-t} - 6e^{-2t} + 2.5e^{-3t}) & (2.5e^{-t} - 4e^{-2t} + 1.5^{-3t}) & (e^{-t} + 2e^{-2t} + e^{-3t}) \\ (-4.5e^{-t} + 12e^{-2t} - 7.5e^{-3t}) & (-2.5e^{2t} + 8e^{-2t} - 4.5e^{-3t}) & (-e^{-t} + 4e^{-2t} - 3e^{-3t}) \\ (-4.5e^{-t} - 3e^{-2t} + 7.5e^{-3t}) & (-2.5e^{-t} - 2e^{-2t} + 4.5e^{-3t}) & (-e^{-t} - e^{-2t} + 3e^{-3t}) \end{bmatrix}$$

Example 6.9. *Consider the state–space representation of a system*

$$\begin{bmatrix} \dot{x}_1 \\ \dot{x}_2 \\ \dot{x}_3 \end{bmatrix} = \begin{bmatrix} 0 & 1 & 0 \\ 0 & 0 & 1 \\ -6 & -11 & -6 \end{bmatrix} \begin{bmatrix} x_1 \\ x_2 \\ x_3 \end{bmatrix} + \begin{bmatrix} 0 \\ 0 \\ 2 \end{bmatrix} u \qquad \text{...(6.77 A)}$$

$$y = \begin{bmatrix} 1 & 0 & 0 \end{bmatrix} \begin{bmatrix} x_1 \\ x_2 \\ x_3 \end{bmatrix} \qquad \text{...(6.77 B)}$$

Change it into DCF (Diagonal Canonical form).

Solution. Comparing equations (6.77) with

$$\dot{\mathbf{x}}(t) = \mathbf{A}\mathbf{x}(t) + \mathbf{B}u(t)$$

$$y(t) = \mathbf{C}\mathbf{x}(t)$$

We have,
$$\mathbf{A} = \begin{bmatrix} 0 & 1 & 0 \\ 0 & 0 & 1 \\ -6 & -11 & -6 \end{bmatrix}$$

$$\mathbf{B} = \begin{bmatrix} 0 \\ 0 \\ 2 \end{bmatrix}$$

and,
$$\mathbf{C} = \begin{bmatrix} 1 & 0 & 0 \end{bmatrix}$$

Characteristic equation of matrix **A** is given by

$$|\lambda\mathbf{I} - \mathbf{A}| = 0$$

Now,
$$\lambda \mathbf{I} - \mathbf{A} = \begin{bmatrix} \lambda & 0 & 0 \\ 0 & \lambda & 0 \\ 0 & 0 & \lambda \end{bmatrix} - \begin{bmatrix} 0 & 1 & 0 \\ 0 & 0 & 1 \\ -6 & -11 & -6 \end{bmatrix} = \begin{bmatrix} \lambda & -1 & 0 \\ 0 & \lambda & -1 \\ 6 & 11 & \lambda+6 \end{bmatrix}$$

$\therefore$
$$|\lambda \mathbf{I} - \mathbf{A}| = 0$$

or,
$$\lambda^3 + 6\lambda^2 + 11\lambda + 6 = 0$$

or,
$$(\lambda+1)\,(\lambda+2)\,(\lambda+3) = 0$$

Therefore, the eigen-values of matrix **A** are,

$$\lambda_1 = -1$$
$$\lambda_2 = -2$$
$$\lambda_3 = -3$$

In the question the matrix **A** is in Bush's form. Thus, in this case the diagonalizing or Modal matrix will be Vander Monde Matrix, given as

$$\mathbf{V} = \begin{bmatrix} 1 & 1 & 1 \\ \lambda_1 & \lambda_2 & \lambda_3 \\ \lambda_1^2 & \lambda_2^2 & \lambda_3^2 \end{bmatrix} = \begin{bmatrix} 1 & 1 & 1 \\ -1 & -2 & -3 \\ 1 & 4 & 9 \end{bmatrix}$$

$\therefore$
$$\mathbf{V}^{-1} = \begin{bmatrix} 3 & 2.5 & 0.5 \\ -3 & -4 & -1 \\ 1 & 1.5 & 0.5 \end{bmatrix}$$

(It is left as an exercise for the readers to calculate $\mathbf{V}^{-1}$ themselves.)

Using similarity transformation $\mathbf{x}(t) = \mathbf{Vz}(t)$ the original state-model modifies to diagonal canonical form

$$\dot{\mathbf{z}}(t) = \Lambda \mathbf{z}(t) + \tilde{\mathbf{B}}\mathbf{u}(t)$$
$$y(t) = \tilde{\mathbf{C}}\mathbf{z}(t)$$

where,
$$\Lambda = \mathbf{V}^{-1}\mathbf{AV}$$

$$= \begin{bmatrix} 3 & 2.5 & 0.5 \\ -3 & -4 & -1 \\ 1 & 1.5 & 0.5 \end{bmatrix} \begin{bmatrix} 0 & 1 & 0 \\ 0 & 0 & 1 \\ -6 & -11 & -6 \end{bmatrix} \begin{bmatrix} 1 & 1 & 1 \\ -1 & -2 & -3 \\ 1 & 4 & 9 \end{bmatrix}$$

$$= \begin{bmatrix} -1 & 0 & 0 \\ 0 & -2 & 0 \\ 0 & 0 & -3 \end{bmatrix}$$

$$\tilde{\mathbf{B}} = \mathbf{V}^{-1}\mathbf{B}$$

$$= \begin{bmatrix} 3 & 2.5 & 0.5 \\ -3 & -4 & -1 \\ 1 & 1.5 & 0.5 \end{bmatrix} \begin{bmatrix} 0 \\ 0 \\ 2 \end{bmatrix} = \begin{bmatrix} 1 \\ -2 \\ 1 \end{bmatrix}$$

$$\tilde{C} = CV$$

$$= [1 \;\; 0 \;\; 0]\begin{bmatrix} 1 & 1 & 1 \\ -1 & -2 & -3 \\ 1 & 4 & 9 \end{bmatrix}$$

$$= [1 \;\; 1 \;\; 1]$$

Now the DCF of state-model can be given as

$$\begin{bmatrix} \dot{z}_1(t) \\ \dot{z}_2(t) \\ \dot{z}_3(t) \end{bmatrix} = \begin{bmatrix} -1 & 0 & 0 \\ 0 & -2 & 0 \\ 0 & 0 & -3 \end{bmatrix}\begin{bmatrix} z_1(t) \\ z_2(t) \\ z_3(t) \end{bmatrix} + \begin{bmatrix} 1 \\ -2 \\ 1 \end{bmatrix} u(t)$$

$$y(t) = [1 \;\; 1 \;\; 1]\begin{bmatrix} z_1(t) \\ z_2(t) \\ z_3(t) \end{bmatrix}$$

Example 6.10. *A system is characterized by the transfer function*

$$\frac{Y(s)}{U(s)} = \frac{2}{s^3 + 6s^2 + 11s + 6}$$

Represent the state-model in Diagonal canonical form (DCF).

Solution.

$$\frac{Y(s)}{U(s)} = \frac{2}{s^3 + 6s^2 + 11s + 6}$$

or,

$$\frac{Y(s)}{U(s)} = \frac{2}{(s+1)(s+2)(s+3)}$$

or,

$$Y(s) = \left[\frac{1}{s+1} - \frac{2}{s+2} + \frac{1}{s+3}\right]U(s) \qquad \text{...(6.78)}$$

Define state-variables as follows

$$X_1(s) = \frac{1}{s+1}U(s)$$

$$X_2(s) = \frac{-2}{s+2}U(s)$$

$$X_3(s) = \frac{1}{s+3}U(s)$$

Then clearly,

$$sX_1(s) = -X_1(s) + U(s)$$

$$sX_2(s) = -2X_2(s) - 2U(s)$$

$$sX_3(s) = -3X_3(s) + U(s)$$

Taking inverse Laplace transform of above equations assuming all initial conditions to be zero, we get

$$\dot{x}_1(t) = -x_1(t) + u(t)$$

$$\dot{x}_2(t) = -2x_2(t) - 2u(t)$$

$$\dot{x}_3(t) = -3x_3(t) + u(t)$$

or, in matrix form,

$$\begin{bmatrix} \dot{x}_1(t) \\ \dot{x}_2(t) \\ \dot{x}_3(t) \end{bmatrix} = \begin{bmatrix} -1 & 0 & 0 \\ 0 & -2 & 0 \\ 0 & 0 & -3 \end{bmatrix} \begin{bmatrix} x_1(t) \\ x_2(t) \\ x_3(t) \end{bmatrix} + \begin{bmatrix} 1 \\ -2 \\ 1 \end{bmatrix} u(t) \quad \text{...(6.79 A)}$$

Now, in terms of state-variables the equations (6.78) can be rewritten as

$$Y(s) = X_1(s) + X_2(s) + X_3(s)$$

Taking inverse Lapalace transform, we get

$$y(t) = x_1(t) + x_2(t) + x_3(t)$$

or, in matrix form

$$y(t) = [1 \;\; 1 \;\; 1] \begin{bmatrix} x_1(t) \\ x_2(t) \\ x_3(t) \end{bmatrix} \quad \text{...(6.79 B)}$$

Equations (6.79) give the desired DCF of state-model.

Alternative Method

The DCF can also be obtained using similarity transformation. For this, convert transfer function into phase-variable form as discussed below:

$$\frac{Y(s)}{U(s)} = \frac{2}{s^3 + 6s^2 + 11s + 6}$$

or,

$$s^3Y(s) = 2U(s) - 6s^2Y(s) - 11sY(s) - 6Y(s) \quad \text{... (6.80)}$$

Define state-variables as

$$X_1(s) = Y(s)$$

$$X_2(s) = sY(s)$$

$$X_3(s) = s^2Y(s)$$

Then clearly,

$$\left.\begin{aligned} sX_1(s) &= sY(s) = X_2(s) \\ sX_2(s) &= s^2Y(s) = X_3(s) \\ sX_3(s) &= s^3Y(s) = 2U(s) - 6s^2Y(s) - 11sY(s) - 6Y(s) \\ &= 2U(s) - 6X_3(s) - 11X_2(s) - 6X_1(s) \end{aligned}\right\} \quad \text{... (6.81)}$$

Taking inverse Laplace transforms of set of equations (6.81); assuming all initial conditions to be zero, we get

$$\dot{x}_1(t) = x_2(t)$$

$$\dot{x}_2(t) = x_3(t)$$

$$\dot{x}_3(t) = 2u(t) - 6x_3(t) - 11x_2(t) - 6x_1(t)$$

or, in matrix form,

$$\begin{bmatrix} \dot{x}_1(t) \\ \dot{x}_2(t) \\ \dot{x}_3(t) \end{bmatrix} = \begin{bmatrix} 0 & 1 & 0 \\ 0 & 0 & 1 \\ -6 & -11 & -6 \end{bmatrix} \begin{bmatrix} x_1(t) \\ x_2(t) \\ x_3(t) \end{bmatrix} + \begin{bmatrix} 0 \\ 0 \\ 2 \end{bmatrix} u(t) \quad \text{...(6.82 A)}$$

Also, $Y(s) = X_1(s)$

Taking inverse Laplace transform, we get

$$y(t) = x_1(t)$$

or,

$$y(t) = [1 \quad 0 \quad 0] \begin{bmatrix} x_1(t) \\ x_2(t) \\ x_3(t) \end{bmatrix} \quad \text{...(6.82 B)}$$

Equation (6.82) give the state-model in phase-variable form.

Further proceeding as in example (6.9), we get the desired answer as

$$\begin{bmatrix} \dot{z}_1(t) \\ \dot{z}_2(t) \\ \dot{z}_3(t) \end{bmatrix} = \begin{bmatrix} -1 & 0 & 0 \\ 0 & -2 & 0 \\ 0 & 0 & -3 \end{bmatrix} \begin{bmatrix} z_1(t) \\ z_2(t) \\ z_3(t) \end{bmatrix} + \begin{bmatrix} 1 \\ -2 \\ 1 \end{bmatrix} u(t) \quad \text{...(6.83 A)}$$

$$y(t) = [1 \quad 1 \quad 1] \begin{bmatrix} z_1(t) \\ z_2(t) \\ z_3(t) \end{bmatrix} \quad \text{...(6.83 B)}$$

Equations (6.83) give the state-model of the system in diagonal canonical form (DCF).

Example 6.11. *Consider the following transfer function*

$$\frac{Y(s)}{U(s)} = \frac{2s^2 + 6s + 7}{(s+1)^2(s+2)}$$

Represent it in Jordan canonical form.

Solution. Notice that the system has repeated roots.

Now,

$$\frac{Y(s)}{U(s)} = \frac{2s^2 + 6s + 7}{(s+1)^2(s+2)}$$

or,

$$\frac{Y(s)}{U(s)} = \frac{3}{(s+1)^2} + \frac{(-1)}{(s+1)} + \frac{3}{(s+2)}$$

or,

$$Y(s) = \frac{3}{(s+1)^2}U(s) + \frac{(-1)}{(s+1)}U(s) + \frac{3}{(s+2)}U(s) \quad \text{...(6.84)}$$

Define state-variables as

$$X_1(s) = \frac{1}{(s+1)^2}U(s)$$

$$X_2(s) = \frac{1}{(s+1)}U(s)$$

$$X_3(s) = \frac{1}{(s+2)}U(s)$$

However, $X_1(s)$ and $X_2(s)$ are related as

$$\frac{X_1(s)}{X_2(s)} = \frac{1}{(s+1)}$$

Then clearly,

$$sX_1(s) = -X_1(s) + X_2(s)$$

$$sX_2(s) = -X_2(s) + U(s)$$

$$sX_3(s) = -2X_3(s) + U(s)$$

and, $$Y(s) = 3X_1(s) - X_2(s) + 3X_3(s)$$

Taking inverse Laplace transforms; assuming all initial conditions to be zero, we get

$$\left.\begin{aligned} \dot{x}_1(t) &= -x_1(t) + x_2(t) \\ \dot{x}_2(t) &= -x_2(t) + u(t) \\ \dot{x}_3(t) &= -2x_3(t) + u(t) \end{aligned}\right\} \text{ State Equations}$$

and, $$y(t) = 3x_1(t) - x_2(t) + 3x_3(t) \text{ ; Output Equation}$$

or, in matrix form,

Jordan Block

$$\begin{bmatrix} \dot{x}_1(t) \\ \dot{x}_2(t) \\ \dot{x}_3(t) \end{bmatrix} = \left[\begin{array}{cc|c} -1 & 1 & 0 \\ 0 & -1 & 0 \\ \hline 0 & 0 & 2 \end{array}\right] \begin{bmatrix} x_1(t) \\ x_2(t) \\ x_3(t) \end{bmatrix} + \begin{bmatrix} 0 \\ 1 \\ 1 \end{bmatrix} u(t) \quad \text{...(6.85 A)}$$

$$y(t) = [3 \;\; -1 \;\; 3] \begin{bmatrix} x_1(t) \\ x_2(t) \\ x_3(t) \end{bmatrix} \quad \text{...(6.85 B)}$$

Equations (6.85) give the state-model of the system in Jordan canonical form.

6.13 STATE-SPACE REPRESENTATION OF DISCRETE-TIME SYSTEMS

The state-space approach for analysis and design of continuous-time systems can be extended for the analysis and design of discrete-time systems. The discrete form of the state-space representation is quite analogous to the continuous form. The general form of state-model for a **multivariable** discrete-time system is

$$\mathbf{x}(k+1) = \mathbf{f}(\mathbf{x}(k), \mathbf{u}(k)) \quad ; \text{ State Equation} \quad \text{...(6.86 A)}$$

$$\mathbf{y}(k) = \mathbf{g}(\mathbf{x}(k), \mathbf{u}(k)) \quad ; \text{ Output Equation} \quad \text{...(6.86 B)}$$

For linear time-invariant system, we may write equations (6.86) as

$$\mathbf{x}(k+1) = \mathbf{A}\mathbf{x}(k) + \mathbf{B}\mathbf{u}(k) \quad ; \quad \text{State Equation} \qquad \text{...(6.87 A)}$$

$$\mathbf{y}(k) = \mathbf{C}\mathbf{x}(k) + \mathbf{D}\mathbf{u}(k) \quad ; \quad \text{Output Equation} \qquad \text{...(6.87 B)}$$

where,

$\mathbf{x}(k) \longrightarrow n \times 1$ State Vector

$\mathbf{u}(k) \longrightarrow m \times 1$ Input Vector

$\mathbf{y}(k) \longrightarrow p \times 1$ Output Vector

$\mathbf{A} \longrightarrow n \times n$ System Matrix

$\mathbf{B} \longrightarrow n \times m$ Input Matrix

$\mathbf{C} \longrightarrow p \times n$ Output Matrix

$\mathbf{D} \longrightarrow p \times m$ Transmission Matrix

The block-diagram representation of the state-model of linear multi-variable time-invariant system is shown in Figure 6.10.

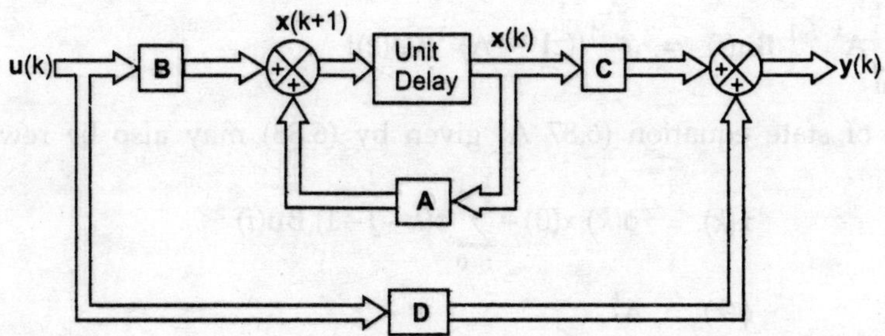

FIGURE 6.10 *Block-diagram representation of the state-model of linear multi-variable time-invariant system.*

6.14 SOLUTION OF STATE-EQUATIONS FOR LINEAR TIME-INVARIANT DISCRETE-TIME SYSTEMS

6.14.1 By use of Recursion Procedure

Consider equation (6.87 A), we may write

$$\mathbf{x}(1) = \mathbf{A}\mathbf{x}(0) + \mathbf{B}\mathbf{u}(0)$$

$$\begin{aligned}\mathbf{x}(2) &= \mathbf{A}\mathbf{x}(1) + \mathbf{B}\mathbf{u}(1)\\ &= \mathbf{A}\{\mathbf{A}\mathbf{x}(0) + \mathbf{B}\mathbf{u}(0)\} + \mathbf{B}\mathbf{u}(1)\\ &= \mathbf{A}^2\mathbf{x}(0) + \mathbf{A}\mathbf{B}\mathbf{u}(0) + \mathbf{B}\mathbf{u}(1)\\ &\vdots \qquad\qquad \vdots \qquad\qquad \vdots\end{aligned}$$

$$\mathbf{x}(k) = \mathbf{A}^k\mathbf{x}(0) + \mathbf{A}^{k-1}\mathbf{B}\mathbf{u}(0) + \mathbf{A}^{k-2}\mathbf{B}\mathbf{u}(1) + \dots + \mathbf{B}\mathbf{u}(k-1)$$

or, we may write

$$\mathbf{x}(k) = \mathbf{A}^k\mathbf{x}(0) + \sum_{i=0}^{k-1}\mathbf{A}^{k-i-1}\,\mathbf{B}\mathbf{u}(i) \qquad \text{... (6.88)}$$

Equation (6.88) gives the **solution** of state equation (6.87 A).

6.14.2 z-Transform Approach to the Solution of Discrete-Time State Equation

Consider equation (5.87 A). Taking the z-transform on both sides of equation (5.87 A), we have

$$z\mathbf{X}(z) - z\mathbf{x}(0) = \mathbf{AX}(z) + \mathbf{BU}(z)$$

or,
$$(z\mathbf{I} - \mathbf{A})\mathbf{X}(z) = z\mathbf{x}(0) + \mathbf{BU}(z)$$

or,
$$\mathbf{X}(z) = (z\mathbf{I} - \mathbf{A})^{-1}z\mathbf{x}(0) + (z\mathbf{I} - \mathbf{A})^{-1}\mathbf{BU}(z) \qquad ...(6.89)$$

Taking inverse z-transform on both sides of equation (6.89), we have

$$\mathbf{x}(k) = \mathcal{Z}^{-1}\{(z\mathbf{I} - \mathbf{A})^{-1}z\}\mathbf{x}(0) + \mathcal{Z}^{-1}\{z\mathbf{I} - \mathbf{A})^{-1}\mathbf{BU}(z)\} \qquad ...(6.90)$$

Equation (6.90) also gives the **Solution** of state equation (6.87 A).

Now, comparing equations (6.88) and (6.90), we have two important results, *i.e.*,

$$\mathbf{A}^k = \mathcal{Z}^{-1}\{(z\mathbf{I} - \mathbf{A})^{-1}z\} \qquad ...(6.91)$$

and,
$$\sum_{i=0}^{k-1} \mathbf{A}^{k-i-1}\,\mathbf{Bu}(i) = \mathcal{Z}^{-1}\{(z\mathbf{I} - \mathbf{A})^{-1}\mathbf{BU}(z)\} \qquad ...(6.92)$$

The solution of state equation (6.87 A) given by (6.88) may also be rewritten as,

$$\mathbf{x}(k) = \phi(k)\,\mathbf{x}(0) + \sum_{i=0}^{k-1} \phi(k-i-1)\,\mathbf{Bu}(i) \qquad ...(6.93)$$

where,
$$\phi(k) = \mathbf{A}^k$$

and is referred to as the **State Transition Matrix** for discrete-time system described by state-model (6.87). The **properties** of state transition matrix, $\phi(k)$ are summarized below:

1. $\phi(0) = \mathbf{I}$
2. $\phi^{-1}(k) = \phi(-k)$
3. $\phi(k, k_0) = \phi(k - k_0) = \mathbf{A}^{(k-k_0)}\ ;\ k > k_0$

Example 6.12. *Determine the solution of the discrete-time state equation,*

$$\mathbf{x}(k+1) = \begin{bmatrix} 0 & 1 \\ -0.16 & -1 \end{bmatrix}\mathbf{x}(k) + \begin{bmatrix} 1 \\ 1 \end{bmatrix}u(k)$$

where, u(k) = 1 for k = 0, 1, 2, ... Assume that initial conditions are

$$\mathbf{x}(0) = \begin{bmatrix} x_1(0) \\ x_2(0) \end{bmatrix} = \begin{bmatrix} 1 \\ -1 \end{bmatrix}$$

Solution. Here,
$$\mathbf{A} = \begin{bmatrix} 0 & 1 \\ -0.16 & -1 \end{bmatrix}$$

$$\therefore \qquad [z\mathbf{I} - \mathbf{A}] = \begin{bmatrix} z & -1 \\ 0.16 & z+1 \end{bmatrix}$$

Now,
$$\text{Adj}\,[z\mathbf{I} - \mathbf{A}] = \begin{bmatrix} z+1 & -0.16 \\ 1 & z \end{bmatrix}^T = \begin{bmatrix} z+1 & 1 \\ -0.16 & z \end{bmatrix}$$

and,
$$|z\mathbf{I} - \mathbf{A}| = z(z+1) + 0.16 = z^2 + z + 0.16 = (z+0.2)(z+0.8)$$

$\therefore$
$$[z\mathbf{I} - \mathbf{A}]^{-1} = \frac{\text{Adj}\,[z\mathbf{I}-\mathbf{A}]}{|z\mathbf{I}-\mathbf{A}|} = \begin{bmatrix} \dfrac{(z+1)}{(z+0.2)(z+0.8)} & \dfrac{1}{(z+0.2)(z+0.8)} \\ \dfrac{-0.16}{(z+0.2)(z+0.8)} & \dfrac{z}{(z+0.2)(z+0.8)} \end{bmatrix}$$

Now,
$$\mathbf{X}(z) = (z\mathbf{I} - \mathbf{A})^{-1} z\mathbf{x}(0) + (z\mathbf{I} - \mathbf{A})^{-1}\mathbf{B}\mathbf{U}(z) = (z\mathbf{I} - \mathbf{A})^{-1}[z\mathbf{x}(0) + \mathbf{B}\mathbf{U}(z)]$$

$\because$
$$\mathbf{U}(z) = \frac{z}{(z-1)}$$

$\therefore$
$$z\mathbf{x}(0) + \mathbf{B}\mathbf{U}(z) = \begin{bmatrix} z \\ -z \end{bmatrix} + \begin{bmatrix} \dfrac{z}{(z-1)} \\ \dfrac{z}{(z-1)} \end{bmatrix} = \begin{bmatrix} \dfrac{z^2}{(z-1)} \\ \dfrac{(-z^2+2z)}{(z-1)} \end{bmatrix}$$

Thus,
$$\mathbf{X}(z) = \begin{bmatrix} \dfrac{(z+1)}{(z+0.2)(z+0.8)} & \dfrac{1}{(z+0.2)(z+0.8)} \\ \dfrac{-0.16}{(z+0.2)(z+0.8)} & \dfrac{z}{(z+0.2)(z+0.8)} \end{bmatrix} \begin{bmatrix} \dfrac{z^2}{(z-1)} \\ \dfrac{(-z^2+2z)}{(z-1)} \end{bmatrix}$$

,
$$= \begin{bmatrix} \dfrac{(z^2+2)z}{(z+0.2)(z+0.8)(z-1)} \\ \dfrac{(-z^2+1.84\,z)z}{(z+0.2)(z+0.8)(z-1)} \end{bmatrix} = \begin{bmatrix} \left(\dfrac{-\frac{17}{6}z}{(z+0.2)} + \dfrac{\frac{22}{9}z}{(z+0.8)} + \dfrac{\frac{25}{18}z}{(z-1)} \right) \\ \left(\dfrac{\frac{3.4}{6}z}{(z+0.2)} + \dfrac{-\frac{17.6}{9}z}{(z+0.8)} + \dfrac{\frac{7}{18}z}{(z-1)} \right) \end{bmatrix}$$

Therefore,
$$\mathbf{x}(k) = \mathcal{Z}^{-1}\,\mathbf{X}(z) = \begin{bmatrix} -\dfrac{17}{6}(-0.2)^k + \dfrac{22}{9}(-0.8)^k + \dfrac{25}{18} \\ \dfrac{3.4}{6}(-0.2)^k - \dfrac{17.6}{9}(-0.8)^k + \dfrac{7}{18} \end{bmatrix}$$

6.15 z-TRANSFER FUNCTION FROM STATE MODEL

Consider the discrete-time state model assuming scalar input, given as

$$\left.\begin{aligned} \mathbf{x}(k+1) &= \mathbf{A}\mathbf{x}(k) + \mathbf{B}u(k) \quad ; \text{ State Equation} \\ y(k) &= \mathbf{C}\mathbf{x}(k) + du(k) \quad ; \text{ Output Equation} \end{aligned}\right\} \qquad \text{...(6.94)}$$

Taking the z-transform on both sides of equation (6.94), we have

$$z\mathbf{X}(z) - z\mathbf{x}(0) = \mathbf{A}\mathbf{X}(z) + \mathbf{B}U(z) \qquad \text{...(6.95 A)}$$

$$Y(z) = \mathbf{C}\mathbf{X}(z) + dU(z) \qquad \text{...(6.95 B)}$$

Consider equation (6.95 A), we have

$$(z\mathbf{I} - \mathbf{A})\mathbf{X}(z) = \mathbf{B}U(z) + z\mathbf{x}(0)$$

Assuming zero initial conditions, *i.e.*, $\mathbf{x}(0) = \mathbf{0}$, we get

$$(z\mathbf{I} - \mathbf{A})\mathbf{X}(z) = \mathbf{B}U(z)$$

or,
$$\mathbf{X}(z) = (z\mathbf{I} - \mathbf{A})^{-1}\,\mathbf{B}U(z) \qquad \text{...(6.96)}$$

Put $\mathbf{X}(z)$ from equation (6.96) in (6.95 B), we obtain

$$Y(z) = \mathbf{C}(z\mathbf{I} - \mathbf{A})^{-1}\mathbf{B}U(z) + dU(z)$$

or,
$$\frac{Y(z)}{U(z)} = \mathbf{C}(z\mathbf{I} - \mathbf{A})^{-1}\mathbf{B} + d$$

or,
$$G(z) = \mathbf{C}(z\mathbf{I} - \mathbf{A})^{-1}\,\mathbf{B} + d \qquad \text{...(6.97)}$$

Equation (6.97) gives the expression of z-transfer function for discrete-time system with scalar input.

PROBLEMS

1. Determine STM when **A** is given by

$$\mathbf{A} = \begin{bmatrix} 0 & 1 & 0 \\ 0 & 0 & 1 \\ 1 & -3 & 3 \end{bmatrix}$$

2. Given $\mathbf{A} = \begin{bmatrix} 0 & 1 \\ 0 & -2 \end{bmatrix}$, compute $e^{\mathbf{A}t}$.

3. Obtain the response $y(t)$ of the system described by

$$\begin{bmatrix} \dot{x}_1 \\ \dot{x}_2 \end{bmatrix} = \begin{bmatrix} -1 & -0.5 \\ 1 & 0 \end{bmatrix}\begin{bmatrix} x_1 \\ x_2 \end{bmatrix} + \begin{bmatrix} 0.5 \\ 0 \end{bmatrix} u$$

$$y = [1 \quad 0]\begin{bmatrix} x_1 \\ x_2 \end{bmatrix}$$

where, $u(t)$ is the unit step occurring at $t = 0$ and $\mathbf{x}^T(0) = [0 \quad 0]$.

4. A system is described by

$$\dot{\mathbf{x}}(t) = \begin{bmatrix} -1 & 1 \\ 0 & -2 \end{bmatrix}\begin{bmatrix} x_1 \\ x_2 \end{bmatrix} + \begin{bmatrix} 1 & 0 & 1 \\ 0 & 1 & 1 \end{bmatrix} u(t)$$

$$y(t) = \begin{bmatrix} 1 & 2 \\ 1 & 0 \\ 1 & 1 \end{bmatrix} \mathbf{x}(t)$$

Obtain the transfer function of the system.

5. Consider a system $\dot{\mathbf{x}}(t) = \mathbf{A}\mathbf{x}(t)$. For this system with $\mathbf{x}(0) = \begin{bmatrix} 1 \\ -2 \end{bmatrix}$, the response is given by

$$\mathbf{x}(t) = \begin{bmatrix} e^{-2t} \\ -2e^{-2t} \end{bmatrix}$$

Determine the system matrix **A** and the state transition matrix.

6. Consider the following matrix

$$\mathbf{A} = \begin{bmatrix} 0 & 1 \\ -6 & -5 \end{bmatrix}$$

Find the STM. Also determine $\mathbf{x}(t)$, given that

$$\mathbf{x}(0) = \begin{bmatrix} 1 \\ 0 \end{bmatrix}$$

7. Given, $\mathbf{A}_1 = \begin{bmatrix} \sigma & 0 \\ 0 & \sigma \end{bmatrix}; \quad \mathbf{A}_2 = \begin{bmatrix} 0 & \omega \\ -\omega & \sigma \end{bmatrix}$

and $\mathbf{A} = \begin{bmatrix} \sigma & \omega \\ -\omega & \sigma \end{bmatrix}$

Compute $e^{\mathbf{A}t}$.

[Hint: $e^{\mathbf{A}t} = e^{(\mathbf{A}_1+\mathbf{A}_2)t} = e^{\mathbf{A}_1 t} \cdot e^{\mathbf{A}_2 t}$ provided $\mathbf{A}_1\mathbf{A}_2 = \mathbf{A}_2\mathbf{A}_1$**]**.

8. Show that the STM for the state equation

$$\begin{bmatrix} \dot{x}_1 \\ \dot{x}_2 \end{bmatrix} = \begin{bmatrix} -3 & 1 \\ 0 & -1 \end{bmatrix}\begin{bmatrix} x_1 \\ x_2 \end{bmatrix} \text{ is } \begin{bmatrix} e^{-3t} & \frac{1}{2}(e^{-t} - e^{-3t}) \\ 0 & -e^{-t} \end{bmatrix}$$

9. A system is described by the following differential equation

$$\dddot{x} + 3\ddot{x} + 4\dot{x} + 4x = u_1 + 3u_2 + 4u_3$$

and outputs are

$$y_1 = 4\dot{x} + 3u_1$$

$$y_2 = \ddot{x} + 4u_2 + u_3$$

Represent the system in state-space.

10. Write the state equations for the circuit shown below :

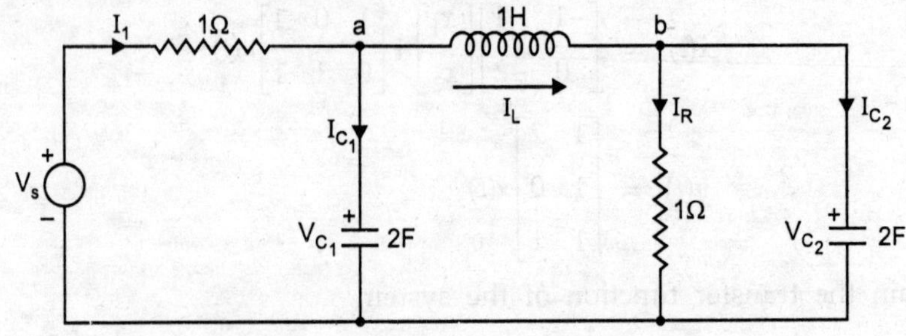

FIGURE 6.11 *Circuit of Problem (10).*

11. Use diagonalization of matrix **A** to determine the time-response of the system :

$$\begin{bmatrix} \dot{x}_1 \\ \dot{x}_2 \end{bmatrix} = \begin{bmatrix} 1 & 1 \\ -6 & -5 \end{bmatrix} \begin{bmatrix} x_1 \\ x_2 \end{bmatrix} + \begin{bmatrix} 1 \\ 0 \end{bmatrix} u$$

and
$$y = \begin{bmatrix} 6 & 1 \end{bmatrix} \begin{bmatrix} x_1 \\ x_2 \end{bmatrix}$$

Given that
$$\begin{bmatrix} x_1(0) \\ x_2(0) \end{bmatrix} = \begin{bmatrix} 1 \\ 0 \end{bmatrix}$$

12. A discrete-time system is described by the difference equation

$$y(k+2) + 5y(k+1) + 6y(k) = u(k)$$
$$y(0) = y(1) = 0;\ T = 1 \text{ sec.}$$

(*a*) Determine the state-model in normal form

(*b*) Compute STM.

(*c*) For input $u(k) = 1;\ k \geq 0$ find the output $y(k)$.

13. Determine the frequency response of the continuous-time system with state-variable description

$$\mathbf{A} = \begin{bmatrix} 2 & -1 \\ 1 & 0 \end{bmatrix};\ \mathbf{B} = \begin{bmatrix} 1 \\ 0 \end{bmatrix};\ \mathbf{C} = \begin{bmatrix} 3 & 1 \end{bmatrix};\ d = 0.$$

Appendix A

MATRIX ALGEBRA

A.1 DEFINITIONS

A **matrix** is an **ordered rectangular array** of elements which may be real or complex numbers, functions or operators. The rectangular array

$$\mathbf{A} \overset{\Delta}{=} \begin{bmatrix} a_{11} & a_{12} & \cdots & a_{1n} \\ a_{21} & a_{22} & \cdots & a_{2n} \\ \vdots & \vdots & \ddots & \vdots \\ a_{m1} & a_{m2} & \cdots & a_{mn} \end{bmatrix}_{m \times n} \quad \text{...(A.1)}$$

is called a matrix of m rows and n columns or simply a matrix of **order** $m \times n$. We shall use **UPPER CASE, BOLD FACE LETTERS** such as **A** to denote **matrices** and corresponding *lower case, light face letters* with subscripts such as a_{ij} to denote its elements.

A.1.1 Square Matrix

If for a matrix $m = n$, *i.e.*, the number of rows is equal to that of columns, the matrix is said to be a **square matrix** of order n.

A.1.2 Row and Column Vectors

A matrix having only one row but n columns, *i.e.*, a matrix of order $1 \times n$ is known as a **row vector.** Similarly, a $m \times 1$ matrix, *i.e.*, a matrix having m rows but only one column is known as a **column vector.**

A.1.3 Diagonal Matrix

A matrix in which **all elements except** the **diagonal elements** are **zero** ($a_{ij} = 0$ for $i \neq j$), is called a **diagonal matrix.**

e.g.,

$$A = \begin{bmatrix} 1 & 0 & 0 \\ 0 & 8 & 0 \\ 0 & 0 & 6 \end{bmatrix} \text{ is a diagonal matrix of order } 3 \times 3.$$

A.1.4 Null (or Zero) Matrix

A matrix in which **all elements** are **zero** ($a_{ij} = 0$ for all i and j) is called a **null** (or **zero**) **matrix** and is denoted by **0**.

e.g.,

$$\mathbf{0} = \begin{bmatrix} 0 & 0 & 0 \\ 0 & 0 & 0 \\ 0 & 0 & 0 \end{bmatrix} \text{ is a null matrix of order } 3 \times 3.$$

A.1.5 Unit (or Identity) Matrix

A **unit matrix,** denoted by **I**, is a **diagonal matrix** with **all diagonal elements** equal to **unity**.

e.g.,

$$\mathbf{I} = \begin{bmatrix} 1 & 0 & 0 & 0 \\ 0 & 1 & 0 & 0 \\ 0 & 0 & 1 & 0 \\ 0 & 0 & 0 & 1 \end{bmatrix} \text{ is a unit matrix of order } 4 \times 4.$$

A.1.6 Determinant of a Square Matrix

The **determinant** of a **square matrix A,** denoted by det **A** or |**A**|, is formed by taking the **determinant** of the elements of the matrix **A**.

e.g.,

$$\text{If} \quad \mathbf{A} = \begin{bmatrix} 1 & 2 & -2 \\ 4 & 6 & 1 \\ 9 & 5 & 3 \end{bmatrix} \text{ then } |\mathbf{A}| = 1\begin{vmatrix} 6 & 1 \\ 5 & 3 \end{vmatrix} - 2\begin{vmatrix} 4 & 1 \\ 9 & 3 \end{vmatrix} + (-2)\begin{vmatrix} 4 & 6 \\ 9 & 5 \end{vmatrix}$$

$$= 1(13) - 2(3) - 2(-34) = 75.$$

A.1.7 Singular and Non-Singular Matrices

A **square matrix** is called **singular** when its **associated determinant** is **zero** and is called **non-singular** when its **associated determinant** is **non-zero.**

A.1.8 Transpose of a Matrix

The **transpose** of matrix **A**, denoted by $\mathbf{A}^T$, is a matrix formed by **interchanging** the rows and columns of **A**.

e.g.,

If $$\mathbf{A} = \begin{bmatrix} 1 & 4 & 2 \\ 3 & 8 & 0 \\ 7 & 5 & 6 \end{bmatrix} \text{ then } \mathbf{A}^T = \begin{bmatrix} 1 & 3 & 7 \\ 4 & 8 & 5 \\ 2 & 0 & 6 \end{bmatrix}$$

It easily follows that,

$$(\mathbf{A}^T)^T = \mathbf{A} \qquad \text{...(A.2)}$$

A.1.9 Symmetric Matrix

A **square matrix** is said to be **symmetric** if it is equal to its transpose ($\mathbf{A}^T = \mathbf{A}$).

e.g.,

$$\mathbf{A} = \begin{bmatrix} 2 & -1 & 3 \\ -1 & 8 & 5 \\ 3 & 5 & 6 \end{bmatrix} \text{ is a symmetric matrix.}$$

A.1.10 Minor

The **minor** M_{ij} of the element a_{ij} of a $n \times n$ matrix **A** is the **determinant** of $(n-1) \times (n-1)$ matrix obtained by **deleting** i^{th} row and j^{th} column of matrix **A**.

A.1.11 Cofactor

The **cofactor** A_{ij} of element a_{ij} of matrix **A** is defined as,

$$A_{ij} = (-1)^{i+j} M_{ij} \qquad \text{...(A.3)}$$

A.1.12 Adjoint Matrix

The **adjoint matrix** of a **square matrix A**, denoted by Adj *A*, is obtained by **replacing** each element a_{ij} of matrix **A** by its **cofactor** A_{ij} and then **transposing** it.

e.g.,

If $$\mathbf{A} = \begin{bmatrix} 1 & 2 & 1 \\ -1 & -4 & -3 \\ -1 & 1 & 3 \end{bmatrix} \text{ then Adj } \mathbf{A} = \begin{bmatrix} +\begin{vmatrix} -4 & -3 \\ 1 & 3 \end{vmatrix} & -\begin{vmatrix} -1 & -3 \\ -1 & 3 \end{vmatrix} & +\begin{vmatrix} -1 & -4 \\ -1 & 1 \end{vmatrix} \\ -\begin{vmatrix} 2 & 1 \\ 1 & 3 \end{vmatrix} & +\begin{vmatrix} 1 & 1 \\ -1 & 3 \end{vmatrix} & -\begin{vmatrix} 1 & 2 \\ -1 & 1 \end{vmatrix} \\ +\begin{vmatrix} 2 & 1 \\ -4 & -3 \end{vmatrix} & -\begin{vmatrix} 1 & 1 \\ -1 & -3 \end{vmatrix} & +\begin{vmatrix} 1 & 2 \\ -1 & -4 \end{vmatrix} \end{bmatrix}^T$$

$$= \begin{bmatrix} -9 & 6 & -5 \\ -5 & 4 & -3 \\ -2 & 2 & -2 \end{bmatrix}^T = \begin{bmatrix} -9 & -5 & -2 \\ 6 & 4 & 2 \\ -5 & -3 & -2 \end{bmatrix}$$

A.1.13 Sub-Matrix

Any matrix obtained from a $m \times n$ matrix **A** by **deleting** some rows and / or some columns is called **sub-matrix** of matrix **A.**

A.1.14 Rank of a Matrix

A **number** r is called the **rank** of a $n \times n$ matrix **A** if it **satisfies** the following conditions:

(*i*) **All** $p \times p$ **sub-matrices** with $p > r$ of matrix **A** have **determinants** equal to **zero,** and

(*ii*) **At least one** $r \times r$ **sub-matrix** of matrix **A** has **non-zero determinant.**

e.g.,

Consider a 3 × 3 matrix,

$$\mathbf{A} = \begin{bmatrix} 3 & 0 & 0 \\ 4 & 1 & -2 \\ 1 & 1 & -2 \end{bmatrix}$$

Since $\begin{vmatrix} 3 & 0 & 0 \\ 4 & 1 & -2 \\ 1 & 1 & -2 \end{vmatrix} = 0$ and $\begin{vmatrix} 3 & 0 \\ 4 & 1 \end{vmatrix} \neq 0$

∴ Rank **A** = 2 while $n = 3$.

A.2 EQUALITY OF MATRICES

Two matrices **A** and **B** are said to be **equal** if they are of the **same order** and **corresponding elements** of the two are **identical,** *i.e.,* if matrices **A** and **B** both are of order $m \times n$ and $a_{ij} = b_{ij}$ for $i = 1, 2, 3, \ldots, m$; $j = 1, 2, 3, \ldots, n$ then **A** = **B.**

A.3 ADDITION OF TWO MATRICES

The **sum** of two matrices of **same order** is found by adding the corresponding elements, *i.e.,*

If $\mathbf{C} = \mathbf{A} + \mathbf{B}$ then $c_{ij} = a_{ij} + b_{ij}$ ...(A.4)

It easily follows that,

$$\mathbf{A} + \mathbf{B} = \mathbf{B} + \mathbf{A} \quad \text{...(A.5)}$$

and, $(\mathbf{A} + \mathbf{B})^T = \mathbf{A}^T + \mathbf{B}^T$

e.g.,

Consider $\mathbf{A} = \begin{bmatrix} 3 & 2 & -1 \\ 7 & -4 & 6 \\ 5 & 11 & 2 \end{bmatrix}$ and $\mathbf{B} = \begin{bmatrix} 2 & -4 & 7 \\ -1 & 3 & 6 \\ 8 & 9 & -4 \end{bmatrix}$

Then, $\mathbf{C} = \mathbf{A} + \mathbf{B} = \begin{bmatrix} 5 & -2 & 6 \\ 6 & -1 & 12 \\ 13 & 20 & -2 \end{bmatrix}$

A.4 MULTIPLICATION OF A MATRIX BY A SCALAR

Multiplication of a matrix by a **scalar** is equivalent to multiplying all its elements by that scalar. Therefore,

If $$\mathbf{A} = \begin{bmatrix} a_{11} & a_{12} & \cdots & a_{1n} \\ a_{21} & a_{22} & \cdots & a_{2n} \\ \vdots & \vdots & \ddots & \vdots \\ a_{m1} & a_{m2} & \cdots & a_{mn} \end{bmatrix} \text{ then } k\mathbf{A} = \begin{bmatrix} ka_{11} & ka_{12} & \cdots & ka_{1n} \\ ka_{21} & ka_{22} & \cdots & ka_{2n} \\ \vdots & \vdots & \ddots & \vdots \\ ka_{m1} & ka_{m2} & \cdots & ka_{mn} \end{bmatrix} \quad \text{...(A.6)}$$

where, k is a scalar.

A.5 MULTIPLICATION OF TWO MATRICES

The **multiplication** of matrices **A** and **B** is defined **only if** the number of columns of matrix **A** is **equal** to the number of rows of matrix **B**. The matrices are then said to be **conformable.** Thus, if matrix **A** is of order $m \times n$ and matrix **B** is of order $n \times p$ then the product **C** = **AB** would be of order $m \times p$.

The elements c_{ij} of matrix **C** are obtained by **multiplying** the elements of the ith row of matrix **A** with the **corresponding** elements of jth column of matrix **B** and then **summing** these **element products**. Mathematically, we can write

$$c_{ij} = \sum_{k=1}^{n} a_{ik} b_{kj} \quad \textit{for } i = 1, 2, 3, \ldots, m \; ; \; j = 1, 2, 3, \ldots, p \quad \text{...(A.7)}$$

e.g.,

Consider $$\mathbf{A} = \begin{bmatrix} a_{11} & a_{12} \\ a_{21} & a_{22} \end{bmatrix}_{2\times 2} \text{ and } \mathbf{B} = \begin{bmatrix} b_{11} & b_{12} & b_{13} \\ b_{21} & b_{22} & b_{23} \end{bmatrix}_{2\times 3}$$

Then, $$\mathbf{C} = \mathbf{AB} = \begin{bmatrix} (a_{11}b_{11} + a_{12}b_{21}) & (a_{11}b_{12} + a_{12}b_{22}) & (a_{11}b_{13} + a_{12}b_{23}) \\ (a_{21}b_{11} + a_{22}b_{21}) & (a_{21}b_{12} + a_{22}b_{22}) & (a_{21}b_{13} + a_{22}b_{23}) \end{bmatrix}_{2\times 3}$$

It is important to note that **matrix multiplication**, in general, is **not commutative,** *i.e.,*

$$\mathbf{AB} \neq \mathbf{BA} \quad \text{...(A.8)}$$

However, the **associative** and **distributive laws hold good** for **matrix multiplication** provided, the appropriate operations are defined.

Associative Law:

$$(\mathbf{AB})\mathbf{C} = \mathbf{A}(\mathbf{BC}) = \mathbf{ABC} \quad \text{...(A.9)}$$

Distributive Law:

$$\mathbf{A}(\mathbf{B} + \mathbf{C}) = \mathbf{AB} + \mathbf{AC} \quad \text{...(A.10)}$$

Multiplication of any matrix by a **unit matrix** results in the **original matrix,** *i.e.,*

$$\mathbf{AI} = \mathbf{IA} = \mathbf{A} \quad \text{...(A.11)}$$

The **transpose** of the **product** of two matrices is the **product** of their **transposes in reverse order**, *i.e.*,

$$(\mathbf{AB})^T = \mathbf{B}^T\ \mathbf{A}^T \qquad \text{...(A.12)}$$

A.6 INVERSE OF A MATRIX

The **inverse** of a **non-singular square matrix** A, denoted by $\mathbf{A}^{-1}$, is defined by the relation

$$\mathbf{A}^{-1}\mathbf{A} = \mathbf{A}\mathbf{A}^{-1} = \mathbf{I} \qquad \text{...(A.13)}$$

and is obtained by using the expression

$$\mathbf{A}^{-1} = \frac{\text{Adj } \mathbf{A}}{\det \mathbf{A}}\ ; \ \det \mathbf{A} \neq 0 \qquad \text{...(A.14)}$$

e.g.,

If $$\mathbf{A} = \begin{bmatrix} 1 & 2 & 1 \\ -1 & -4 & -3 \\ -1 & 1 & 3 \end{bmatrix}$$

Then, $$\text{Adj } \mathbf{A} = \begin{bmatrix} -9 & 6 & -5 \\ -5 & 4 & -3 \\ -2 & 2 & -2 \end{bmatrix}^T = \begin{bmatrix} -9 & -5 & -2 \\ 6 & 4 & 2 \\ -5 & -3 & -2 \end{bmatrix}$$

and, $$\det \mathbf{A} = \begin{vmatrix} 1 & 2 & 1 \\ -1 & -4 & -3 \\ 1 & 1 & 3 \end{vmatrix} = -2$$

$$\mathbf{A}^{-1} = \frac{\text{Adj } \mathbf{A}}{\det \mathbf{A}} = \begin{bmatrix} 4.5 & 2.5 & 1 \\ -3 & -2 & -1 \\ 2.5 & 1.5 & 1 \end{bmatrix}$$

The following are the **important properties** characterizing the **inverse**:

$$(\mathbf{AB})^{-1} = \mathbf{B}^{-1}\mathbf{A}^{-1} \qquad \text{...(A.15)}$$

$$(\mathbf{A}^{-1})^T = (\mathbf{A}^T)^{-1} \qquad \text{...(A.16)}$$

$$(\mathbf{A}^{-1})^{-1} = \mathbf{A} \qquad \text{...(A.17)}$$

Appendix B

Fourier Transform Pairs

Continuous-Time Signal $f(t)$	*Laplace Transform* $F(jw)$
$\sum_{n=-\infty}^{\infty} F_n\, e^{jn\omega_0 t}$; $\omega_0 = \frac{2\pi}{T_0}$	$2\pi \sum_{n=-\infty}^{\infty} F_n\, \delta(\omega - n\omega_0)$
$e^{j\omega_0 t}$	$2\pi\ \delta(\omega - \omega_0)$
$e^{-at}\ u(t)$; $a > 0$	$\frac{1}{a + j\omega}$
$e^{at}\ u(-t)$; $a > 0$	$\frac{1}{a - j\omega}$
$e^{-a\|t\|}$; $a > 0$	$\frac{2a}{a^2 + \omega^2}$
$\delta(t)$	1
1	$2\pi\ \delta(\omega)$
$\delta_{T_0}(t)$	$\omega_0 \sum_{n=-\infty}^{\infty} \delta(\omega - n\omega_0)$
$u(t)$	$\pi\ \delta(\omega) + \frac{1}{j\omega}$
$\text{sgn}(t)$	$\frac{2}{j\omega}$

$\begin{cases} 1 & ; \quad -T_0 \le t \le T_0 \\ 0 & ; \quad \text{Otherwise} \end{cases}$	$\left(\frac{2}{\omega}\right) \sin \omega T_0$
$\left(\frac{1}{\pi t}\right) \sin W_0 t$	$\begin{cases} 1 & ; \quad -W_0 \le \omega \le W_0 \\ 0 & ; \quad \text{Otherwise} \end{cases}$
$\sin \omega_0 t$	$\frac{\pi}{j} \delta(\omega - \omega_0) - \frac{\pi}{j} \delta(\omega + \omega_0)$
$\cos \omega_0 t$	$\pi\, \delta(\omega - \omega_0) + \pi\, \delta(\omega + \omega_0)$

Appendix C

LAPLACE TRANSFORM PAIRS

Continuous-Time Signal $f(t)$	*Laplace Transform* $F(s)$	*Region of Convergence* ROC
$\delta(t)$	1	Entire s-plane
$u(t)$	$\frac{1}{s}$	$\mathcal{Re}(s) > 0$
$-u(-t)$	$\frac{1}{s}$	$\mathcal{Re}(s) < 0$
$r(t)$	$\frac{1}{s^2}$	$\mathcal{Re}(s) > 0$
$e^{-at}\, u(t)$	$\frac{1}{s+a}$	$\mathcal{Re}(s) > -a$
$-\, e^{-at}\, u(-t)$	$\frac{1}{s+a}$	$\mathcal{Re}(s) < -a$
$\frac{t^{n-1}}{(n-1)\,!}\, u(t)$	$\frac{1}{s^n}$	$\mathcal{Re}(s) > 0$
$-\frac{t^{n-1}}{(n-1)\,!}\, u(-t)$	$\frac{1}{s^n}$	$\mathcal{Re}(s) < 0$
$e^{-at}\,\frac{t^{n-1}}{(n-1)\,!}\, u(t)$	$\frac{1}{(s+a)^n}$	$\mathcal{Re}(s) > -a$

$-e^{-at}\dfrac{t^{n-1}}{(n-1)!}u(-t)$	$\dfrac{1}{(s+a)^n}$	$\mathscr{Re}(s) < -a$
$t^n\ u(t)$	$\dfrac{n!}{s^{n+1}}$	$\mathscr{Re}(s) > 0$
$-t^n\ u(-t)$	$\dfrac{n!}{s^{n+1}}$	$\mathscr{Re}(s) < 0$
$[\sin\ \omega t]\ u(t)$	$\dfrac{\omega}{s^2+\omega^2}$	$\mathscr{Re}(s) > 0$
$[\cos\ \omega t]\ u(t)$	$\dfrac{s}{s^2+\omega^2}$	$\mathscr{Re}(s) > 0$
$[e^{-at}\ \sin\ \omega t]\ u(t)$	$\dfrac{\omega}{(s+a)^2+\omega^2}$	$\mathscr{Re}(s) > -a$
$[e^{-at}\ \cos\ \omega t]\ u(t)$	$\dfrac{s}{(s+a)^2+\omega^2}$	$\mathscr{Re}(s) > -a$
$[\sinh\ \omega t]\ u(t)$	$\dfrac{\omega}{s^2-\omega^2}$	$\mathscr{Re}(s) > 0$
$[\cosh\ \omega t]\ u(t)$	$\dfrac{s}{s^2-\omega^2}$	$\mathscr{Re}(s) > 0$
$[e^{-at}\ \sinh\ \omega t]\ u(t)$	$\dfrac{\omega}{(s+a)^2-\omega^2}$	$\mathscr{Re}(s) > -a$
$[e^{-at}\ \cosh\ \omega t]\ u(t)$	$\dfrac{s}{(s+a)^2-\omega^2}$	$\mathscr{Re}(s) > -a$

Appendix D

z-Transform Pairs

Discrete-Time Signal $f(k)$	*z-Transform* $F(z)$	*Region of Convergence ROC*
$\delta(k)$	1	Entire z-plane
$u(k)$	$\frac{z}{z-1}$	$\lvert z \rvert > 1$
$u(k-1)$	$\frac{1}{z-1}$	$\lvert z \rvert > 1$
$u(-k-1)$	$\frac{1}{z-1}$	$\lvert z \rvert < 1$
$r(k)$	$\frac{z}{(z-1)^2}$	$\lvert z \rvert > 1$
$a^k u(k)$	$\frac{z}{z-a}$	$\lvert z \rvert > \lvert a \rvert$
$-a^k u(-k-1)$	$\frac{z}{z-a}$	$\lvert z \rvert < \lvert a \rvert$
$a^{k-1} u(k-1)$	$\frac{1}{(z-a)}$	$\lvert z \rvert > \lvert a \rvert$
$-a^{k-1} u(-k)$	$\frac{1}{(z-a)}$	$\lvert z \rvert < \lvert a \rvert$

$k\,a^{k-1}\,u(k-1)$	$\dfrac{z}{(z-a)^2}$	$\lvert z\rvert > \lvert a\rvert$
$(k+1)a^k\,u(k)$	$\dfrac{z^2}{(z-a)^2}$	$\lvert z\rvert > \lvert a\rvert$
$-(k+1)a^k\,u(-k-1)$	$\dfrac{z^2}{(z-a)^2}$	$\lvert z\rvert < \lvert a\rvert$
$(k-1)a^{k-2}\,u(k-2)$	$\dfrac{1}{(z-a)^2}$	$\lvert z\rvert > \lvert a\rvert$
$-(k-1)a^{k-2}\,u(-k+1)$	$\dfrac{1}{(z-a)^2}$	$\lvert z\rvert < \lvert a\rvert$
$\dfrac{1}{2\,!}(k+1)(k+2)a^k\,u(k)$	$\dfrac{z^3}{(z-a)^3}$	$\lvert z\rvert > \lvert a\rvert$
$-\dfrac{1}{2\,!}(k+1)(k+2)a^k\,u(-k-1)$	$\dfrac{z^3}{(z-a)^3}$	$\lvert z\rvert < \lvert a\rvert$
$\dfrac{1}{2\,!}(k-2)(k-1)a^{k-3}\,u(k-3)$	$\dfrac{1}{(z-a)^3}$	$\lvert z\rvert > \lvert a\rvert$
$-\dfrac{1}{2\,!}(k-2)(k-1)a^{k-3}\,u(-k+2)$	$\dfrac{1}{(z-a)^3}$	$\lvert z\rvert < \lvert a\rvert$
$[\sin\,\omega k]\;u(k)$	$\dfrac{z\sin\omega}{z^2-2z\cos\omega+1}$	$\lvert z\rvert > 1$
$[\cos\,\omega k]\;u(k)$	$\dfrac{z^2-z\cos\omega}{z^2-2z\cos\omega+1}$	$\lvert z\rvert > 1$

Appendix E

ANSWERS TO PROBLEMS

CHAPTER–1

1. $f(k) = u(k) - u(k-10)$
2. (*a*) Non-linear
   (*b*) Causal
3. Non-linear
4. Non-linear
5. Unstable
6. Time-varying
7. Time-invariant
8. 4
9. $\frac{A}{T_0}[t\,u(t)-(t-T_0)\,u(t-T_0)-T_0\,u(t-T_0)]$
10. $f(t) = u(t) - 2u(t-T_0) + 2u(t-2T_0) - 2u(t-3T_0) + ...$
11. $f(t) = u(t) - 3u(t-1) + 3u(t-2) - 3u(t-3) + ...$

CHAPTER–2

1. $f(t)=\frac{A}{\pi}+\frac{A}{2}\sin\omega_0 t-\frac{2A}{3\pi}\cos 2\omega_0 t-\frac{2A}{15\pi}\cos 4\omega_0 t+\ ......\ ;\ -\infty<t<+\infty$
2. $a_0=\frac{A}{2}\ ;\ a_n=0\ ;\ b_n=-\frac{A}{n\pi}$ *and* $F_n=-j\frac{A}{2\pi n}$
3. $f(t)=\frac{A}{4}-\frac{2A}{\pi^2}\cos\omega_0 t+\frac{A}{\pi}\sin\omega_0 t-\frac{A}{2\pi}\sin 2\omega_0 t-\frac{2A}{9\pi^2}\cos 3\omega_0 t+\frac{A}{3\pi}\sin 3\omega_0 t-......$
4. $a_0=\frac{A}{2}\ ;\ a_n=-\frac{4A}{n^2\pi^2}\ ;\ b_n=0$

5. $f(t) = 0.5 + \frac{12}{\pi^2}\cos \pi t + \frac{12}{9\pi^2}\cos 3\pi t + \frac{12}{25\pi^2}\cos 5\pi t + \ldots\ldots$

CHAPTER–3

1. $F(j\omega) = j\left(\frac{2}{\omega}\cos\omega - \frac{2}{\omega^2}\sin\omega\right)$

2. $F(j\omega) = \frac{1}{(a+j\omega)^2}$

3. $F(j\omega) = \frac{2}{\omega-10}\sin[\pi(\omega-10)]$

4. $G(j\omega) = \frac{j\omega+2}{j\omega+1}$; $g(t) = \delta(t) + e^{-t}u(t)$

5. (a) $y(t) = \frac{1}{b-a}[e^{-at}u(t) - e^{-bt}u(t)]$ (b) $y(t) = t\, e^{-at}\, u(t)$

6. (a) $f(t) = \frac{\sin[\pi(t+1)]}{\pi(t+1)} + \frac{\sin[\pi(t-1)]}{\pi(t-1)}$ (b) $f(t) = [e^{-t} - 2e^{-2t}]u(t)$

7. $y(t) = \left[\frac{1}{4}e^{-t} + \frac{1}{2}t\, e^{-t} - \frac{1}{4}e^{-3t}\right]u(t)$

CHAPTER–4

1. $F(s) = \frac{e^{-5s}}{s}$ ROC: $\mathcal{R}e(s) > 0$

2. (a) $F(s) = -\frac{1}{s^2(s-3)}$ (b) $F(s) = -\frac{e^{-2(s+1)}}{(s+1)^2}$

3. $f(0^+) = 0$ and $f(\infty) = -1$

4. $F(s) = \frac{A}{s^2 T_0}[1-(1+sT_0)\, e^{-sT_0}]$

5. $F(s) = \frac{5\, e^{-2s}}{s^2}[1 - e^{-2s} + 2s\, e^{-2s}]$

6. $F(s) = \frac{1}{s}\left[\frac{1-e^{-sT_0}}{1+e^{-sT_0}}\right]$

7. $\frac{e^{-as}}{s^2}$

8. $F(s) = \frac{1}{s}\left[\frac{1-2e^{-s}}{1+e^{-s}}\right]$

9. $y(t) = e^{-2t}\, u(t) - 3e^{-5t}\, u(t)$

10. $f(t) = -2e^{t}u(-t) + 2e^{-t}\cos(t)u(t) - 4e^{-t}\sin(t)u(t)$

11. $f(t) = 30[e^{-t}\sin t]\, u(t)$

12. $f(t) = 2e^{t}\, u(t) + 2e^{-t}\cos(t)\, u(t) - 4e^{-t}\sin(t)\, u(t)$

13. $g(t) = \delta(t) + 2e^{-3t}\, u(t) + e^{-2t}\, u(t)$

14. (a) $g(t) = 2e^{-3t}\, u(t) - e^{2t}\, u(-t)$ (b) $g(t) = 2e^{-3t}\, u(t) + e^{2t}\, u(t)$; No

15. $i(t) = -10^3\ e^{-20t}\ u(t)$

16. $i(t) = \frac{E}{R}\cos\left(\frac{t}{\sqrt{LC}}\right) u(t)$

17. $i(t) = 0.5\ u(t-2)\ [1 - e^{-2(t-2)}]u(t)$

18. $i(t) = 60\ e^{-2\times10^5\ t}\ u(t)$

19. $v(t) = j\ 2.887\ (e^{(0.5-j\sqrt{0.75})\ t} - e^{(0.5+j\sqrt{0.75})\ t})\ u(t).$

CHAPTER–5

1. (a) $F(z) = \frac{z(z+1)}{(z-1)^3}$ (b) $F(z) = \frac{z(z+a)}{(z-a)^3}$

   (c) $F(z) = \frac{z \sinh\beta}{z^2 - 2z\cosh\beta + 1}$ (d) $F(z) = \frac{z(z-\cosh\beta)}{z^2 - 2z\cosh\beta + 1}$

2. (a) $F(z) = \frac{z}{z - e^{-aT}}$ (b) $F(z) = \frac{Tz\,e^{-aT}}{(z - e^{-aT})^2}$

   (c) $F(z) = \frac{ze^{-aT}\ \sin\omega T}{e^{2aT}z^2 - 2ze^{-aT}\ \cos\omega T + 1}$ (d) $F(z) = \frac{z(z-2+T)}{(z-1)^2}$

3. (a) $f(k) = 0.541\ [1 - (-0.85)^k]$; $k = 0, 1, 2, ...$

   (b) $f(k) = \left[\frac{1}{2}(1)^k - 2(2)^k + \frac{3}{2}(3)^k\right] u(k)$

   (c) $f(k) = [0.96(1)^k + (1.29)^k\ (0.96\ \cos\ 51.3k - 0.8 \sin\ 51.3k)]u(k)$

4. (a) $G(z) = \frac{z\sin\omega T}{z^2 - 2z\cos\omega T + 1}$ (b) $G(z) = \frac{ze^{-\alpha T}\ \sin\omega T}{z^2 - 2ze^{-\alpha T}\ \cos\omega T + e^{-2\alpha T}}$

   (c) $G(z) = \frac{z(z-\cos\omega T)}{z^2 - 2z\cos\omega T + 1}$ (d) $G(z) = \frac{z^2 - ze^{-\alpha T}\ \cos\omega T}{z^2 - 2ze^{-\alpha T}\ \cos\omega T + e^{-2\alpha T}}$

5. $Y(z) = \left(\frac{z}{z-0.5}\right)^2$

6. $f(k) = \cos\left(\frac{k\pi}{2}\right)$; $k = 0, 1, 2, ...$

7. $y(k) = \left[\frac{1}{6} - \frac{3}{2}(-1)^k + \frac{7}{3}(-2)^k\right] u(k)$

8. $y(k) = \left[2\ (1)^k - 2(0.5\sqrt{2})^k\ \cos\left(\frac{k\pi}{4}\right)\right] u(k).$

CHAPTER–6

1. $$e^{\mathbf{A}t} = \begin{bmatrix} \left(e^t - te^t + \frac{1}{2}t^2e^t\right) & (te^t - t^2e^t) & \left(\frac{1}{2}t^2e^t\right) \\ \left(\frac{1}{2}t^2e^t\right) & (e^t - te^t - t^2e^t) & \left(te^t + \frac{1}{2}t^2e^t\right) \\ \left(te^t + \frac{1}{2}t^2e^t\right) & (-3te^t - t^2e^t) & \left(e^t + 2te^t + \frac{1}{2}t^2e^t\right) \end{bmatrix}$$

2. $e^{\mathbf{A}t} = \begin{bmatrix} 1 & \frac{1}{2}(1-e^{-2t}) \\ 0 & e^{-2t} \end{bmatrix}$

3. $y(t) = e^{-0.5t} \sin 0.5t$

4. $G(s) = \dfrac{s^3 - 3s^2 - 9s + 2}{s^2 + 3s + 2}$.

5. $\mathbf{A} = \begin{bmatrix} 0 & 1 \\ -2 & -3 \end{bmatrix}$; $e^{\mathbf{A}t} = \begin{bmatrix} (2e^{t} - e^{-2t}) & (e^{-t} - e^{-2t}) \\ (-2e^{-t} + 2e^{-2t}) & (-e^{-t} + 2e^{-2t}) \end{bmatrix}$

6. $e^{\mathbf{A}t} = \begin{bmatrix} (3e^{-2t} - 2e^{-3t}) & (e^{-2t} - e^{-3t}) \\ (-6e^{-2t} + 6e^{-3t}) & (3e^{-3t} - e^{2t}) \end{bmatrix}$; $\mathbf{x}(t) = \begin{bmatrix} (3e^{-2t} - 2e^{-3t}) \\ (-6e^{-2t} + 6e^{-3t}) \end{bmatrix}$

7. $e^{\mathbf{A}t} = \begin{bmatrix} e^{\sigma t} \cos \omega t & e^{\sigma t} \sin \omega t \\ -e^{\sigma t} \sin \omega t & e^{\sigma t} \cos \omega t \end{bmatrix}$

9. $\begin{bmatrix} \dot{x}_1 \\ \dot{x}_2 \\ \dot{x}_3 \end{bmatrix} = \begin{bmatrix} 0 & 1 & 0 \\ 0 & 0 & 1 \\ -4 & -4 & -3 \end{bmatrix} \begin{bmatrix} x_1 \\ x_2 \\ x_3 \end{bmatrix} + \begin{bmatrix} 0 & 0 & 0 \\ 0 & 0 & 0 \\ 1 & 3 & 4 \end{bmatrix} \begin{bmatrix} u_1 \\ u_2 \\ u_3 \end{bmatrix}$

$\begin{bmatrix} y_1 \\ y_2 \end{bmatrix} = \begin{bmatrix} 0 & 4 & 0 \\ 0 & 0 & 1 \end{bmatrix} \begin{bmatrix} x_1 \\ x_2 \\ x_3 \end{bmatrix} + \begin{bmatrix} 3 & 0 & 0 \\ 0 & 4 & 1 \end{bmatrix} \begin{bmatrix} u_1 \\ u_2 \\ u_3 \end{bmatrix}$

10. $\begin{bmatrix} \dot{V}_{C_1} \\ \dot{V}_{C_2} \\ \dot{I}_L \end{bmatrix} = \begin{bmatrix} -1/2 & 0 & -1/2 \\ 0 & -1/2 & 1/2 \\ 1 & -1 & 0 \end{bmatrix} \begin{bmatrix} V_{C_1} \\ V_{C_2} \\ I_L \end{bmatrix} + \begin{bmatrix} 1/2 \\ 0 \\ 0 \end{bmatrix} [V_s]$

11. $\begin{bmatrix} x_1 \\ x_2 \end{bmatrix} = \begin{bmatrix} \frac{1}{6} + \frac{5}{2}e^{-2t} - \frac{5}{3}e^{-3t} \\ -5e^{-2t} + 5e^{-3t} \end{bmatrix}$

$y = 1 + 10e^{-2t} - 5e^{-3t}$

12. (*a*) $\mathbf{A} = \begin{bmatrix} -2 & 0 \\ 0 & -3 \end{bmatrix}$; $\mathbf{B} = \begin{bmatrix} 1 \\ 1 \end{bmatrix}$ and $\mathbf{C} = [1 \quad -1]$

(*b*) $STM = \mathbf{A}^k = \begin{bmatrix} (-2)^k & 0 \\ 0 & (-3)^k \end{bmatrix}$

(*c*) $y(k) = \frac{1}{4}(-3)^k - \frac{1}{3}(-2)^k + \frac{1}{12}$

13. $G(j\omega) = \dfrac{3j\omega + 1}{(j\omega)^2 - 2j\omega + 1}$.

INDEX

A

B

C

D

E

F

G

H

I

J

L

M

N

O

P

R

S

T

U

V

Z